MATHEMATICAL LOGIC

MATHEMATICAL LOGIC

BY

Willard Van Orman Quine

REVISED EDITION

Cambridge
HARVARD UNIVERSITY PRESS

SEVENTH PRINTING OF THE REVISED EDITION, 1974

LIBRARY OF CONGRESS CATALOG CARD NUMBER 51-7541
ISBN 0-674-55450-7

PRINTED IN THE UNITED STATES OF AMERICA

PREFACE

Of that which receives precise formulation in mathematical logic, an important part is already vaguely present as a basic ingredient of daily discourse. The passage from non-mathematical, non-philosophical common sense to the first technicalities of mathematical logic is thus but a step, quickly taken. Once within the field, moreover, one need not travel to its farther end to reach a frontier; the field is itself a frontier, and investigators are active over much of its length. Even within an introductory exposition there is room for novelties which may not be devoid of interest to the specialist. Textbook and treatise can hence be combined within the same covers in mathematical logic as in few other fields. Such is the intended status of the present book.

The material presented is substantially that covered in my course Mathematics 19 at Harvard. I have undertaken to present it in such a way as to presuppose neither previous acquaintance with mathematical logic nor any special training in mathematics or philosophy. But the book is intended for the serious reader; there are sections which demand close study. Rigor has not, in general, been consciously compromised in favor of perspicuity.

Chapter I deals with the *truth-functional composition* of statements: the modification of statements by denial, and the formation of compounds by such connectives as 'and', 'or', 'if-then'. These are reduced to the single connective of joint denial, 'neither-nor', following Sheffer. The contrast is emphasized between use of expressions and discourse about expressions, and the controversy over implication is considered in the light of this distinction. A metamathematical, or syntactical, notation is introduced to facilitate discourse about statements and other expressions; and the principles of statement composition are expounded in these terms, without recourse to the so-called propositional variables '*p*', '*q*', etc. of earlier works. The property of *tautology* — logical truth of the sort which depends only on truth-functional composition — receives a new formulation in metamathematical terms, and a series of tautologous forms is adduced. The deductive method is dispensed with, at the level of tautology, in favor of the method of tabular calculation.

Chapter II treats of *quantification:* the formal correspondent of the idioms involving 'all' and 'some.' It is in connection with quantification that the *variable* makes its first appearance; and accordingly some attention is devoted to clarifying the nature of the variable, and exhibiting its analogy to the pronoun of ordinary language. Quantification theory, like the preceding part of logic, is expounded within the medium of metamathematics; its presentation in any other medium appears disadvantageous, indeed, because of subtleties having to do with the so-called bound and free occurrences of variables. Here as in Chapter I, methods of calculation are deemed preferable to methods of deduction; but a compromise now has to be struck, because of the fact that no method of calculation is adequate to the truths of quantification theory. The compromise consists in specifying an infinite set of so-called *axioms of quantification* by a method of calculation, and then allowing the theorems to proceed from these by one deductive rule of the simplest sort: *modus ponens.* A further departure, which recommends itself as more intuitive than the usual procedure, is this: the axioms of quantification are specified in such a way that the theorems come to include only *statements* in the strict sense — formulæ containing no "free" variables, no variables beyond those used in quantification.

In Chapter III the connective 'ϵ' of *membership* emerges, and therewith classes. This connective, taken as supplementary to quantification and joint denial, proves to afford an adequate basis for the definition of the notions of logic and those of arithmetic and derivative disciplines as well. An essential step in the series of definitions is the introduction of *abstraction* — the notation whereby classes are specified through conditions on their members. Its introduction is accomplished by contextual definition. On the basis of abstraction in turn the device of *description* is introduced, which corresponds to the idiom 'the entity such that . . .'. It now becomes possible to construe *names* in general, abstract and concrete alike, as mere shorthand which is eliminable at will from all discourse. Besides the economy which it affords, this procedure proves to have the value of effecting a cleavage between syntactical questions of meaningfulness and factual questions of existence.

In Chapter IV the axioms of quantification are supplemented with a new infinity of axioms, providing the essential properties of membership and derivative notions; but no rule of inference is needed beyond the original *modus ponens*. Identity and unit classes are defined, and likewise the notions of the ordinary algebra of classes; and a train of theorems is derived.

The logical antinomies, e.g. Russell's, are avoided by a method which is less restrictive than past methods such as the theory of types. The method turns, like von Neumann's, on construing certain entities as incapable of membership; but the entities so construed are far scarcer than under von Neumann's theory. Besides making for algorithmic facility, this liberalization turns out to render the existence of infinite classes demonstrable without special postulation.

In Chapter V *relations* are defined on the basis of class theory, by the Wiener-Kuratowski method, and the usual concepts of relation theory are introduced and investigated. *Functions*, in turn, are defined on the basis of relation theory. Analogues of the device of abstraction which was used in class theory reappear in connection with relations and functions. The definition of *natural number* comes in Chapter VI; a subsequent definition then introduces the operation of raising a relation to a numerical power, which operation is found to provide a simple and uniform means of defining the various operations of arithmetic. Certain of the basic theorems of arithmetic are proved, and the standard constructions which lead from this level into higher reaches of quantitative mathematics are briefly outlined.

In Chapter VII we turn to the formalization of the metamathematical or syntactical machinery involved in discourse *about* a formalism such as presented in the foregoing chapters. Gödel's theorem regarding the incompletability of logic and arithmetic is derived along novel lines, and its scope is somewhat extended.

CAMBRIDGE, MASS.
April 7, 1940.

ACKNOWLEDGMENTS

Grateful acknowledgment is due Harvard University for having expedited my work on this book by the grant of a free semester (1938–39). I am grateful further to the American Philosophical Society for funds covering assistance in preparing the manuscript. To the editor of *Technology Review* thanks are due for permission to weave passages of my "Relations and Reason" into the last few paragraphs of the Introduction.

I am indebted to Professors H. S. Leonard, C. I. Lewis, and E. V. Huntington and Mr. H. N. Goodman for criticisms of an early draft of Chapter I. In connection with subsequent chapters at least eleven of my past and present students in Mathematics 19 have obliged me with corrections or other helpful suggestions: Professor R. L. Korgen, Dr. G. W. Brown, Miss L. D. Steinhardt, and Messrs. G. D. W. Berry, B. Friedman, I. Kaplansky, W. H. Kruskal, M. J. Norris, R. M. Ravven, K. R. Symon, and, notably in connection with Chapter VII, M. G. Wurtele. I want to thank Mr. Friedman, Mr. and Mrs. Berry, Mr. F. W. Peel, and my wife for the pains they have taken in preparing the manuscript for the press, and Mr. Friedman again for expert assistance with the bibliographical annotations and index.

PREFACE TO THE REVISED EDITION

The revisions made in this edition have affected half the pages, mostly in small ways. Some thirty pages have undergone major changes.

The prime mover of the most important revision was Rosser, who discovered, shortly after the first edition appeared, that the axioms of class theory in the middle of the book were contradictory. On learning of his discovery I arranged with the publisher to paste a corrigendum slip into the remaining stock of the book, indicating a makeshift repair of the system. The text of this slip was also inserted in the second printing (1947). Lately, however, Wang has devised a better repair, admirably suited both to the spirit of the original system and to most of the details of the original exposition. It has hence been possible in the present edition to put the text to rights and leave no scar. The revisions thus entailed have affected pages 157–160, 162, 163, 166, 193, 238, and 305.

Another revision stems from Berry, who showed how a minor modification of the concept of closure (page 79) could yield a reduction in the axioms of quantification. Berry's idea was mentioned in the second printing (page 89), but in the present edition it has been directly incorporated, at the cost of reworking pages 79–95 and 300–301 and changing reference numbers and quantifiers in a hundred pages.

Over the decade since the first edition appeared, two major difficulties for readers have become manifest. One is the distinction between theorems formally deduced and metatheorems informally established. The other is the reasoning behind Gödel's theorem of incompletability. These two difficulties have been alleviated in the present edition, the one by the addition of an appendix and the other by the insertion of some supplementary exposition (pages 307–312).

In the treatment of real numbers two subtle errors, discovered by Mr. Ralph M. Krause and Dr. Wang, have been eradicated by revising pages 273–275. Other emendations and additions on miscellaneous topics have been made on pages 25, 81, 89, 144, 177f, 191, 202, 208f, 229f, 292, 299, and 318.

CAMBRIDGE, MASS.
January 28, 1951.

CONTENTS

PAGE

Preface . . . v
Acknowledgments . . . viii
Preface to Revised Edition . . . ix
Introduction . . . 1

CHAPTER ONE. *Statements* . . . 9
1. Conjunction, Alternation, and Denial . . . 11
2. The Conditional . . . 14
3. Iterated Composition . . . 18
4. Use versus Mention . . . 23
5. Statements about Statements . . . 27
6. Quasi-Quotation . . . 33
7. Parentheses and Dots . . . 37
8. Reduction to Three Primitives . . . 42
9. Reduction to One Primitive . . . 45
10. Tautology . . . 50
11. Selected Tautologous Forms . . . 55

CHAPTER TWO. *Quantification* . . . 63
12. The Quantifier . . . 65
13. Formulæ . . . 71
14. Bondage, Freedom, Closure . . . 76
15. Axioms of Quantification . . . 80
16. Theorems . . . 85
17. Metatheorems . . . 89
18. Substitutivity of the Biconditional . . . 96
19. Existential Quantification . . . 101
20. Distribution of Quantifiers . . . 105
21. Alphabetic Variance . . . 109

CHAPTER THREE. *Terms* . . . 117
22. Class and Member . . . 119
23. Logical Formulæ . . . 123

PAGE

24. Abstraction 128
25. Identity 134
26. Abstraction Resumed 140
27. Descriptions and Names 146

CHAPTER FOUR. *Extended Theory of Classes* . . . 153
28. Stratification 155
29. Further Axioms of Membership 160
30. Substitutivity of Identity 167
31. Substitution for Variables 170
32. Further Consequences 175
33. Logical Product, Sum, Complement . . . 179
34. Inclusion 185
35. Unit Classes 189

CHAPTER FIVE. *Relations* 195
36. Pairs and Relations 197
37. Abstraction of Relations 202
38. Converse, Image, Relative Product . . . 208
39. The Ancestral 215
40. Functions 221
41. Abstraction of Functions 225
42. Identity and Membership as Relations . . 229

CHAPTER SIX. *Number* 235
43. Zero, One, Successor 237
44. Natural Numbers 241
45. Counter Sets 246
46. Finite and Infinite 250
47. Powers of Relations 253
48. Arithmetical Sum, Product, Power . . . 259
49. Familiar Identities of Arithmetic . . . 262
50. Ratios 266
51. Real Numbers 271
52. Further Extensions 275

CHAPTER SEVEN. *Syntax* 281
53. Formality 283

PAGE

54. The Syntactical Primitive 287
55. Protosyntax 291
56. Formula and Matrix Defined 295
57. Axioms of Quantification Defined 299
58. Theorem Defined 302
59. Protosyntax Self-Applied 306
60. Incompleteness 310

APPENDIX. *Theorem versus Metatheorem* 319

List of Definitions 323

List of Theorems and Metatheorems 325

Bibliographical References 331

Index of Proper Names 339

Index of Subjects 341

MATHEMATICAL LOGIC

INTRODUCTION

MATHEMATICAL logic differs from the traditional formal logic so markedly in method, and so far surpasses it in power and subtlety, as to be generally and not unjustifiably regarded as a new science. Its crude beginnings are placed with George Boole, in the middle of the last century. Fragments foreshadowing mathematical logic date back much farther than Boole — as far back indeed as Leibniz; but it was from Boole onward through Peirce, Schröder, Frege, Peano, Whitehead, Russell, and their successors that mathematical logic underwent continuous development and reached the estate of a reputable department of knowledge.

The traditional formal logic, dating in its essentials from Aristotle, is nevertheless the direct progenitor of mathematical logic. The striking differences between the two must not be allowed to obscure the fact that they are both "logic" in the strictest sense of the word. They both have, vaguely speaking, the same subject matter. Just what that subject matter is, it is not easy to say; the usual characterizations of logic as "the science of necessary inference", "the science of forms", etc., are scarcely informative enough to be taken as answers.

But if we shift our attention from subject matter to vocabulary, it is easy to draw a superficial distinction between the truths of logic and true statements of other kinds. A logically true statement has this peculiarity: basic particles such as 'is', 'not', 'and', 'or', 'unless', 'if', 'then', 'neither', 'nor', 'some', 'all', etc. occur in the statement in such a way that the statement is true independently of its other ingredients. Thus, consider the classical example:

(1) If every man is mortal and Socrates is a man then Socrates is mortal.

Not only is this statement true, but it is true independently of the constituents 'man', 'mortal', and 'Socrates'; no alteration of these words is capable of turning the statement into a falsehood. Any other statement of the form:

(2) If every — is — and — is a — then — is —

is equally true, so long merely as the first and fourth blanks are filled alike, and the second and last, and the third and fifth. A still simpler logical truth is:

(3) Socrates is mortal or Socrates is not mortal;

alteration of 'Socrates' and 'mortal' is incapable of making the statement false.

A word may be said to occur *essentially* in a statement if replacement of the word by another is capable of turning the statement into a falsehood.[1] When this is not the case, the word may be said to occur *vacuously*. Thus the words 'Socrates' and 'man' occur essentially in the statement 'Socrates is a man', since the statements 'Bucephalus is a man' and 'Socrates is a horse' are false; on the other hand 'Socrates' and 'mortal' occur vacuously in (3), and 'Socrates', 'man', and 'mortal' occur vacuously in (1). The logical truths, then, are describable as those truths in which only the basic particles alluded to earlier occur essentially.

Those particles may be said to constitute the *logical vocabulary*. They are basic to all discourse. If we were to undertake e.g. to specify a geological vocabulary, comprising the words which occur essentially in the truths of geology, we should have to include in it not only such words as 'moraine', 'fault', etc., but also the whole of the logical vocabulary; and similarly for any other discipline. Accordingly the truths of logic may be reckoned trivially among the truths of geology, and among the truths of economics, and so on. This lends some sense to the dictum that logic has a universal subject matter, and is the common denominator of the special sciences.

The words comprising the logical vocabulary can be substantially reduced, for we can paraphrase some of them with help of the rest.

[1] For a somewhat more careful formulation see my essay "Truth by Convention," pp. 93 ff.

The sense of 'unless', e.g., is adequately conveyed by 'or'; again the joint use of 'neither-nor' and 'not', in the fashion 'neither not . . . nor not . . .', provides an adequate paraphrase of 'and'. When this sort of reduction is carried to the limit, the logical vocabulary comes to comprise only 'is', 'neither-nor', and a device corresponding roughly to the word 'every', together with a certain scheme of pronouns auxiliary to the latter device. (Cf. §§ 9, 12, 22–23.)

What is ordinarily classified under mathematical logic or traditional formal logic includes, indeed, not only logical truths in the proposed sense — truths involving just the logical vocabulary essentially — but also statements *about* such truths. It is customary to include within logic not only such statements as (1) and (3), but also such statements as this:

(4) Any statement of the form (2) is true so long as the first and fourth blanks are filled alike, and the second and last, and the third and fifth.

The present book, in particular, consists in large part of such statements as (4). We must thus distinguish two senses of logic, a broader and a narrower; logic in the narrower sense comprises those truths which contain only the so-called logical vocabulary essentially, while logic in the broader sense includes both logic in the narrower sense and discourse about it. Discourse of the latter kind is classifiable, in large part at least, under the head of *formal syntax* (cf. Ch. VII). Over the years the term 'logic' has of course been applied also to a vast range of other topics, encroaching upon rhetoric, psychology, epistemology, metaphysics; but I shall not attempt to find a unifying principle among these far-flung applications of the term.

Preparatory to the foregoing characterization of logical truths, we took the *general* notion of truth for granted. We simply distinguished logical truths from others, by specifying what words occur essentially in the logical truths. Now the general notion of truth, central as it is to baffling problems of philosophy, may appear rather too big a thing to take for granted. Repudiation of "truth with a capital 'T'" is a favorite way, indeed, of professing alignment with the hard-headed. But in point of fact there is no denying that we know what it means to say that a given statement is true — absolutely True — just as clearly as we understand the given statement itself. The circumstances under which the statement:

(5) Jones smokes

would be said to be true, e.g., are precisely the circumstances under which Jones himself would be said to smoke. Truth *of the statement* (5) is no more mysterious than the notions of Jones and smoking. Applied to any statement *S*, the word 'true' is no obscurer than the obscurest word in the statement *S* itself; for to say that *S* is true is simply to say *S*.[1] This would seem to condone the present expository recourse to the general notion of truth, however the subtle problems which that notion involves may fare.

To determine the truth of the statement (5), it is not enough to inspect the statement; we must also observe Jones. With most other statements the case is similar. Thus when we say that it is the business of the geologist to find out what statements are geological truths, we do not commit him to a sedentary study of statements; we call upon him to spend a good part of his time inspecting bluffs and craters. But logic, indeed mathematics generally, is otherwise. The truth of (1) and (3), e.g., is recognizable on inspection merely of the statements themselves. Confronted with any logical truth or indeed any true statement of mathematics, no matter how complex, we recognize its truth if at all merely by inspecting the statement and reflecting or calculating; observation of craters, test tubes, or human behavior is of no avail. Insofar then as logical truth is discernible at all, standards of logical truth can be formulated in terms merely of more or less complex notational features of statements; and correspondingly for mathematics generally. Thus it is that the logician and mathematician talk *about* statements so much more than the geologist does; and thus it is that logic in a broad sense is commonly taken to include discourse about the statements of logic in the narrower

[1] For an instructive elaboration of this theme see §1 of Tarski's "Wahrheitsbegriff."

sense, whereas no such bifurcation is ordinarily imposed upon geology.

If standards of logical or mathematical truth are to be formulated in terms merely of the observable features of statements, a first important step is revision and schematization of the language in such fashion as to put the relevant features of statements into the simplest possible form. Revision in this direction has long since taken place in mathematics; perhaps the most basic departures from ordinary language are the use of parentheses to indicate grouping, and the use of variables for those purposes of cross-reference for which ordinary language uses pronouns (cf. § 12). Through emulation of mathematics, a similar revision took place in logic from Boole onward; this is indeed so obvious a point of contrast between the old logic and the new that the latter is often called "symbolic logic". But far more radical linguistic revisions are wanted if such projects as the formulation of truth criteria are to be expedited to the fullest. It is helpful to reduce the notions of logic and mathematics to a minimum, by defining some in terms of others; for this reduces the variety of statements which the truth criterion must cover. Reduction of this kind can, it turns out, be carried to great lengths. Mention has already been made of the meager array of devices which suffices for logic: analogues of 'is', 'neither-nor', 'every', and pronouns. What is more surprising, these devices prove adequate not only to logic but to pure mathematics generally (cf. §§ 23, 52); mathematics reduces to logic.

Having given the language of logic the most economical and schematic form we can, we might next hope to devise some routine test which, applied merely to the notational patterns of statements, will distinguish always between those which are logically true and those which are not. But this is a good deal to hope, particularly in view of the reducibility of mathematics in general to logic. Every mathematical problem would become soluble by a mechanical procedure — even the celebrated Fermat problem, which has resisted solution for three centuries. Publication of proofs in mathematics would never again be necessary; results would simply be stated subject to mechanical check on the part of the reader.

Diffident of so bold a project, we might try to formulate some less powerful notational criterion of logical and mathematical truth: a criterion whose fulfillment by any given statement is discernible only by luck rather than by an infallible routine test. Such, indeed, is the character of mathematical proof; a proof once discovered can be mechanically checked, but the actual discovery of the proof is a hit and miss matter. (Cf. §§ 16, 55.) Our present more modest objective, then, can take the form of an explicit formulation of the notion of proof, or theorem, such as will involve reference only to the notational patterns of statements. But recent developments indicate, actually, that an exhaustive criterion of logical or mathematical truth along even these more modest lines is impossible. Given any rules of proof which do not actually lead to falsehoods, there will be mathematical truths which cannot be proved by the rules (cf. § 60). There will always be some demonstrably indemonstrable mathematical truths. We must content ourselves with a version of "theorem" which covers only one or another important portion of the truths of logic and mathematics.

The fact remains that insofar as logical or mathematical truth can be detected at all, the standards can be formulated explicitly as criteria hinging upon the notational patterns of statements. Such truths of logic and mathematics as are not obvious are recognized, if at all, by proof; and the considerations which go to make up a proof do admit of explicit formulation in terms of notational features of statements. Such formulation is not only possible, but highly desirable for the increased insight which it promises. The very discovery that there must always be indemonstrable mathematical truths is a discovery which could not have been made without analyzing the notion of "proof" or "theorem" in terms explicitly of notational features of statements.

The described developments are by no means the whole motivation of the modern refinements in logic. The reduction of the notions of mathematics and logic to a minimum, for example, is prompted by much more than the special purpose mentioned earlier. It is indeed an end in itself, for it reveals just what few notions are finally presupposed by the whole of mathematics. Through this process of paraphrasing some notions in terms of others, moreover, and finally paraphrasing all in terms of the chosen few, the several mathematical and logical notions of course undergo illuminating analysis.

Schematization of the logical language is of significance also in more practical spheres. Logic has its practical use in inference from premisses which are not logical truths to conclusions which are not logical truths. Logic countenances such inference when the conditional statement 'If ... then ...' connecting premiss with conclusion is itself logically true (like (1) above); and it is in this way that logical truth links up with extra-logical concerns. Precisely the analogous account holds with regard to applications of mathematics generally; the tremendous utility of mathematical techniques in natural science turns simply on the importance of discerning mathematical truths of the form 'If . . . then . . .' whose component parts are statements of natural science. Unlike the numerical branches of mathematics, however, logic itself has traditionally figured in natural science only tacitly and at a pretty rudimentary level of inference; it has played fully as subordinate a rôle as arithmetic might have played in the days of Roman numerals. But when we schematize logic along the lines of modern mathematics (retaining convenient abbreviations, of course, over and above the minimum logical vocabulary), we have a tool which is about as effective as arithmetic and derivative branches of mathematics — and which far surpasses these latter in scope of applicability.

Where number is irrelevant, regimented mathematical technique has hitherto tended to be lacking. Thus it is that the progress of natural science has depended so largely upon the discernment of measurable quantity of one sort or another. Measurement consists in correlating our subject matter with the series of real numbers; and such correlations are desirable because, once they are set up, all the well-worked theory of numerical mathematics lies ready at hand as a tool for our further reasoning. But no science can rest entirely on measurement, and many scientific investigations are quite out of reach of that device. To the scientist longing for non-quantitative techniques, then, mathematical logic brings hope. It provides explicit techniques for manipulating the most basic ingredients of discourse. Its yield for science may be expected to consist also in a contribution of rigor and clarity — a sharpening of the concepts of science. Such sharpening of concepts should serve both to disclose hitherto hidden consequences of given scientific hypotheses, and to obviate subtle errors which may stand in the way of scientific progress.

Mathematical logic has been applied,[1] but the most important applications are surely still to come. The usefulness of a theory is not to be measured solely in terms of the application of prefabricated techniques to preformulated problems; we must allow the applicational needs themselves, rather, to play their part in motivating further elaborations of theory. The history of mathematics has consisted to an important degree in such give and take between theory and application. Much of the promise of mathematical logic for science lies in its potentialities as a basis from which to construct subsidiary techniques of unforeseen kinds in response to special needs.

[1] See e.g. Berkeley, Carnap (*Aufbau*), Dittrich, Hempel and Oppenheim, Hull, Quine ("Relations and Reason"), Russell ("Order in Time"), Shannon, Stamm, Whitehead ("Mathematical Concepts," "Théorie rélationniste"), Wiener ("Synthetic Logic," "Measurement"), Woodger, Zwicky.

CHAPTER ONE

STATEMENTS

§1. *Conjunction, Alternation, and Denial*

MATHEMATICAL logic at its most elementary level deals with *statements*, or declarative sentences, and with ways of compounding them into further statements. One of these modes of composition, known to logicians as *conjunction*, consists in joining two statements by 'and' — or, in the notation of mathematical logic, by the dot '.'. 'Some are born great and some achieve greatness', or 'some are born great . some achieve greatness', is the conjunction of 'some are born great' and 'some achieve greatness'. A conjunction is true if both component statements are true; otherwise false. To determine the *truth value* (truth or falsehood) of a conjunction, therefore, it is sufficient to know the truth values of the components.

A mode of statement composition will be said to be *truth-functional* if the truth value of the compound is determined in all cases by the truth values of the components. Conjunction is thus truth-functional. So are all other modes of statement composition required for mathematical logic, or mathematics generally. There is reason to believe that none but truth-functional modes of statement composition are needed in any discourse, mathematical or otherwise; but this is a controversial question. (Cf. § 5.)

Each such mode of composition is exhaustively describable by a *truth table* — a table specifying the truth value of the compound for each assignment of truth values to the components. Thus the table for conjunction is as follows:

1st component	*2d component*	*Conjunction*
T	T	T
F	T	F
T	F	F
F	F	F

It tells us that the conjunction is true when both components are true, false when the first component is false and the second true, and false likewise in the two remaining cases.

A second mode of composition, no less familiar than conjunction, is *alternation* — composition of statements by means of the connective 'or'. Reflection upon ordinary usage of 'or' reveals that an alternation is true if just one of its components is true, and false if they are both false. This gives us three-quarters of a truth table for alternation:

1st component	*2d component*	*Alternation*
T	T	
F	T	T
T	F	T
F	F	F

But the top row is indicated less clearly by usage. We must decide whether 'or' is to be construed in an *exclusive* sense, corresponding to the Latin 'aut', or in an *inclusive* sense, corresponding to the Latin 'vel'. When 'or' is used in the exclusive sense, the compound is regarded as true only if exactly one of the two components is true; joint truth of the components falsifies the compound. An 'or'-compound in this sense can be expressed more clearly by adding the words 'but not both'. When 'or' is used in the inclusive sense, on the other hand, the compound is regarded as true if at least one of the components is true; joint truth of the components verifies the compound. An 'or'-compound in this sense can be expressed more clearly by adding the words 'or both', or by admitting the barbarism 'and/or'. The blank in our table receives an 'F' or a 'T' according as 'or' is construed in the exclusive or the inclusive sense.

Common usage can be adduced in support of either course. If a witness guesses that either the steering gear was loose or the driver was drunk, and it is found that the steering gear was loose and the driver was drunk, we do not regard the witness as mistaken; to this extent usage supports the inclusive 'or'. At the same time the prevalent use of the expressions 'or both' and 'and/or' is a presumption in favor of the exclusive interpretation, since otherwise these expressions would always be superfluous.

In mathematical logic the ambiguity of ordinary usage is resolved by adopting a special symbol 'v', suggestive of 'vel', to take the place of 'or' in the inclusive sense. Alternation is identified with this usage; thus the truth table of alternation becomes:

1st component	*2d component*	*Alternation*
T	T	T
F	T	T
T	F	T
F	F	F

The exclusive use of 'or' is not frequent enough in technical developments to warrant a special name and symbol.

The connectives 'and' and 'or' are used in ordinary discourse not only between statements but between nouns, verbs, prepositions, indeed nearly all parts of speech. It is only as connectives of statements, however, that they receive the symbolic rendering '.' and 'v'. The other uses do not fall under the head of conjunction and alternation.

Conjunction and alternation are *binary*, in that they combine statements two at a time. But *denial*, to which we now turn, is *singulary:*[1] it is a method merely of elaborating a single statement to form a new statement. It has the following truth table, which is limited to two rows because there is just one component.

Component	*Denial*
T	F
F	T

The denial is true if and only if the original statement is false.

The method of forming the denial in ordinary language is irregular. Sometimes 'not' is attached to the main verb; thus the denial of 'Jones is away' is 'Jones is not away'. But if the verb

[1] The series of adjectives 'binary', 'ternary', 'quaternary', 'quinary', . . . leaves mathematicians in a quandary when $n = 1$. It is customary to stammer out some such makeshift as 'unary' or 'uninary' or 'unitary'. But the proper word is apparent if we reflect that the series of Latin distributives 'bini', 'terni', 'quaterni', 'quini', . . . begins with 'singuli'.

is governed by 'sometimes' or 'always', the denial is formed rather by substituting 'never' or 'not always'. If the statement is a conjunction or alternation and hence has no main verb, then denial is accomplished by one or another periphrasis.

In mathematical logic the denial of a statement is formed by prefixing the tilde '$\sim$', which is a modified 'n' and is conveniently read 'not'. Thus 'Jones is not away' gives way to '$\sim$ Jones is away'. Over the ordinary use of 'not' this notation has two advantages: it leaves the internal constitution of the affected statement intact, and it applies immediately to statements of any form and complexity.

The first branch of mathematical logic to reach maturity was the so-called algebra of logic, which was founded by Boole (1847) and improved by Jevons, Peirce, and Schröder. This algebra, while regarded primarily as an algebra of classes (cf. § 33), was recognized as amenable to reinterpretation as a theory governing conjunction, alternation, and denial. Truth tables, and the graphic method of calculation which they provide (cf. § 3), came only in 1920–21 (Łukasiewicz, Post, Wittgenstein); but much the same technique in non-tabular form was made known by Peirce (3.387 f) in 1885, and is indeed implicit in Boole's "general rule of development" (*Laws of Thought*, p. 75 f). The signs '.', '$\vee$', and '$\sim$' are from Whitehead and Russell.

§ 2. *The Conditional*

ANOTHER binary connective, and a particularly important one, is 'if-then'. A statement thus compounded is known as a *conditional* (or *hypothetical*). The first of its two component statements, placed between 'if' and 'then', is called the *antecedent* (or *protasis* or *hypothesis*) of the conditional; the second component, placed after 'then', is called the *consequent* (or *apodasis* or *thesis*).

Any conditional with true antecedent and false consequent is false. The conditional 'If Jones was here then the glove is his' is clearly disproved when it turns out that Jones was here and the glove is not his. But none of the other three combinations of truth values is adequate to disproving the conditional. In case Jones was not here, the conditional stipulates nothing regarding the ownership of the glove; and in case Jones was here and the glove is his, the conditional is supported. Are we then to construe a conditional as false in the one case and true in the other three?

Antecedent	*Consequent*	*Conditional*
T	T	T
F	T	T
T	F	F
F	F	T

The mode of composition described in the table constitutes the nearest truth-functional approximation to the conditional of ordinary discourse, and may be called the *truth-functional conditional.* All truth-functional conditionals with false antecedents and all with true consequents are true. Only those are false which have true antecedents and false consequents. If we construe 'if-then' in this sense, the following become true:

(1) If France is in Europe then the sea is salt,
(2) If France is in Australia then the sea is salt,
(3) If France is in Australia then the sea is sweet.

Symbolically the truth-functional conditional is rendered by the connective '⊃', thus:

(4) Jones was here ⊃ the glove is his.

(4) is true in case Jones was not here, true also in case the glove is his, and false just in case he was here and the glove is not his. It is true whenever at least one of the two statements '∼ Jones was here' and 'the glove is his' is true, and false in the case where 'Jones was here' and '∼ the glove is his' are both true. (4) can hence be paraphrased in terms of denial and alternation thus:

(5) ∼ Jones was here ∨ the glove is his,

or in terms of denial and conjunction thus:

(6) ∼ (Jones was here . ∼ the glove is his).

The parentheses in (6) indicate the grouping, as usual in mathematics. They indicate that the denial sign applies to the whole conjunction. When there is no such indication, a denial sign always applies to the shortest statement following it; thus (5) is understood as:

$$(\sim \text{Jones was here}) \vee \text{the glove is his}$$

rather than:

$$\sim (\text{Jones was here} \vee \text{the glove is his}).$$

A statement of the form '—— ⊃ ——' is conveniently pronounced in the usual fashion 'If —— then ——';[1] but we have now to consider to what extent this conforms to the ordinary usage of 'if-then'. Let us examine the latter more closely. The subjunctive mood is ordinarily used in a conditional when the speaker believes the antecedent to be false, and the indicative mood when he wants to leave the antecedent unprejudiced. Usage of the subjunctive conditional is certainly at variance with the above table; for if all conditionals with false antecedents are true, no subjunctive conditional is worth affirming. The subjunctive conditional is in fact not directly identifiable with any truth-functional mode of composition, but calls for a more elaborate analysis. (Cf. § 5.) This task is important for the philosophy of science, for the subjunctive conditional is implicit in those terms of disposition or potentiality with which the natural sciences abound: 'soluble', 'malleable', 'hard', 'sensitive', 'intelligent', etc. The task is not one, however, which need be accomplished within the framework of mathematical logic. Mathematics makes no use of the subjunctive conditional; the indicative form suffices. It is hence useful to shelve the former and treat the 'if-then' of the indicative conditional as a connective distinct from 'if-then' subjunctively used. It remains to consider whether the above table gives an acceptable version of the indicative conditional.[2]

[1] The reading 'implies', for '⊃', is to be avoided (cf. § 5). Its popularity probably arises from the fact that the 'if-then' reading calls for an inconvenient excursion into the context; 'then' supplants '⊃', but 'if' has to be inserted in an earlier position. Note however that we can avoid this inconvenience by reading '⊃', at will, simply as 'only if'. A little reflection shows that 'If — then —' can always be given the alternative rendering '— only if —', without disturbing the order of the components. Whereas '— if —' is of course the reverse of 'If — then —', insertion of 'only' has the peculiar effect of restoring the normal order of antecedent and consequent.

[2] Not only the present section but also § 5 is relevant to the controversy which

One who affirms an indicative conditional is ordinarily uncertain as to the truth values of both antecedent and consequent. He believes it is not the case that the first is true and the second false, but ordinarily he does not know whether they are both true, or both false, or first false and second true. It is for this reason, perhaps, that the truth of (1)–(3) seems strange; not because these would be false under ordinary usage, but because they would be idle or senseless. Indeed, since usage conforms to the third line of the table, and usage lapses as soon as a case is precisely located elsewhere in the table, there is no clear conflict between the table and the indicative conditional of ordinary usage. The table departs from usage only in committing itself on points where usage is lacking. This much departure is necessary in any complete formulation — any formulation which explains the conditional for every pair of statements.

What the truth table adds, in thus deciding the cases beyond the range of ordinary usage, is essentially theoretical; no supplementary practical use of 'if-then' is thereby prescribed. In practice, even in the light of the truth table, one would naturally not bother to affirm a conditional if he were in position to affirm the consequent outright or to deny the antecedent — any more than one bothers to affirm an alternation when he knows which component is true. We say:

(7) If Jones has malaria then he needs quinine

because we know about malaria but are in doubt both of Jones's ailment and of his need of quinine. Thus only those conditionals are worth affirming which follow from some manner of relevance between antecedent and consequent — some manner of law, perhaps, connecting the matters which these two component statements describe. Such connection underlies the useful application of the truth-functional conditional without participating in the meaning of that notion. The situation is similar, indeed, in the case of alternation.

A conditional thus inferred from a general law must be dis-

---

has arisen over this point. Those acquainted with the controversy should keep in mind that in the present section nothing is said of implication.

tinguished from the law itself, which commonly bears the outward form of a conditional in turn:

(8) If one has malaria then he needs quinine.

Certainly (8) asserts a connection between malaria and quinine which transcends any truth table; but actually the statement is elliptical, and analysis deprives it of the form of a pure conditional. It is short for some such statement as this:

(9) Whatever person be selected, if he has malaria then he needs quinine.

This statement is of an important kind which will be taken up in Chapter II under the head of *quantification*. Its initial part confers universality; the remainder has no separate status as a statement at all, for it contains pronouns which refer back to the initial part and lose all meaning on isolation. (9) is not itself a conditional, but has the effect of simultaneous affirmation of a vast array of conditionals. Each of these separate conditionals, e.g. (7), can still be construed in accordance with the truth table; the strong interconnection between malaria and quinine which (9) conveys is not to be laid to the conditional mode of composition but to the initial indication of universality.

The truth-functional version of the conditional goes back to Philo of Megara. It was revived in modern logic by Frege (1879) and Peirce (1885). The conditional sign '$\supset$' was used by Gergonne as early as 1816, though not in the truth-functional sense. The appropriateness of the truth-functional version was vigorously debated in ancient times (cf. Peirce, 3.441 ff; Łukasiewicz, "Zur Geschichte", p. 116), and has become a current topic of controversy as well. The issue has been clouded, however, by failure to distinguish clearly between the conditional and implication (cf. § 5)

§ 3. *Iterated Composition*

WHEN truth-functional composition is applied in iteration to form a complex statement, e.g.:

(1) Germany will withdraw ∨ (France will invade . England will mobilize),

(2) (Jones was here ⊃ the glove is his) . (the glove is his ⊃ Jones was here),

(3) (Jones came . ~ (Smith stayed . ~ Robinson left)) v Robinson left,

we can always calculate the net import of the result by truth tables. We can determine systematically what truth values on the part of the ultimate components would make the whole true and what ones would make the whole false.

This is accomplished, for example in the case of (2), by assigning all possible combinations of truth values to the ultimate components 'Jones was here' and 'the glove is his' and then deriving the truth value of (2) for each of these assignments by consulting the truth tables of the modes of composition involved. The process can conveniently be carried out by a tabular construction whose

(*Jones was here*	⊃	*the glove is his*)	.	(*the glove is his*	⊃	*Jones was here*)
T	T	T	T	T	T	T
F	T	T	F	T	F	F
T	F	F	F	F	T	T
F	T	F	T	F	T	F

TABLE 1

outcome is Table 1. The first and third columns, standing beneath 'Jones was here' and 'the glove is his', exhaust the four combinations of truth values; they are the familiar columns used in earlier tables. These columns are repeated under the repeated occurrences of 'Jones was here' and 'the glove is his'. Each of the three remaining columns stands beneath a statement connective, and lists the truth values, for the four cases, of the compound produced by that connective. Thus, the second column indicates the truth value of the conditional:

Jones was here ⊃ the glove is his

for the successive cases; the four entries in the column were found by inspecting the two adjacent columns and taking the appropriate values for the conditional according to the conditional table (§ 2). The next to last column, indicating the value of:

The glove is his $\supset$ Jones was here

for the successive cases, was obtained similarly from its adjacent columns. Finally, the middle column indicates the value of the whole conjunction (2) for the successive cases; the four entries here were found by inspecting the columns belonging to the two immediate components of the conjunction, namely the two conditionals, and taking the appropriate values for the whole according to the conjunction table (§ 1).

The completed Table 1 tells us, then, in its middle column, that (2) is true in just the first and last cases: true if Jones was here and the glove is his, true if Jones was not here and the glove is not his, and false otherwise.

The form of iterated composition illustrated in (2) is, as it happens, a particularly important one; for it corresponds to the idiom 'if and only if'. Symbolically it is usually rendered in abbreviated fashion by the connective '$\equiv$'; thus (2) becomes:

Jones was here $\equiv$ the glove is his.

This binary mode of composition may be called the *biconditional*, more specifically the truth-functional biconditional, for it is the conjunction of opposite truth-functional conditionals. It has Table 2 as its truth table, for we saw that (2) is true just in case

1st component	*2d component*	*Biconditional*
T	T	T
F	T	F
T	F	F
F	F	T

TABLE 2

the components are both true or both false. The biconditional is true just in case its components are alike in truth value; false just in case one component is true and the other false. This mode of composition is related to the 'if and only if' of ordinary usage precisely as the truth-functional conditional is related to the ordinary usage of 'if-then' (cf. § 2).

1st component	*2d component*	*3d component*
T	T	T
F	T	T
T	F	T
F	F	T
T	T	F
F	T	F
T	F	F
F	F	F

TABLE 3

Let us now analyze (3). Here there are three ultimate components, 'Jones came', 'Smith stayed', and 'Robinson left', instead of just two. Our first task, then, is to exhaust the ways of assigning truth values to three components. A convenient systematic method is shown in Table 3. The upper half of this array was formed by setting down the familiar array for two components and adjoining a third column composed uniformly of 'T'; the lower half was formed similarly using 'F'. The array for four components would be formed from this array for three components by the same method, and so on. Obviously this scheme always exhausts the combinations. In general, we see that the array for n components will have 2^n rows; and that 'T' and 'F' will always be alternated simply in the first column, in pairs in the second column, in fours in the third column, and so on.

We have now to calculate the truth value of (3) for each of the eight ways of assigning truth values to the three ultimate components. The construction, analogous to that of Table 1, has Table 4 as its outcome. The columns under 'Jones came', 'Smith stayed', and 'Robinson left' are the columns of Table 3, with repetition for the repeated component 'Robinson left'. The sixth column indicates the truth values of the denial '∼ Robinson left' for the eight cases; its entries are derived from those in the seventh column by means of the denial table. The fifth column indicates the values of:

Smith stayed . ∼Robinson left;

it is derived from the adjacent columns by the conjunction table.

(*Jones came*	.	~(*Smith stayed*		.	~*Robinson*	*left*))	∨	*Robinson left*
T	T	T	T	F	F	T	T	T
F	F	T	T	F	F	T	T	T
T	T	T	F	F	F	T	T	T
F	F	T	F	F	F	T	T	T
T	F	F	T	T	T	F	F	F
F	F	F	T	T	T	F	F	F
T	T	T	F	F	T	F	T	F
F	F	T	F	F	T	F	F	F

TABLE 4

The third column indicates the values of the denial of this conjunction; it is derived from the fifth column by the denial table. The second column indicates the values of:

Jones came . ~(Smith stayed . ~Robinson left);

it is derived from the adjacent columns by the conjunction table. Finally, the eighth column indicates the values of the whole alternation (3); it is derived from the second and last columns by the alternation table.

Table 4, thus constructed, tells us in its eighth column that (3) is true in all cases except the fifth, sixth, and last. Thus (3) is false if Jones came, Smith stayed, and Robinson did not leave; false also if Jones did not come, Smith stayed, and Robinson did not leave; false also if Jones did not come, Smith did not stay, and Robinson did not leave; but true in all other cases.

In Table 5 the method is applied to a case where there is just one ultimate component. Here there are just two rows, exhausting the truth possibilities for the single component. The third column shows that the whole is true if Jones was not here, otherwise false.

~	*Jones was here*	≡	(*Jones was here*	⊃	*Jones was here*)
F	T	F	T	T	T
T	F	T	F	T	F

TABLE 5

§4. *Use versus Mention*

IN THE literature on the logic of statements, and in other foundational studies of mathematics as well, confusion and controversy have resulted from failure to distinguish clearly between an object and its name. Ordinarily the failure to maintain this distinction is not to be attributed to any close resemblance between the object and the name, even if the object happens to be a name in turn; for even the discrimination between one name and another is a visual operation of an elementary kind. The trouble comes rather in forgetting that a statement about an object must contain a name of the object rather than the object itself. If the object is a man or a city, physical circumstances prevent the error of using it instead of its name; when the object is a name or other expression in turn, however, the error is easily committed.

As an illustration of the essential distinction, consider these three statements:

(1) Boston is populous,
(2) Boston is disyllabic,
(3) 'Boston' is disyllabic.

The first two are incompatible, and indeed (1) is true and (2) false. Boston is a city rather than a word, and whereas a city may be populous, only a word is disyllabic. To say that the place-name in question is disyllabic we must use, not that name itself, but a name of it. The name of a name or other expression is commonly formed by putting the named expression in single quotation marks; the whole, called a *quotation*, denotes its interior. This device is used in (3), which, like (1), is true. (3) contains a name of the disyllabic word in question, just as (1) contains a name of the populous city in question. (3) is *about* a word which (1) *contains;* and (1) is about no word at all, but a city. In (1) the place-name is *used*, and in this way the city is *mentioned;* in (3) a quotation is used, and the place-name is mentioned. We mention x by using a name of x; and a statement about x contains a name of x.[1]

[1] By these considerations, the first sentence of the present paragraph might be

The foregoing treatment of (1)–(3) is itself replete with mention of expressions, yet free from quotations. These were avoided by circumlocution. As an exercise in quotation marks, however, it may be useful now to add a few comments involving them. 'Boston is populous' is about Boston and contains 'Boston'; ''Boston' is disyllabic' is about 'Boston' and contains ''Boston''. ''Boston'' designates 'Boston', which in turn designates Boston. To mention Boston we use 'Boston' or a synonym, and to mention 'Boston' we use ''Boston'' or a synonym. ''Boston'' contains six letters and just one pair of quotation marks; 'Boston' contains six letters and no quotation marks; and Boston contains some 800,000 people.

Such examples as (3), or:

(4) 'Boston' has six letters,
(5) 'Boston' is a noun,
(6) 'Boston' occurs in Walt Whitman's *Chants Democratic*,

must not be thought of as exhausting the kinds of things that can be said about an expression. These four statements ascribe properties to 'Boston' which might be classed respectively as phonetic, morphological, grammatical, and literary; roughly speaking, they have nothing to do with meaning. But an expression which has use in language will also have *semantic* properties, or properties which arise from the meaning of the expression. Such properties are ascribed to 'Boston' by these statements:

(7) 'Boston' designates Boston,
(8) 'Boston' designates a populous city,
(9) 'Boston' designates the capital of Massachusetts,
(10) 'Boston' is synonymous with 'the capital of Massachusetts'.

(7)–(10) are just as genuinely statements about 'Boston' as are (3)–(6), and omission of the quotation marks from (7)–(10) would give results no less objectionable than (2); for it is only

criticized for failure to enclose the whole statements (1)–(3) in quotation marks. But it is clearer to avoid quotation, in such cases, by agreeing to regard the colon as equivalent to quotation when followed by displayed text (text centered in a new line).

expressions, not places, that can designate or be synonymous. A statement about an expression may depend for its verification upon considerations of sound or shape or literary locale, or even upon considerations of population or other extra-linguistic matters of fact with which the expression is indirectly connected by its use; but so long as the statement is about the expression it must contain a name of the expression.

Lack of care in thus distinguishing the name from the named is common in mathematical writings. The following passage, from a widely used textbook on the differential calculus, is fairly typical:

> The expression $D_x y \Delta x$ is called the *differential* of the function and is denoted by dy:
>
> $$dy = D_x y \Delta x.$$

The third line of this passage, an equation, is apparently supposed to reproduce the sense of the first two lines. But actually, whereas the equation says that the entities dy and $D_x y \Delta x$ (whatever these may be) are the same, the preceding two lines say rather that the one is a name of the other. And the first line of the passage involves further difficulties; taken literally it implies that the exhibited expression '$D_x y \Delta x$' constitutes a name of some other, unexhibited *expression* which is known as a differential. But all these difficulties can be removed by a slight rephrasing of the passage: drop the first two words and change 'and is denoted by' to 'or briefly'.

Expository confusions of this sort have persisted because, in most directions of mathematical inquiry, they have not made themselves felt as a practical obstacle. They do give rise to minor perplexities, indeed, even at the level of elementary arithmetic. A student of arithmetic may wonder, e.g., how 6 can be the denominator of $\frac{4}{6}$ and not of $\frac{2}{3}$ when $\frac{4}{6}$ is $\frac{2}{3}$; this puzzle arises from failure to observe that it is the *fractions* '$\frac{4}{6}$' and '$\frac{2}{3}$' that have denominators, whereas it is the designated *ratios* $\frac{4}{6}$ and $\frac{2}{3}$ that are identical. But it is primarily in mathematical logic that carelessness over these distinctions is found to have its more serious effects. At the level of the logic of statements, one effect is obliteration of the distinction between predicates of statements and composition of statements — a distinction which will be considered in the next section.

Scrupulous use of quotation marks is the main practical measure against confusing objects with their names. But it has already been suggested that this particular method of naming expressions is not theoretically essential. E.g., using elaborately descriptive names of 'Boston', we might paraphrase (3) in either of the following ways:

> The word composed successively of the second, fifteenth, nineteenth, twentieth, fifteenth, and fourteenth letters of the alphabet is disyllabic.
>
> The 4354th word of *Chants Democratic* is disyllabic.

Quotation is the more graphic and convenient method, but it has a certain anomalous feature which calls for special caution: from the standpoint of logical analysis each whole quotation must be regarded as a single word or sign, whose parts count for no more than serifs or syllables. A quotation is not a *description*, but a *hieroglyph;* it designates its object not by describing it in terms of other objects, but by picturing it. The meaning of the whole does not depend upon the meanings of the constituent words. The personal name buried within the first word of the statement:

(11) 'Cicero' has six letters,

e.g., is logically no more germane to the statement than is the verb 'let' which is buried within the last word. Otherwise, indeed, the identity of Tully with Cicero would allow us to interchange these personal names, in the context of quotation marks as in any other context; we could thus argue from the truth (11) to the falsehood:

'Tully' has six letters.

Frege seems to have been the first logician to recognize the importance of scrupulous use of quotation marks for avoidance of confusion between use and mention of expressions (cf. *Grundgesetze*, vol. 1, p. 4); but unfortunately his counsel and good example in this regard went unheeded by other logicians for some thirty years. For further discussion of this topic see Carnap, *Syntax*, pp. 153-160. Concerning the necessity of treating a whole quotation as a single sign, see also § 1 of Tarski's "Wahrheitsbegriff".

§ 5. *Statements about Statements*

TO SAY that a city or a word has a given property, e.g. populousness or disyllabism, we attach the appropriate predicate to a name of the city or word in question (cf. § 4). To say that a statement has a given property, e.g. the phonetic property of being a hexameter or the semantic property of truth or falsehood, we attach the appropriate predicate to a name of the statement in question — not to the statement itself. Thus, to attribute truth to:

(1) Jones is ill

we write:

(2) 'Jones is ill' is true,

and to attribute falsehood we write:

(3) 'Jones is ill' is false.

Equivalently, we may write:

(4) (1) is true,
(5) (1) is false;

but never:

(6) Jones is ill is true,
(7) Jones is ill is false,

on the analogy of:

(8) $\sim$ Jones is ill.

(2)–(5) are about the statement (1), but (8) is not; it, like (1), is about Jones. 'Is true' and 'is false' attach to names of statements precisely because, unlike '$\sim$', they are predicates by means of which we speak *about* statements. Whereas statement connectives ('$\sim$', '$.$', '$\vee$', '$\supset$', '$\equiv$') attach to statements to form statements, a predicate is an expression which attaches to names to form statements. Grammar alone is enough to condemn (6) and (7), since each occurrence of 'is' should have a noun as subject. Confusion over this matter results in the view that the suffix 'is true' is vacuous, and that the suffix 'is false' is the English translation of the prefix '$\sim$'; the view, in other words, that (6) is equivalent to (1) and (7) to (8).

In order to say that two objects stand in a given relation, e.g. hate, or remoteness, one puts an appropriate binary predicate (transitive verb) between names of the objects thus: 'Roosevelt hates Hitler', 'Berlin is far from Washington'. To say that two statements stand in a given relation, whether the phonetic relation of rhyming or the semantic relation of implication, we put the appropriate binary predicate between names of the statements — not between the statements themselves. We may write:

(9) 'All men are mortal' implies 'all white men are mortal',
(10) The third statement of the book implies the seventh,

but never:

(11) All men are mortal implies all white men are mortal

on the analogy of:

(12) If all men are mortal then all white men are mortal,
(13) All men are mortal $\supset$ all white men are mortal.

The verb 'implies' belongs between names of statements precisely because, unlike '$\supset$' or 'if-then', it expresses a relation between statements; it is a binary predicate by means of which we talk *about* statements. (9) and (10) are about statements, while (12) and (13) are about men.

The relation of implication in one fairly natural sense of the term, viz. *logical implication*, is readily described with help of the auxiliary notion of *logical truth*. A statement is logically true if it is not only true but remains true when all but its logical skeleton is varied at will; in other words, if it is true and contains only logical expressions essentially, any others vacuously (cf. Introduction). Now one statement may be said logically to imply another when the truth-functional conditional which has the one statement as antecedent and the other as consequent is logically true. Thus (9), so construed, is equivalent to:

(13) is logically true.

A trivial analogue, *material implication*, may be said to hold whenever the truth-functional conditional which has the one statement as antecedent and the other as consequent is true. Thus one statement materially implies another provided merely that the first is false or the second true. This relation is so broad as not to deserve the name of implication at all except by analogy. But — and this is the point usually missed — 'materially implies' is still a binary predicate, not a binary statement connective. It stands to '$\supset$' precisely as 'is false' stands to '$\sim$'. Insertion of the connective '$\supset$' between statements as in (13) amounts to inserting the verb 'materially implies', not between the statements themselves as in (11), but between their names as in (9).

With a few trivial exceptions such as material implication, any relation between statements will depend on something more than the truth values of the statements related. Such is the case, e.g., with the phonetic relation of rhyming. The same holds for the semantic relation of logical implication described above, and for any other relation which has (unlike material implication) a serious claim to the name of implication. Such relations are quite consonant with a policy of shunning non-truth-functional modes of statement composition (cf. § 1), since a relation of statements is not a mode of statement composition. On this account, the policy of admitting none but truth-functional modes of statement composition is not so restrictive as might have at first appeared; what could be accomplished by a subjunctive conditional or other non-truth-functional mode of statement composition can commonly be accomplished just as well by talking *about* the statements in question, thus using an implication relation or some other strong relation of statements instead of the strong mode of statement composition. Instead of saying:

> If Perth were 400 miles from Omaha then Perth would be in America

one might say:

> 'Perth is 400 miles from Omaha' implies 'Perth is in America',

in some appropriate sense of implication.

Much of what has been said regarding implication applies equally to other semantic relations of statements, e.g. *equivalence* and *compatibility*. Statements are logically equivalent when they logically imply each other, and logically compatible when one does not imply the other's denial. Or, what comes to the same thing, statements are logically equivalent when the biconditional formed from them is logically true, and they are logically compatible except when the conjunction formed from them is logically false, i.e., except when the denial of the conjunction is logically true. Trivial analogues, material equivalence and compatibility, are similarly determined: statements are materially equivalent when they materially imply each other, and materially compatible when one does not materially imply the other's denial. Or, what comes to the same thing, statements are materially equivalent whenever their biconditional is true, and materially compatible except when their conjunction is false, hence whenever their conjunction is true. Material equivalence is agreement in truth value, and material compatibility is joint truth. Equivalence and compatibility, even in this degenerate sense, must be distinguished from the biconditional and conjunction; insertion of '$\equiv$' or '.' between statements amounts to inserting 'is materially equivalent to' or 'is materially compatible with', not between the statements themselves, but between their names.

Note that 'is true', 'is false', 'implies', 'is equivalent to', etc. do not admit of iterated application as do the statement connectives. The expressions which '$\sim$', '.', '$\supset$', etc. govern and the expressions which they produce are homogeneously statements; the expressions produced can hence be so governed in turn, and thus we obtain statements of the forms:

$$\sim\sim \text{——}, \qquad \text{——} \supset (\text{——} \supset \text{——}), \qquad (\sim \text{——} \,.\, \text{——}) \supset \sim \text{——},$$

etc. But 'is false', 'implies', etc. govern names and produce statements; hence the expressions produced cannot be so governed in turn. Formally the predicates 'is false' and 'implies' resemble the predicates 'is negative' and '$\leqslant$' of arithmetic rather than the statement connectives '$\sim$' and '$\supset$'. Just as it is true for all numbers x, y, z that

$$(x \leqslant y \,.\, y \leqslant z) \supset x \leqslant z$$

and $(y \text{ is negative} \,.\, x \leqslant y) \supset x \text{ is negative},$

so it is true for all statements ϕ, ψ, χ that

$$(\phi \text{ implies } \psi \,.\, \psi \text{ implies } \chi) \supset \phi \text{ implies } \chi$$

and $(\psi \text{ is false} \,.\, \phi \text{ implies } \psi) \supset \phi \text{ is false};$

on the other hand the contexts:

$$(\phi \text{ implies } \psi \,.\, \psi \text{ implies } \chi) \text{ implies } (\phi \text{ implies } \chi)$$

and: $(\psi \text{ is false} \,.\, \phi \text{ implies } \psi) \text{ implies } (\phi \text{ is false})$

make no more sense than:

$$(x \leqslant y \,.\, y \leqslant z) \leqslant (x \leqslant z),$$
$$(y \text{ is negative} \,.\, x \leqslant y) \leqslant (x \text{ is negative}).$$

In Whitehead and Russell's exposition and terminology the distinction between predicate and statement connective is blurred. The notation '— ⊃ —' is explained indiscriminately in the sense of the truth-functional conditional and in the sense of material implication. It is translated not only thus:

(14) If — then —

but also thus:

(15) If — is true then — is true,

(16) — is false or — is true,

(17) — implies —.

Similarly '≡' is explained both in the sense of the truth-functional biconditional and in the sense of material equivalence, and '∼' is explained both in the sense of denial and in the sense of falsehood. The authors even adopt 'implication' or 'material implication' as their regular terminology in connection with '⊃', and 'equivalence' in connection with '≡'. Actually, as we have seen, the blanks in (14) admit only statements whereas those in (15)–(17) admit only names of statements. In the construction of examples, indeed, grammatical sense leads Whitehead and Russell to fill the blanks of '— is true', '— is false', and '— implies —' with quotations rather than statements;[1] but the distinction is straightway obliterated in the discussion.

Once having noted the discrepancy between (14) and the other proposed translations of '— ⊃ —', one need not delay in making his choice. In all technical developments the expressions which Whitehead and Russell adjoin to the sign

[1] "'*x* wrote Waverly' is true" (vol. 1, p. 68); "'the author . . . was a poet' is false" (ibid.); "'Socrates is a man' implies 'Socrates is mortal'" (ibid., pp. 20, 138).

'⊃' have the form of statements rather than names. The mere fact of its iteration indeed, e.g. in the manner

$$— ⊃ (— ⊃ —),$$

is enough to determine the sign as a statement connective rather than a predicate about statements. In short, the versions (15)–(17) do not operate in Whitehead and Russell's work beyond the level of unfortunate exposition and nomenclature. The English idiom which '— ⊃ —' supplants in practice is not (15), (16), or (17), but (14). The case is similar with '≡' and '∼'.

On the topic of implication Whitehead and Russell have many critics, who rightly object that the trivial relation of material implication expressed in (16) is too weak to constitute a satisfactory version of (17). But it is seldom observed that this objection does not condemn the truth-functional conditional '— ⊃ —' as a version of 'if — then —'. Lewis, Smith, and others have undertaken systematic revision of '⊃' with a view to preserving just the properties appropriate to a satisfactory relation of implication; but what the resulting systems describe are actually modes of statement composition — revised conditionals of a non-truth-functional sort — rather than implication relations between statements.

If we were willing to reconstrue statements as names of some sort of entities, we might take implication as a relation between those entities rather than between the statements themselves; and correspondingly for equivalence, compatibility, etc. This procedure would dissolve the distinction between material implication and the truth-functional conditional, and likewise between other sorts of implication and other sorts of conditionals. 'Implies' would come to enjoy simultaneously the status of a binary predicate and the status of a binary statement connective. Expressions such as (11) would be legitimized; and so also would the iterated use of implication, characteristic of Lewis and Smith. For thus construing statements as names some slight support can be adduced, indeed, by appeal to substantive clauses. The statement 'All men are mortal' might be held to designate that abstract entity, whatever it is, which we ordinarily designate by the substantive 'that all men are mortal'. A deterring consideration, however, is the obscurity of these alleged entities. What are they like? and under what circumstances may the entities designated by two statements be said to be the same or different entities? Certain entities which are perhaps less obscure than these but no less abstract will indeed be countenanced at a later point (§ 22), viz. classes or properties, if only through ignorance of how to get on without them; but entities designated by statements are happily dispensable.[1] It thus seems well to adhere to the common-sense view that statements are not names at all, though they may contain names along with verbs and adverbs and the rest. A statement remains meaningful, but meaningful by virtue of its structure together with the meanings of the constituent names and other words; its meaningfulness does not consist in its being a name of something.

Conceding that 'implies' belongs between names of statements as in (9), rather than between statements, one might still urge that such a relation of implication produces a derivative mode of composition of the statements themselves — namely,

---

[1] See my "Ontological Remarks," "Logistical Approach," "Designation."

a mode which consists notationally of compounding the statements by means of 'implies' *and* the two pairs of quotation marks.[1] If implication is construed as going beyond questions of truth value, this derivative mode of statement composition will not be truth-functional. Implication thus construed would then seem, after all, to interfere with a policy of admitting none but truth-functional modes of statement composition. By the same argument, indeed, a purely morphological or phonetic relation such as containing or rhyming would interfere similarly. Actually, however, derivation of modes of statement composition from relations in the suggested fashion involves abuse of quotation. The statements buried in the quotations in (9) cannot be treated in turn as constituents of (9), for a quotation figures as a single irreducible word. Similar abuse of quotation was seen in § 4 to lead from ' 'Cicero' has six letters' to ' 'Tully' has six letters'.[2]

These latter remarks serve only to show that we can construe implication as going beyond questions of truth value without *thereby* committing ourselves to any form of conditional which goes beyond questions of truth value. The need for some such strong form of conditional might still be urged on other grounds. Certainly not all uses of the subjunctive conditional submit to the easy method of paraphrase illustrated in the case of Perth and America. When this fails we may look to other devices, e.g. Carnap's method of reduction sentences ("Testability," pp. 439–453); but if any really useful cases prove to resist all such methods of analysis, then we shall perhaps have to choose pragmatically between the usefulness of those cases and the convenience and clarity of the truth functional kind of statement composition. Mathematics itself gives rise to no such recalcitrant cases; and any which seem to arise beyond the bounds of mathematics should be critically regarded.[3]

§ 6. *Quasi-Quotation*

IN DISCUSSING the modes of statement composition we are having continually to talk of expressions. Quotation suffices for the mention of any specific expression, such as '$\vee$' or '$\equiv$' or 'Jones is away', but is not available when we want to speak generally of an unspecified expression of such and such kind. On such occasions use has been made of general locutions such as 'a conditional', 'the first component', etc.; and more difficult cases have been managed indirectly by introducing a blank '——' from time to

[1] Analogous reasoning appears in Huntington's "Note on a Recent Set", p. 11.
[2] See also Tarski, "Wahrheitsbegriff," § 1.
[3] For further discussion and references see Carnap, *Syntax*, §§ 67–71.

time. But the developments to follow call for a more elastic method of referring to unspecified expressions.

For the beginnings of such a method, the use of letters in algebra provides us with an adequate model. In algebra 'x', 'y', etc. are used as names of unspecified numbers; we may suppose them replaced by names of any specific numbers we choose. Analogously, Greek letters other than 'ϵ', 'ι', 'λ'[1] will now be used as names of unspecified *expressions;* we may suppose them replaced by names (e.g. quotations) of any specific expressions we choose.

A discussion of numbers may, for example, begin thus:

(1) Let x be a factor of y.

Throughout the discussion thus prefaced, we are to think of x and y as any specific numbers we like which satisfy the condition (1) — say the numbers 5 and 15, or 4 and 32. We are to think of the letters 'x' and 'y' as if they were names of the numbers 5 and 15, or names of the numbers 4 and 32, etc. We are to imagine the letters 'x' and 'y' replaced by the numerals (expressions) '5' and '15', or by '4' and '32', etc.

Similarly a discussion of expressions might begin thus:

(2) Let μ be part of ν.

Throughout the discussion thus prefaced, we are to think of μ and ν as any specific expressions we like which satisfy the condition (2) — say the expressions 'York' and 'New York', or '3' and '32'. We are to think of the letters 'μ' and 'ν' as if they were names of the expressions 'York' and 'New York', or names of the expressions '3' and '32', etc. We are to imagine the letters 'μ' and 'ν' replaced by the quotations ''York'' and ''New York'', or by ''3'' and ''32'', etc.

The reader is urged to compare the above short paragraph with the preceding one, word by word; also to review § 4. Roughly speaking, the letters 'x', 'y', etc. may be described as ambiguous numerals, ambiguous names of numbers, variables ambiguously designating numbers, or, in the usual technical phrase, variables taking numbers as their values. Correspondingly the letters

[1] These three letters are reserved for later purposes; cf. §§ 22, 35, 41.

'μ', 'ν', etc. may be roughly described as ambiguous quotations, ambiguous names of expressions, variables ambiguously designating expressions, or variables taking expressions as their values. This does not mean simply that 'μ' and 'ν' take the place of expressions, or are replaceable by expressions, for this is true of 'x' and 'y' as well. Rather, the letters 'μ' and 'ν' take the place of quotations or other names of expressions, just as 'x' and 'y' take the place of numerals or other names of numbers.

Occasionally Greek letters will be used with accents or subscripts attached: 'μ'', 'μ''', 'μ_1', 'μ_2', 'μ_n', etc. Such variants may be regarded simply as so many further Greek letters. Three Greek letters, 'ϕ', 'ψ', and 'χ', together with their accented and subscripted variants, will be limited in their use to those cases where the expression designated is intended to be a statement. They serve as names of unspecified statements, and are replaceable by statement quotations or other names of specific statements.[1]

There is need also of a convenient way of speaking of specific contexts of unspecified expressions: speaking, e.g., of the result of enclosing the unspecified expression μ in parentheses, or the result of joining the unspecified statements ϕ and ψ in that order by the sign '$\equiv$'. Note that quotation is not available here. The quotations:

$$\text{'}(\mu)\text{'}, \qquad \text{'}\phi \equiv \psi\text{'}$$

designate only the specific expressions therein depicted, containing the specific Greek letters 'μ', 'ϕ', and 'ψ'. Reference to the intended contexts of the unspecified expressions μ, ϕ, and ψ will be accomplished by a new notation of *corners*, thus:

(3) $\ulcorner(\mu)\urcorner$, $\ulcorner\phi \equiv \psi\urcorner$.

Because of the close relationship which it bears to quotation, this device may be called *quasi-quotation*. It amounts to quoting the constant contextual backgrounds, '( )' and ' ≡ ', and imagining the unspecified expressions μ, ϕ, and ψ written in the blanks. If in particular we take the expression 'Jones' as μ,

[1] The three letters have indeed already appeared in this use in § 5, where the sense intended was apparent.

'Jones is away' as ϕ, and 'Smith is ill' as ψ, then $\ulcorner(\mu)\urcorner$ is '(Jones)' and $\ulcorner\phi \equiv \psi\urcorner$ is 'Jones is away $\equiv$ Smith is ill'.

The quasi-quotations (3) are synonymous with the following verbal descriptions:

The result of writing '(' and then μ and then ')',
The result of writing ϕ and then '$\equiv$' and then ψ;

or, equivalently:

The result of putting μ in the blank of '(   )',
The result of putting ϕ and ψ in the respective blanks of ' $\equiv$ ';

or, equivalently:

The result of putting μ for 'μ' in '(μ)',
The result of putting ϕ for 'ϕ' and ψ for 'ψ' in '$\phi \equiv \psi$'.

We may translate any quasi-quotation:

$$\ulcorner\text{——}\urcorner$$

into words in corresponding fashion:

The result of putting μ for 'μ', ν for 'ν', . . . , ϕ for 'ϕ', ψ for 'ψ', . . . in '——'.

Described in another way: a quasi-quotation designates that (unspecified) expression which is obtained from the contents of the corners by replacing the Greek letters (other than 'ϵ', 'ι', 'λ') by the (unspecified) expressions which they designate.

When a Greek letter stands alone in corners, quasi-quotation is vacuous: $\ulcorner\mu\urcorner$ is μ. For, by the foregoing general description, $\ulcorner\mu\urcorner$ is the result of putting μ for 'μ' in 'μ'; $\ulcorner\mu\urcorner$ is what the letter 'μ' becomes when that letter itself is replaced by the (unspecified) expression μ; in other words, $\ulcorner\mu\urcorner$ is simply that expression μ.

Quasi-quotation would have been convenient at earlier points, but was withheld for fear of obscuring fundamentals with excess machinery. Now, however, it may be a useful exercise to recapitulate some sample points from §§ 1–5 in terms of this device. A conjunction $\ulcorner\phi \,.\, \psi\urcorner$ is true just in case ϕ and ψ are both true, and an alternation $\ulcorner\phi \vee \psi\urcorner$ is false just in case ϕ and ψ are both false. A conditional $\ulcorner\phi \supset \psi\urcorner$ is true if ϕ is false or ψ true, and false if ϕ is true and ψ false. A biconditional $\ulcorner\phi \equiv \psi\urcorner$ is true just in case ϕ and ψ are alike in truth value. A denial $\ulcorner{\sim}\,\phi\urcorner$ is true just in case ϕ is false. ϕ logically implies ψ or is logically equivalent to ψ according as $\ulcorner\phi \supset \psi\urcorner$ or $\ulcorner\phi \equiv \psi\urcorner$ is logically true, and ϕ materially implies ψ or is materially equivalent to ψ according as $\ulcorner\phi \supset \psi\urcorner$ or $\ulcorner\phi \equiv \psi\urcorner$ is true.

§ 7. *Parentheses and Dots*

PARENTHESES, taken for granted thus far as an auxiliary notation, are most simply construed as forming integral parts of the binary connectives. The notations of conjunction, alternation, the conditional, and the biconditional will no longer be regarded as $\ulcorner\phi \,.\, \psi\urcorner$, $\ulcorner\phi \vee \psi\urcorner$, $\ulcorner\phi \supset \psi\urcorner$, and $\ulcorner\phi \equiv \psi\urcorner$, but as $\ulcorner(\phi \,.\, \psi)\urcorner$, $\ulcorner(\phi \vee \psi)\urcorner$, $\ulcorner(\phi \supset \psi)\urcorner$, and $\ulcorner(\phi \equiv \psi)\urcorner$. But the notation of denial remains simply $\ulcorner{\sim}\,\phi\urcorner$. This formulation yields just the usage of parentheses which we have hitherto followed, except in one respect: a conjunction, alternation, conditional, or biconditional comes now to bear an outside pair of parentheses even when it stands apart from any further symbolic context. Such outside parentheses have hitherto been omitted.

When the binary modes of composition are thus construed, no auxiliary technique of grouping is needed. The syntactical simplicity thus gained proves useful in certain abstract studies (e.g. § 10; also later chapters). In applications, however, such simplicity is less important than facility of reading; and excess of parentheses is a hindrance, for we have to count them off in pairs to know which ones are mates. It is hence convenient in practice to omit the outside parentheses as hitherto, and furthermore to suppress most of the remaining parentheses in favor of a more graphic notation of *dots*.

Parentheses mark the outer limits of a binary compound; dots determine those same limits less directly. Roughly speaking, a group of dots placed beside '$\vee$', '$\supset$', or '$\equiv$' indicates that the component on that side has its other end at the nearest larger group of dots, if any, and otherwise at the limit of the whole symbolic context. E.g., since '≡' in

$$\ulcorner \phi \vee : \psi \supset \chi \,.\!\equiv \phi \,.:\supset.\, \psi \supset \phi \urcorner$$

has just one dot prefixed, the first component of the biconditional begins at the last previous group of two or more dots — hence, as it happens, at the group of two. Again, since '≡' has no dots suffixed, the second component of the biconditional stops as soon as it meets any dots at all — hence at the group of three. The components of the biconditional are thus $\ulcorner \psi \supset \chi \urcorner$ and ϕ. The alternation has $\ulcorner \psi \supset \chi \,.\!\equiv \phi \urcorner$ as second component, for this runs from the pair of dots to the first larger group. The conditional corresponding to the second occurrence of '⊃' has $\ulcorner \phi \vee : \psi \supset \chi \,.\!\equiv \phi \urcorner$ in its entirety as antecedent, since the group of dots prefixed to the second occurrence of '⊃' exceeds any previous group. Similarly that conditional has $\ulcorner \psi \supset \phi \urcorner$ as consequent. The whole would appear in terms of parentheses as

$$\ulcorner((\phi \vee ((\psi \supset \chi) \equiv \phi)) \supset (\psi \supset \phi))\urcorner.$$

Dots serve in this fashion to reinforce the connectives '∨', '⊃', and '≡'. In the case of conjunction, already represented by a dot, such reinforcement will be accomplished simply by using a group of dots instead of the one; thus the conjunction of ϕ with $\ulcorner \psi \vee \chi \,.\!\supset \phi \urcorner$ appears as $\ulcorner \phi : \psi \vee \chi \,.\!\supset \phi \urcorner$. Use of dots for conjunction is distinguishable from the other uses by the absence of any adjacent connective '∨', '⊃', or '≡'. A group of *n* dots standing as a conjunction sign will be regarded as indicating a smaller break than that indicated by a group of *n* dots alongside '∨', '⊃', or '≡', though of course a greater break than that indicated by *n*–1 dots. Thus, in

$$\ulcorner \phi \,.\, \psi \,.\!\supset \chi : \psi \vee \phi :\equiv \chi \urcorner$$

the conjunction indicated by the single dot has ϕ and ψ as components; the conjunction indicated by the pair of dots has $\ulcorner \phi \,.\, \psi \,.\!\supset \chi \urcorner$ and $\ulcorner \psi \vee \phi \urcorner$ as components; the conditional has $\ulcorner \phi \,.\, \psi \urcorner$ and χ as components; and the biconditional has $\ulcorner \phi \,.\, \psi \,.\!\supset \chi : \psi \vee \phi \urcorner$ and χ. The whole would appear in terms of parentheses as:

$$\ulcorner((((\phi \,.\, \psi) \supset \chi) \,.\, (\psi \vee \phi)) \equiv \chi)\urcorner.$$

A place occupied by dots will be called a *joint;* also, if '∨', '⊃', or '≡' lacks dots on one side, the narrow blank place on that side will be called a joint. One joint will be said to be *looser* than another if it contains more dots than the other, or if it contains just as many dots as the other and is adjacent to '∨', '⊃', or '≡' while the other is not. Now the way to read dots can be summed up thus, subject to revisions in the immediate sequel: (I) *The second component of a conjunction, alternation, conditional, or biconditional ends at the first subsequent joint, if any, which is looser than the one at its beginning; otherwise it runs to the end of the whole symbolic context. The first component begins at the last previous joint, if any, which is looser than the one at its end; otherwise it begins at the beginning of the whole context.*

Allowance must be made now for the survival of parentheses in the midst of the dot notation, for it is convenient to be able to mix the two notations at will. After a denial sign, indeed, a compound will always retain its outer parentheses, even in dot-strewn contexts. E.g., we write $\ulcorner\sim(\phi \supset. \psi \vee \chi)\urcorner$; never $\ulcorner\sim : \phi \supset. \psi \vee \chi\urcorner$. Dots may or may not be used to reinforce a binary connective, but they will never be used to reinforce '∼'. Now the presence of parentheses obliges us to qualify (I) so that joints sealed up in parentheses will not be made to affect the grouping of outside text. E.g., the pair of dots in

$$\ulcorner\phi \equiv. \psi \vee \sim(\phi \supset. \psi \,.\, \chi :\supset \chi)\urcorner$$

must not be made to terminate the second component of the biconditional; that component is supposed to run to the end of the whole expression. Similarly for the second component of the alternation. (I) can be suitably amended by adding the obvious requirement that a component must contain no dangling parenthesis; it must contain equal numbers of left- and right-hand parentheses.

The presence of parentheses also has another effect: if a component of a binary compound has its one end inside some parenthesized expression, it may be bounded at its other end by the limit of that parenthesized expression rather than by a joint. In the case of a second component, such an ending will be forced by any right-hand parenthesis whose mate did not intervene — hence by any right-hand parenthesis which may turn up, so long as the component already contains equal numbers of left- and right-hand parentheses. The case of a first component is analogous.

In view of these considerations, (I) gives way to the following final formulation. (II) *The components of a conjunction, alternation, conditional, or biconditional are as short as these conditions allow: each component contains just as many left- as right-hand parentheses; the second component ends either at the end of the whole symbolic context or else just before a right-hand parenthesis or a joint looser than the one at its beginning; and the first component begins either at the beginning of the whole context or else just after a left-hand parenthesis or a joint looser than the one at its end.*

Note that (II) compels a denial sign to apply, as usual, to the shortest statement following it. It compels interpretation of $\ulcorner \sim \phi \vee \psi \urcorner$ as the alternation of $\ulcorner \sim \phi \urcorner$ and ψ, interpretation of $\ulcorner \sim (\phi \supset . \psi \vee \chi) \equiv \chi \urcorner$ as the biconditional of $\ulcorner \sim (\phi \supset . \psi \vee \chi) \urcorner$ and χ, and so on. For, according to (II), the first component of the alternation or biconditional must begin just after a joint or a left-hand parenthesis or at the beginning of the whole context; it cannot begin just after '$\sim$'.

As appears from the examples, the use of dots proceeds without regard to the complexity of the unspecified statements denoted by the Greek letters. We write '$\ulcorner \phi \supset . \psi \vee \chi \urcorner$', not '$\ulcorner \phi \mathbin{.:\supset .} \psi \vee \chi \urcorner$', even though ϕ be thought of in particular as a statement complex enough to contain say a group of two dots in turn. This practice can be justified by thinking of the statements indicated by Greek letters as constructed always with parentheses rather than dots, to the extent at least of bearing their outermost parentheses.

The parenthesis notation formulated at the beginning of the section is retained, for the sake of theoretical developments, as the "official" notation; dots enter only as shorthand. This unofficial use of the dot notation has its justification in the fact that the parenthesis notation can easily be restored whenever we like. To accomplish this in any given case we have only to put parentheses around each expression μ which satisfies the following conditions

(i) and (ii), and then drop all dots except single dots for conjunction.

(i) μ is a component of a conjunction, alternation, conditional, or biconditional, according to (II), or else it exhausts the whole symbolic context;

(ii) μ contains a binary connective which is unparenthesized in μ; i.e., one which is preceded in μ by no more left- than right-hand parentheses.

Note now, in preparation for one further notational departure, that the grouping in an iterated conjunction is always immaterial; $\ulcorner\phi \,.\, \psi : \chi\urcorner$ and $\ulcorner\phi : \psi \,.\, \chi\urcorner$, for example, are equivalent. Any iterated conjunction of ϕ_1, ϕ_2, . . . , and ϕ_n, regardless of the grouping, is true in just one case: the case where ϕ_1, ϕ_2, . . . , and ϕ_n are all true. For, a conjunction is true just in case both its immediate components are true; either component, if in turn a conjunction, is true just in case both its components are true; and so on. The grouping in an iterated alternation is likewise immaterial; every iterated alternation of ϕ_1, ϕ_2, . . . , and ϕ_n is true in the same cases — all cases except the one where ϕ_1, ϕ_2 . . . , and ϕ_n are all false. For, an alternation is false just in case both components are false; either component, if in turn an alternation, is false just in case both its components are false; and so on. Because of these circumstances it is sometimes convenient to use uniform connectives throughout the iteration — thus writing

$$\ulcorner\phi_1 \,.\, \phi_2 \,.\, \phi_3 \,.\, \ldots \,.\, \phi_n\urcorner, \quad \ulcorner\phi_1 \vee \phi_2 \vee \phi_3 \vee \ldots \vee \phi_n\urcorner,$$

also

$$\ulcorner\phi \supset \phi' \;.\!\vee\!.\; \psi \supset \psi' \;.\!\vee\!.\; \chi \supset \chi'\urcorner,$$

etc., without further indication of grouping. One or another arbitrary indication of grouping must be restored, however, before analyzing such a compound along the lines prescribed in the foregoing paragraphs.

The indication of grouping by means of dots dates from Peano. In details the usage varies from author to author. Curry's usage ("Dots as Brackets") is neater than that set forth above, but not as easy to read. Since parentheses are so much simpler in theory, I have preferred to retain them as the "official" notation and then to frame dot conventions with an eye only to perspicuity and approximate conformity to prevalent usage. The syntactical simplicity of the "official" notation could have been further enhanced, indeed, by writing binary compounds in the fashion ⌜$\vee \phi \psi$⌝, ⌜$\supset \phi \psi$⌝, etc. instead of ⌜$(\phi \vee \psi)$⌝, ⌜$(\phi \supset \psi)$⌝, etc.; under this rearrangement, as Lukasiewicz has remarked ("Untersuchungen," p. 31, reprint p. 3), grouping is unambiguously discernible without help of parentheses or other punctuation.

§ 8. *Reduction to Three Primitives*

THE TRUTH tables of singulary and binary modes of composition have the forms depicted respectively in Tables 1 and 2. Each way of putting 'T's and 'F's for the question marks gives us a different

Component	*Compound*
T	?
F	?

TABLE 1

1st component	*2d component*	*Compound*
T	T	?
F	T	?
T	F	?
F	F	?

TABLE 2

mode of composition. If we proposed to take up one or another *ternary* mode of composition, or mode of combining statements three at a time, our table would have a column for each of the three components and a column for the compound. There would be eight rows, exhausting the possible cases of truth and falsehood on the part of three components (cf. § 3). In general, the truth table for an n-ary mode of composition appears as in Table 3; the successive rows exhaust the 2^n possible cases of truth and false-

	1st component	*2d component*	*3d component*	. . .	*nth component*	*Compound*
	ϕ_1	ϕ_2	ϕ_3	. . .	ϕ_n	
Case 1	T	T	T	. . .	T	?
Case 2	F	T	T	. . .	T	?
Case 3	T	F	T	. . .	T	?
Case 4	F	F	T	. . .	T	?
Case 5	T	T	F	. . .	T	?
.	.	.	.	. . .	.	.
.	.	.	.	. . .	.	.
.	.	.	.	. . .	.	.
Case 2^n	F	F	F	. . .	F	?

TABLE 3

hood on the part of n components $\phi_1, \phi_2, \ldots, \phi_n$ (cf. § 3), and the last column specifies the truth value of the compound for the successive cases. Each of the 2^{2^n} ways of replacing the question marks by 'T' or 'F' determines a different n-ary mode of composition.

Of the 2^{2^1} (= 4) singulary modes thus determined, we have considered only one: denial. Of the 2^{2^2} (= 16) binary modes we have considered only four. The 2^{2^3} (= 256) ternary modes, the 2^{2^4} (= 65,536) quaternary ones, and so on, all remain untouched. Actually, however, the few modes already considered are all we ever need; for it will now be shown that *every mode of composition describable by a truth table is translatable into terms of denial, conjunction, and alternation.*[1] In the next section a still more striking reduction will be made.

Case 1 of Table 3 is the case where $\phi_1, \phi_2, \phi_3, \ldots$, and ϕ_n are all true; in other words, it is the case where

$$\ulcorner \phi_1 \cdot \phi_2 \cdot \phi_3 \cdot \ldots \cdot \phi_n \urcorner$$

is true (cf. § 7). Let us call this conjunction χ_1. Again, Case 2 is the case where ϕ_1 is false and $\phi_2, \phi_3, \ldots$, and ϕ_n are true; in other words, it is the case where $\ulcorner \sim \phi_1 \urcorner, \phi_2, \phi_3, \ldots$, and ϕ_n are all true; in other words, it is the case where

$$\ulcorner \sim \phi_1 \cdot \phi_2 \cdot \phi_3 \cdot \ldots \cdot \phi_n \urcorner$$

is true. Let us call this conjunction χ_2. Similarly Case 3 is the case where

$$\ulcorner \phi_1 \cdot \sim \phi_2 \cdot \phi_3 \cdot \ldots \cdot \phi_n \urcorner$$

is true; let us call this χ_3. Case 4 is the case where

$$\ulcorner \sim \phi_1 \cdot \sim \phi_2 \cdot \phi_3 \cdot \ldots \cdot \phi_n \urcorner$$

is true; let us call this χ_4. In this fashion Case i, for each i from 1 to 2^n, is allotted a statement χ_i which is true in just that case and is composed of $\phi_1, \phi_2, \phi_3, \ldots, \phi_n$ by means of nothing more than conjunction and denial. (In the trivial situation where n is 1, conjunction of course drops out; χ_1 is ϕ_1 and χ_2 is $\ulcorner \sim \phi_1 \urcorner$.)

Now every truth table is formed from Table 3 (for appropriate n) by assigning 'T' and 'F' to the successive cases, in place of the

[1] This was first shown by Post.

question marks. But if it is formed by assigning 'T' to just one case, say Case i, and 'F' to the rest, then it is the table for χ_i; for χ_i is true in Case i and false in the others. If on the other hand it is formed by assigning 'T' to many cases, say Cases i, j, k, . . . , then it is the table for

$$\ulcorner \chi_i \vee \chi_j \vee \chi_k \vee \ldots \urcorner;$$

for this alternation is true wherever any of χ_i, χ_j, χ_k, . . . is true (cf. § 7), hence in Cases i, j, k, ... If finally the table is formed by assigning 'F' to all cases, then it is the table for any trivial conjunction such as

$$\ulcorner \phi_1 \,.\, \sim\phi_1 \,.\, \phi_2 \,.\, \ldots \,.\, \phi_n \urcorner;$$

for there is no case where ϕ_1 and $\ulcorner \sim \phi_1 \urcorner$ are both true, hence no case where this conjunction is true. Now since

$$\chi_i, \qquad \ulcorner \chi_i \vee \chi_j \vee \chi_k \vee \ldots \urcorner, \qquad \ulcorner \phi_1 \,.\, \sim\phi_1 \,.\, \phi_2 \,.\, \ldots \,.\, \phi_n \urcorner$$

are all compounded of ϕ_1, ϕ_2, . . . , ϕ_n by means only of denial, conjunction, and alternation, we see that every truth table describes a mode of composition involving nothing beyond those three. All truth-functional modes of composition are thus translatable into terms of denial, conjunction, and alternation. We can take these as the sole basic or *primitive* truth-functional modes of composition, and construe all others merely as results of repeated application of the primitive modes.

Given the truth table of any mode of composition, the method of translating that mode into terms of the three primitives is apparent from the foregoing proof; we have only to form the conjunction χ_i, or the alternation of the conjunctions χ_i, χ_j, χ_k, . . . , corresponding to the case or cases marked with 'T' in the table. If there is no 'T', we resort to the trivial device of forming a conjunction containing ϕ_1 and $\ulcorner \sim \phi_1 \urcorner$. As examples, consider the conditional and the biconditional. Here n is 2; χ_1, χ_2, χ_3, and χ_4 become respectively $\ulcorner \phi_1 \,.\, \phi_2 \urcorner$, $\ulcorner \sim \phi_1 \,.\, \phi_2 \urcorner$, $\ulcorner \phi_1 \,.\, \sim \phi_2 \urcorner$, and $\ulcorner \sim \phi_1 \,.\, \sim \phi_2 \urcorner$, or better $\ulcorner \phi \,.\, \psi \urcorner$, $\ulcorner \sim \phi \,.\, \psi \urcorner$, $\ulcorner \phi \,.\, \sim \psi \urcorner$, and $\ulcorner \sim \phi \,.\, \sim \psi \urcorner$. Now the conditional has in its table three 'T's, marking Cases 1, 2, and 4; hence $\ulcorner \phi \supset \psi \urcorner$ is expressible as $\ulcorner \chi_1 \vee \chi_2 \vee \chi_4 \urcorner$, i.e.,

$$\ulcorner \phi \,.\, \psi \,.\!\vee\!.\, \sim \phi \,.\, \psi \,.\!\vee\!.\, \sim \phi \,.\, \sim \psi \urcorner.$$

Again, the biconditional has in its table two 'T's, marking Cases 1 and 4; hence ⌜$\phi \equiv \psi$⌝ is expressible as ⌜$\chi_1 \vee \chi_4$⌝, i.e.,

$$⌜\phi \,.\, \psi \,.\!\vee\!.\, \sim \phi \,.\, \sim \psi⌝.$$

The method always gives us a translation into terms of denial, conjunction, and alternation, but it does not always happen to give the shortest. For example, ⌜$\phi \supset \psi$⌝ admits also of the shorter translations ⌜$\sim \phi \vee \psi$⌝ and ⌜$\sim (\phi \,.\, \sim \psi)$⌝, as observed in § 2.

It is trivial but perhaps instructive to apply the method also to the singulary modes of composition. Here χ_1 and χ_2 are simply ϕ_1 and ⌜$\sim \phi_1$⌝; or, dropping the subscript, ϕ and ⌜$\sim \phi$⌝. Now one of the four singulary modes assigns 'T' to Cases 1 and 2 alike; it is thus expressible as ⌜$\chi_1 \vee \chi_2$⌝, or ⌜$\phi \vee \sim \phi$⌝. Another assigns 'T' to Case 1 alone; hence it is expressible as χ_1, or simply ϕ. A third assigns 'T' to Case 2 alone; hence it is expressible as χ_2, or ⌜$\sim \phi$⌝. The remaining singulary mode assigns 'F' to both cases; hence, by the device usual in the absence of 'T's, it is expressible as ⌜$\phi \,.\, \sim \phi$⌝.

§9. *Reduction to One Primitive*

WE NOW turn to the binary mode of statement composition embodied in the connective 'neither-nor'. This will be called *joint denial*, and rendered symbolically in the notation ⌜$(\phi \downarrow \psi)$⌝. A joint denial, e.g. 'Neither is Jones away nor is Smith ill', symbolically:

(Jones is away ↓ Smith is ill),

is true just in case its components are both false. This gives us the following table:

ϕ	ψ	⌜$(\phi \downarrow \psi)$⌝
T	T	F
F	T	F
T	F	F
F	F	T

Since ⌜$(\phi \downarrow \psi)$⌝ is true just in the case where ⌜$\sim \phi$⌝ and ⌜$\sim \psi$⌝ are both true, it is translatable into terms of denial and conjunction as $\ulcorner(\sim\phi \,.\, \sim\psi)\urcorner$. Again, since it is true just in the case where $\ulcorner(\phi \vee \psi)\urcorner$ is false, it is translatable into terms of denial and alternation as $\ulcorner\sim(\phi \vee \psi)\urcorner$. The fact that $\ulcorner(\phi \downarrow \psi)\urcorner$ denies $\ulcorner(\phi \vee \psi)\urcorner$ is reflected in ordinary language, indeed, by the cancellatory 'n' which turns 'either-or' into 'neither-nor'. The vertical mark in '$\downarrow$' may be thought of as having the same cancellatory effect upon the sign '$\vee$'; it is like the vertical mark in the inequality sign '$\neq$' of arithmetic.

With denial, conjunction, and alternation at hand, a special notation for joint denial is superfluous. But it is introduced for the sake of translation in the opposite direction; for it turns out that denial, conjunction, and alternation can all be expressed in terms of joint denial alone — $\ulcorner\sim\phi\urcorner$ as $\ulcorner(\phi \downarrow \phi)\urcorner$, $\ulcorner(\phi \,.\, \psi)\urcorner$ as $\ulcorner((\phi \downarrow \phi) \downarrow (\psi \downarrow \psi))\urcorner$, and $\ulcorner(\phi \vee \psi)\urcorner$ as $\ulcorner((\phi \downarrow \psi) \downarrow (\phi \downarrow \psi))\urcorner$. This is seen by constructing tables according to the method of § 3:

$\ulcorner(\phi \downarrow \phi)\urcorner$
T F T
F T F

$\ulcorner((\phi \downarrow \phi)$	$\downarrow$	$(\psi \downarrow \psi))\urcorner$
T F T	T	T F T
F T F	F	T F T
T F T	F	F T F
F T F	F	F T F

$\ulcorner((\phi \downarrow \psi)$	$\downarrow$	$(\phi \downarrow \psi))\urcorner$
T F T	T	T F T
F F T	T	F F T
T F F	T	T F F
F T F	F	F T F

Comparison of the middle columns of these tables with the tables of denial, conjunction, and alternation shows that $\ulcorner(\phi \downarrow \phi)\urcorner$ is true in just the case where $\ulcorner\sim\phi\urcorner$ is true, that $\ulcorner((\phi \downarrow \phi) \downarrow (\psi \downarrow \psi))\urcorner$ is true in just the case where $\ulcorner(\phi \,.\, \psi)\urcorner$ is true, and that $\ulcorner((\phi \downarrow \psi) \downarrow (\phi \downarrow \psi))\urcorner$ is true in just the cases where $\ulcorner(\phi \vee \psi)\urcorner$ is true.

Since all truth-functional modes of composition are translatable into terms of denial, conjunction, and alternation (cf. § 8), the translation in turn of these three into terms of joint denial shows that all the truth-functional modes of composition are translatable into terms of joint denial alone. Iterated application of this one mode of composition suffices for expressing anything that could be expressed by any of the others. We can take joint denial as the sole primitive truth-functional mode of composition, and construe all the others merely as results of applying this primitive mode in iteration.

In theoretical developments (e.g. §§ 10, 18) it is convenient to be able to treat '↓' or 'neither-nor' thus as the sole truth-functional connective. But in applications it is convenient to have the signs '∼', '.', 'v', '⊃', and '≡' as well, for the brevity and clarity which they afford. Adoption of '↓' as sole connective demands continual use of compounds so cumbersome, indeed, that conventions of shorthand reducing them to manageable length would become a practical necessity.

Between the theoretical advantage of a single connective and the practical advantage of a multiplicity, this idea of shorthand effects a complete reconciliation. By way of basic or primitive notation we may adopt the form of notation $\ulcorner(\phi \downarrow \psi)\urcorner$; but we can still introduce the convenient forms of notation $\ulcorner\sim \phi\urcorner$, $\ulcorner(\phi \,.\, \psi)\urcorner$, $\ulcorner(\phi \vee \psi)\urcorner$, etc., simply as shorthand abbreviations for the appropriate complexes involving '↓'. Thus, where ϕ is any statement, $\ulcorner\sim \phi\urcorner$ will be construed as an abbreviation of $\ulcorner(\phi \downarrow \phi)\urcorner$. This convention of abbreviation will be referred to as D1.

D1. $\ulcorner\sim \phi\urcorner$ *for* $\ulcorner(\phi \downarrow \phi)\urcorner$.

Similarly, where ϕ and ψ are any statements, $\ulcorner(\phi \,.\, \psi)\urcorner$ and $\ulcorner(\phi \vee \psi)\urcorner$ will be regarded as abbreviations of $\ulcorner((\phi \downarrow \phi) \downarrow (\psi \downarrow \psi))\urcorner$ and $\ulcorner((\phi \downarrow \psi) \downarrow (\phi \downarrow \psi))\urcorner$. More briefly, we may explain $\ulcorner(\phi \,.\, \psi)\urcorner$ and $\ulcorner(\phi \vee \psi)\urcorner$ as abbreviations of $\ulcorner(\sim \phi \downarrow \sim \psi)\urcorner$ and $\ulcorner\sim (\phi \downarrow \psi)\urcorner$; for, D1 explains $\ulcorner(\sim \phi \downarrow \sim \psi)\urcorner$ and $\ulcorner\sim (\phi \downarrow \psi)\urcorner$ as abbreviations in turn of $\ulcorner((\phi \downarrow \phi) \downarrow (\psi \downarrow \psi))\urcorner$ and $\ulcorner((\phi \downarrow \psi) \downarrow (\phi \downarrow \psi))\urcorner$. Our next two conventions of abbreviation are hence as follows:

D2. $\ulcorner(\phi \,.\, \psi)\urcorner$ *for* $\ulcorner(\sim \phi \downarrow \sim \psi)\urcorner$,

D3. $\ulcorner(\phi \vee \psi)\urcorner$ *for* $\ulcorner\sim (\phi \downarrow \psi)\urcorner$.

Such conventions of abbreviation are called formal definitions; hence the use of 'D' in numbering them. To define a sign formally is to adopt it as shorthand for some form of notation already at hand. If the sign has a preconceived meaning, as in the present instances, and the definition suits that meaning, then the definition amounts to an elimination: it shows that the sign is dispensable in favor of those occurring in the definition. To define a sign is to show how to avoid it.

The relationships utilized for the definitions D1–3 were shown by truth tables to follow from the meanings hitherto assigned to '$\sim$', '.', 'v', and '$\downarrow$'. But they are readily checked also on the linguistic level. 'Jones is not away' has a prolix but obviously accurate equivalent in 'Neither is Jones away nor is Jones away'; so has 'Jones is away and Smith is ill' in 'Neither is Jones not away nor is Smith not ill'; and 'Jones is away or Smith is ill' amounts obviously to the denial of 'Neither is Jones away nor is Smith ill'.

Definitions of $\ulcorner(\phi \supset \psi)\urcorner$ and $\ulcorner(\phi \equiv \psi)\urcorner$ are apparent from earlier observations (§§ 2, 3):

D4. $\ulcorner(\phi \supset \psi)\urcorner$ *for* $\ulcorner(\sim \phi \vee \psi)\urcorner$,
D5. $\ulcorner(\phi \equiv \psi)\urcorner$ *for* $\ulcorner((\phi \supset \psi) \,.\, (\psi \supset \phi))\urcorner$.

$\ulcorner(\phi \supset \psi)\urcorner$ and $\ulcorner(\phi \equiv \psi)\urcorner$ are defined ultimately in terms of joint denial, by virtue of these two definitions together with their predecessors. D4, for example, defines $\ulcorner(\phi \supset \psi)\urcorner$ as $\ulcorner\sim (\sim \phi \downarrow \psi)\urcorner$ in view of D3, and hence as

$$\ulcorner(((\phi \downarrow \phi) \downarrow \psi) \downarrow ((\phi \downarrow \phi) \downarrow \psi))\urcorner$$

in view of D1.

As stated earlier (§ 7), indications of grouping in an iterated conjunction or alternation will often be suppressed in view of the fact that differences in this respect are immaterial to truth value. Let us now give this procedure the status of definitional abbreviation, by construing the ungrouped conjunction or alternation as shorthand for a conjunction or alternation grouped in some one arbitrary fashion. The following definitions will serve.

D6. $\ulcorner(\phi_1 \,.\, \phi_2 \,.\, \phi_3)\urcorner$ *for* $\ulcorner((\phi_1 \,.\, \phi_2) \,.\, \phi_3)\urcorner$,
$\ulcorner(\phi_1 \,.\, \phi_2 \,.\, \phi_3 \,.\, \phi_4)\urcorner$ *for* $\ulcorner(((\phi_1 \,.\, \phi_2) \,.\, \phi_3) \,.\, \phi_4)\urcorner$, *etc.*
D7. $\ulcorner(\phi_1 \vee \phi_2 \vee \phi_3)\urcorner$ *for* $\ulcorner((\phi_1 \vee \phi_2) \vee \phi_3)\urcorner$,
$\ulcorner(\phi_1 \vee \phi_2 \vee \phi_3 \vee \phi_4)\urcorner$ *for* $\ulcorner(((\phi_1 \vee \phi_2) \vee \phi_3) \vee \phi_4)\urcorner$, *etc.*

Over the shorthand introduced by D1–7, further shorthand is superimposed by the technique of using dots instead of parentheses (§ 7). That technique will not be formulated in a numbered definition, but it has the same status.

Joint denial is not the only truth-functional mode of composition which is adequate to definition of all the rest. The same is true of *alternative denial*, which is described by the following truth table:

ϕ	ψ	$\ulcorner(\phi \mid \psi)\urcorner$
T	T	F
F	T	T
T	F	T
F	F	T

Just as $\ulcorner(\phi \downarrow \psi)\urcorner$ is equivalent to $\ulcorner(\sim \phi \,.\, \sim \psi)\urcorner$ and to $\ulcorner\sim (\phi \vee \psi)\urcorner$, so $\ulcorner(\phi \mid \psi)\urcorner$ is equivalent to $\ulcorner(\sim \phi \vee \sim \psi)\urcorner$ and to $\ulcorner\sim (\phi \,.\, \psi)\urcorner$. Now denial, conjunction, and alternation can all be translated into terms of this one mode of composition, as follows:[1] $\ulcorner\sim \phi\urcorner$ as $\ulcorner(\phi \mid \phi)\urcorner$, $\ulcorner(\phi \,.\, \psi)\urcorner$ as $\ulcorner((\phi \mid \psi) \mid (\phi \mid \psi))\urcorner$, and $\ulcorner(\phi \vee \psi)\urcorner$ as $\ulcorner((\phi \mid \phi) \mid (\psi \mid \psi))\urcorner$. This can be checked by tables just as was done in the case of joint denial. Instead of adopting joint denial as our primitive mode of composition, then, we might just as well have chosen alternative denial. D1 would then give way to:

D1′. $\ulcorner\sim \phi\urcorner$ *for* $\ulcorner(\phi \mid \phi)\urcorner$.

Similarly $\ulcorner(\phi \,.\, \psi)\urcorner$ and $\ulcorner(\phi \vee \psi)\urcorner$ would be introduced as abbreviations of $\ulcorner((\phi \mid \psi) \mid (\phi \mid \psi))\urcorner$ and $\ulcorner((\phi \mid \phi) \mid (\psi \mid \psi))\urcorner$. Simplified by application of the abbreviation D1′, these definitions appear as follows:

D2′. $\ulcorner(\phi \,.\, \psi)\urcorner$ *for* $\ulcorner\sim (\phi \mid \psi)\urcorner$,
D3′. $\ulcorner(\phi \vee \psi)\urcorner$ *for* $\ulcorner(\sim \phi \mid \sim \psi)\urcorner$.

Choice between this course and the previous one is a matter of indifference. But there is no third course affording equal economy; it can be shown that the truth-functional modes of composition are not all reducible to any binary one other than joint and alternative denial.[2]

[1] The definability of denial, conjunction, and alternation in terms of joint denial was first pointed out by Sheffer in 1913; and similarly for alternative denial. The adequacy of joint denial was known to Peirce in 1880, and both facts were known to him in 1902; but his notes on the subject remained unpublished until 1933 (4.12, 4.264).

[2] Cf. Żyliński.

§ 10. *Tautology*

THUS FAR in our study of the truth-functional modes of composition we have devoted our attention to concepts and notations. We turn now to principles, logical truths. The study of the logical truths at this level proves to be a trivial one, and will not detain us long.

Statements which are true by virtue solely of the truth-functional modes of composition will be called *tautologous;* they involve only the truth-functional notations essentially, all else vacuously (cf. Introduction). If we think of all such notations as merely abbreviating iterated joint denials, according to D1–5 (§ 9), then we may describe a tautologous statement as one which involves only '↓' (better, '(↓   )') essentially. All tautologous statements are logically true, but not all logically true statements are tautologous. The method of showing a statement to be tautologous consists merely of constructing a table under it in the usual way (§ 3) and observing that the column under the main connective is composed entirely of 'T's. When this is the case we know that the statement is true by virtue solely of its truth-functional structure and independently of the ultimate constituent statements.

For example, where ϕ, ψ, and χ are any statements, $\ulcorner \phi \supset \psi \,.\, \psi \supset \chi \,.\!\vee\!.\, \chi \supset \phi \urcorner$ is found to be tautologous by constructing Table 1 and observing its eighth column; $\ulcorner \phi \supset \psi \,.\!\supset \phi :\equiv \phi \urcorner$ is found tautologous

$\ulcorner\phi$	$\supset$	ψ	.	ψ	$\supset$	χ	.v.	χ	$\supset$	$\phi\urcorner$
T	T	T	T	T	T	T	T	T	T	T
F	T	T	T	T	T	T	T	T	F	F
T	F	F	F	F	T	T	T	T	T	T
F	T	F	T	F	T	T	T	T	F	F
T	T	T	F	T	F	F	T	F	T	T
F	T	T	F	T	F	F	T	F	T	F
T	F	F	F	F	T	F	T	F	T	T
F	T	F	T	F	T	F	T	F	T	F

TABLE 1

$\ulcorner\phi$	$\supset$	ψ	$.\supset$	ϕ	$:\equiv$	$\phi\urcorner$
T	T	T	T	T	T	T
F	T	T	F	F	T	F
T	F	F	T	T	T	T
F	T	F	F	F	T	F

TABLE 2

$\ulcorner\phi$	$\vee$	$\sim$	$\phi\urcorner$
T	T	F	T
F	T	T	F

TABLE 3

$\ulcorner\sim$	$(\phi$	.	$\sim$	$\phi)\urcorner$
T	T	F	F	T
T	F	F	T	F

TABLE 4

by constructing Table 2 and observing its sixth column; $\ulcorner\phi \vee \sim\phi\urcorner$ is found tautologous by constructing Table 3 and observing its second column; and $\ulcorner\sim(\phi \,.\, \sim\phi)\urcorner$ is found tautologous by constructing Table 4 and observing its first column.

$\ulcorner\phi \vee \sim\phi\urcorner$ illustrates the *law of the excluded middle*, which is commonly phrased as saying that every statement is true or false; and $\ulcorner\sim(\phi \,.\, \sim\phi)\urcorner$ illustrates the *law of contradiction* — no statement is both true and false. These two laws are not to be identified with $\ulcorner\phi \vee \sim\phi\urcorner$ and $\ulcorner\sim(\phi \,.\, \sim\phi)\urcorner$ themselves; the latter are unspecified statements pending specification of ϕ, and they become the minor truths 'Jones is ill ∨ ∼Jones is ill' and '∼ (Jones is ill . ∼ Jones is ill)' when ϕ is specified as 'Jones is ill'. But the law of the excluded middle may be formulated as saying that $\ulcorner\phi \vee \sim\phi\urcorner$ is true for every statement ϕ, and the law of contradiction correspondingly.

For any statement ϕ, the compounds $\ulcorner\phi \vee \sim\phi\urcorner$ and $\ulcorner\sim(\phi \,.\, \sim\phi)\urcorner$ are obviously and trivially true. Many longer tautologous statements are equally obvious, e.g.

$$\ulcorner\phi \equiv \psi \,.\!\equiv\!.\, \psi \equiv \phi\urcorner, \qquad \ulcorner\phi \supset \psi \,.\, \psi \supset \chi \,.\!\supset\!.\, \phi \supset \chi\urcorner.$$

This is not, however, characteristic of all; the ones established in Tables 1 and 2, e.g., are far from obvious. But if tautologous statements are not in general obvious, at least they are essentially trivial, in that they can be proved by the simple mechanical expedient of truth tables.

When the truth tables produce a solid column of 'T's under the main connective of a statement, we know the statement is tautologous. But note that the appearance of an 'F' in the column is not, conversely, a proof that the statement is not tautologous. A table for $\ulcorner\chi \supset \phi \,.\!\equiv \phi\urcorner$, e.g., indicates falsehood in the fourth case:

$$\begin{array}{ccccc} \ulcorner\chi & \supset & \phi \,. & \equiv & \phi\urcorner \\ \mathrm{F} & \mathrm{T} & \mathrm{F} & \mathrm{F} & \mathrm{F} \end{array}$$

Yet, as Table 2 shows, $\ulcorner\chi \supset \phi \,.\!\equiv \phi\urcorner$ will still be tautologous if χ happens to be $\ulcorner\phi \supset \psi\urcorner$. In general, though the truth-table test may give a negative result when a statement is analyzed into components $\phi_1, \phi_2, \ldots, \phi_n$, analysis into shorter components $\phi'_1, \phi'_2, \ldots, \phi'_n$, may still show the statement to be tautologous. The occurrence of an 'F' in the column under the main connective does not disprove that a statement is tautologous, unless we know that we have analyzed the statement into its *ultimate truth-functional components* and thus laid bare the whole of its truth-functional structure.

Precise formulation of this notion of ultimate truth-functional components is facilitated by taking joint denial as our primitive truth-functional mode of statement composition and thinking of '$\sim$', '.', 'v', '$\supset$', and '$\equiv$' as merely an unofficial shorthand eliminable by D1–5.[1] The *immediate truth-functional components* of ϕ, now, are to be understood as *the* ψ *and* χ (if such there be) *whose joint denial* $\ulcorner(\psi \downarrow \chi)\urcorner$ *is* ϕ. The *truth-functional components* of ϕ, more generally, are to be understood as comprising ϕ *itself, together with its immediate truth-functional components* (if any), *together with the immediate truth-functional components of those immediate truth-functional components* (if any), *and so on*. The ultimate truth-functional components of ϕ, finally, are those truth-functional components of ϕ which are not joint denials in turn.

Now a tautologous statement is describable as one which proves true by the truth-table method under every assignment of truth values to its ultimate truth-functional components. This description of tautology is more explicit and mechanical than the earlier description in terms of essential occurrence, but the truth-table method to which it refers could do with more explicit characterization in turn.

As we have seen, there are 2^n possible distributions of truth values to the ultimate truth-functional components $\phi_1, \phi_2, \ldots, \phi_n$ of ϕ; the first distribution marks all as true, the second marks ϕ_1 false and the rest true, the third marks ϕ_2 false and the rest true, the fourth marks ϕ_1 and ϕ_2 false and the rest true, and so on. Consultation of the truth tables of '$\downarrow$', '$\sim$', '.', 'v', '$\supset$', and '$\equiv$' enables us, under each one of these 2^n distributions, to derive truth values step by step for more and more complex truth-functional components of ϕ until finally ϕ itself is marked as true or as false.

[1] The ensuing developments would proceed analogously if alternative denial were taken as primitive instead of joint denial.

For each i from 1 to 2^n, thus, there is a definite set of truth-functional components of ϕ which will be marked true; let us call this set the ith *truth set* of ϕ. It comprises those of $\phi_1, \phi_2, \ldots, \phi_n$ which are marked true under the ith distribution of truth values, together with all those more complex components of ϕ which come derivatively to be marked true by consultation of the tables of '$\downarrow$', '$\sim$', '.', 'v', '$\supset$', and '$\equiv$'. Described in terms of the tabular constructions, as of Tables 1–4, the ith truth set comprises those statements which are marked 'T' in the ith row. The third row of Table 2 shows, e.g., that the third truth set of $\ulcorner \phi \supset \psi .\supset \phi :\equiv \phi \urcorner$ embraces ϕ, $\ulcorner \phi \supset \psi .\supset \phi \urcorner$, and $\ulcorner \phi \supset \psi .\supset \phi :\equiv \phi \urcorner$ (if we suppose that ϕ and ψ here are the ultimate truth-functional components).

Towards a more rigorous characterization of this notion of truth set, let us take '$\downarrow$' again as our primitive connective and think of '$\sim$', '.', 'v', '$\supset$', and '$\equiv$' as mere shorthand. Now a truth set S of ϕ will include, to begin with, anywhere from all to none of the ultimate truth-functional components $\phi_1, \phi_2, \ldots, \phi_n$ of ϕ (depending on whether S is the first, second, . . . , or 2^nth truth set). From among the truth-functional components of ϕ having the form $\ulcorner \phi_i \downarrow \phi_j \urcorner$, next, we decide which ones accrue to S by consulting the joint-denial table:

$\ulcorner \phi_i$	$\downarrow$	$\phi_j \urcorner$
T	F	T
F	F	T
T	F	F
F	T	F

In view of this table, we accord $\ulcorner \phi_i \downarrow \phi_j \urcorner$ to S if and only if S includes neither ϕ_i nor ϕ_j. Similarly for the succeeding levels of complexity: any truth-functional component $\ulcorner (\phi_i \downarrow \phi_j) \downarrow \phi_k \urcorner$ or $\ulcorner \phi_k \downarrow (\phi_i \downarrow \phi_j) \urcorner$ of ϕ will belong to S if and only if neither of the parts $\ulcorner \phi_i \downarrow \phi_j \urcorner$ and ϕ_k belongs to S; likewise any truth-functional component $\ulcorner (\phi_i \downarrow \phi_j) \downarrow (\phi_k \downarrow \phi_l) \urcorner$ of ϕ will belong to S if and only if neither of the parts $\ulcorner (\phi_i \downarrow \phi_j) \urcorner$ and $\ulcorner (\phi_k \downarrow \phi_l) \urcorner$ belongs to S; and so on. In general, (I) *any joint denial* $\ulcorner \psi \downarrow \chi \urcorner$ *which is a truth-functional component of* ϕ *will belong to* S *if and only if neither* ψ *nor* χ *belongs to* S. From among the ultimate truth-functional components $\phi_1, \phi_2, \ldots, \phi_n$ of ϕ, a truth set S may contain any arbitrary selection (there being a different truth set for each such selection); but once this part of S is fixed, the rest of the membership of S is fixed by (I). Thus a *truth set of* ϕ is describable simply as *any set* S *of truth-functional components of* ϕ *which conforms to* (I).

To say that ϕ proves true by the truth-table method under each assignment of truth values to its ultimate truth-functional components is merely to say that ϕ belongs to each of its truth sets, since the successive truth sets of ϕ comprise those components of ϕ which are marked 'T' in the successive rows of the truth table. We thus arrive at the following formulation of the notion of tautology: *ϕ is tautologous if and only if it belongs to each of its truth sets.*

When a statement is written out in full, with all the abbreviations '$\sim$', '.', 'v', '$\supset$', and '$\equiv$' eliminated in favor of '$\downarrow$', the writing of rows of 'T's and 'F's under the statement in usual truth-table fashion is merely a way of recording the successive truth sets. Thus it is that solid 'T's, in the column corresponding to the whole statement, establish tautology. In practice, of course, we do not eliminate the abbreviations '$\sim$', '.', 'v', '$\supset$', and '$\equiv$'; in constructing the truth table of a given statement we thus consult not only the table of '$\downarrow$' but also those of '$\sim$', '.', etc. This is just a shortcut, based on the consideration that the tables of '$\sim$', '.', etc. record the net results of expanding '$\sim$', '.', etc. into terms of '$\downarrow$' and applying the table of '$\downarrow$' (cf. § 9).

Often a statement is not given explicitly, but is described only incompletely in terms of unspecified components designated by Greek letters 'ϕ', 'ψ', etc. Such is the situation confronting us in Tables 1–4. Under these circumstances we do not know whether ϕ, ψ, etc. are ultimate components; consequently, as remarked earlier, the presence of an 'F' in the main column of the table will *not* rule out the possibility that the statement is tautologous. It is important to note, however, that the opposite sort of reasoning does hold: when the main column exhibits 'T' throughout we can conclude that the statement is tautologous regardless of whether the Greek letters refer to ultimate components. Consider, e.g., Table 1. The 'T' in the first row and main column shows that the whole statement will belong to *all* truth sets which contain ϕ, ψ, and χ; the 'T' in the second row and main column shows that the whole statement will belong to all truth sets which lack ϕ and contain ψ and χ; and so on. Table 1 shows therefore that the whole statement belongs to all its truth sets, regardless of what the nature and number of its ultimate truth-functional components may be. The table shows that any statement of the form $\ulcorner \phi \supset \psi \,.\, \psi \supset \chi \,.\!\vee\!.\, \chi \supset \phi \urcorner$ is tautologous, regardless of the complexity of ϕ, ψ, and χ.

The term 'tautology' is taken from Wittgenstein. The present notion of tautologous statements, as those true by virtue solely of truth-functional composition, seems to agree with his usage; he contrives to make the term cover truths which involve also quantification, but this is consequent only upon an effort to explain quantification as a sort of infinite mode of truth-functional composition. A broader use of the term 'tautologous' has arisen in subsequent literature, because of Wittgenstein's doctrine that all mathematics and logic is tautologous. This doctrine was intended by Wittgenstein as a thesis, not as a definition of tautology; and indeed it is a difficult thesis to defend. But some who do not maintain the thesis in any such form, and who regard the inferences of logic and mathematics as "merely verbal transformations" or "disguised repetitions" only in some much broader sense, have been led thus to transfer the term 'tautologous' to this broader sense. It is not clear just what this broader sense is (cf. my "Truth by Convention"); but, whatever it is, there is already a term of long standing ready at hand for it — Kant's term 'analytic'. Hence, following a suggestion of Carnap's, I am confining the term 'tautologous' to the narrower sense — though in abstraction from Wittgenstein's theories.

A rule amounting substantially to the tabular test of tautology was set forth by Peirce (3.387 f), and is indeed implicit in Boole's general rule of development. The rigorous formulation of tautology presented above is new. Another formulation, equally strict but more complex, has been offered by Tarski ("Untersuchungen," Def. 4).

§ 11. *Selected Tautologous Forms*

THERE IS no end to the variety of ways in which statements can be compounded by truth-functional connectives to produce tautology. From among this infinity of forms, certain samples were considered in the preceding section. Further samples, all of them conditionals or biconditionals, will now be recorded and commented upon.

Where ϕ is any statement, the statements

(1) $\ulcorner \phi \equiv \phi \urcorner$,

(2) $\ulcorner \phi \equiv \sim\sim \phi \urcorner$,

(3) $\ulcorner \phi \equiv . \phi . \phi \urcorner$,

(4) $\ulcorner \phi \equiv . \phi \vee \phi \urcorner$

are tautologous.[1] This can be verified by constructing four two-row tables, in the manner of Tables 3 and 4 of § 10. The correlative conditionals $\ulcorner \phi \supset \phi \urcorner$, $\ulcorner \phi \supset \sim\sim \phi \urcorner$, $\ulcorner \sim\sim \phi \supset \phi \urcorner$, etc. are of course likewise tautologous.

A binary mode of statement composition, rendered say by the connective κ, is said to be *reflexive* if $\ulcorner \phi \, \kappa \, \phi \urcorner$ is true for all statements ϕ, and *idempotent* if $\ulcorner \phi \equiv . \phi \, \kappa \, \phi \urcorner$ is true for all statements ϕ. The tautology of $\ulcorner \phi \equiv \phi \urcorner$ and $\ulcorner \phi \supset \phi \urcorner$ thus shows the reflexivity of the biconditional and the conditional; and (3) and (4) reflect the idempotence of conjunction and alternation. The *law of double denial*, according to which consecutive denial signs cancel, is reflected in (2).

The property of tautology provides derivative relations of *tautologous implication* and *equivalence* which are narrower than logical implication and equivalence (cf. § 5). ϕ may be said to imply ψ tautologously when $\ulcorner \phi \supset \psi \urcorner$ is tautologous; and ϕ may be said to be tautologously equivalent to ψ when $\ulcorner \phi \equiv \psi \urcorner$ is tautologous. The antecedent of a tautologous conditional tautologously implies the consequent, and the two sides of a tautologous biconditional are tautologously equivalent. From the tautologous forms listed in (1)–(4), thus, we see that any statement is tautologously equivalent to itself, to its double denial, to its self-conjunction, and to its self-alternation.

Where ϕ and ψ are any statements, the conditionals

(5) $\ulcorner \phi . \psi . \supset \phi \urcorner$,

(6) $\ulcorner \phi \supset . \phi \vee \psi \urcorner$

are tautologous. This can be verified by construction of four-row tables, in the manner of Table 2 of § 10. Conjunctions tautolo-

[1] Note incidentally that (2)–(4) are merely different abbreviations, by D1–3, of one and the same form:

$$\ulcorner \phi \equiv ((\phi \downarrow \phi) \downarrow (\phi \downarrow \phi))\urcorner.$$

gously imply their components, and alternations are tautologously implied by them.

A mode of statement composition is said to be *commutative* if $\ulcorner\phi\ \kappa\ \psi\ .\!\equiv\!.\ \psi\ \kappa\ \phi\urcorner$ is true for all statements ϕ and ψ. Conjunction, alternation, and the biconditional are commutative; for, where ϕ and ψ are any statements, construction of tables shows that

(7) $\ulcorner\phi\ .\ \psi\ .\!\equiv\!.\ \psi\ .\ \phi\urcorner$,
(8) $\ulcorner\phi \vee \psi\ .\!\equiv\!.\ \psi \vee \phi\urcorner$,
(9) $\ulcorner\phi \equiv \psi\ .\!\equiv\!.\ \psi \equiv \phi\urcorner$

are tautologous.

DeMorgan's law, according to which a denied conjunction is equivalent to the alternation of the denials and a denied alternation is equivalent to the conjunction of the denials, is reflected in the following two forms of tautology.

(10) $\ulcorner{\sim}(\phi\ .\ \psi) \equiv.\ {\sim}\phi \vee {\sim}\psi\urcorner$.
(11) $\ulcorner{\sim}(\phi \vee \psi) \equiv.\ {\sim}\phi\ .\ {\sim}\psi\urcorner$.

In view of D4, (11) has:

(12) $\ulcorner{\sim}(\phi \supset \psi) \equiv.\ \phi\ .\ {\sim}\psi\urcorner$

as a minor variant. The three forms (10)–(12) show how to break up a denied conjunction, alternation, or conditional. The tautologous form

(13) $\ulcorner{\sim}(\phi \equiv \psi) \equiv.\ \phi \equiv {\sim}\psi\urcorner$

does the same for the biconditional.

The *law of transposition*, according to which the antecedent and consequent of a conditional may be switched provided that each is changed to the extent of attaching or dropping an initial denial sign, is reflected in the tautologous forms

(14) $\ulcorner\phi \supset \psi\ .\!\equiv\!.\ {\sim}\psi \supset {\sim}\phi\urcorner$,
(15) $\ulcorner{\sim}\phi \supset \psi\ .\!\equiv\!.\ {\sim}\psi \supset \phi\urcorner$,
(16) $\ulcorner\phi \supset {\sim}\psi\ .\!\equiv\!.\ \psi \supset {\sim}\phi\urcorner$.

Note that (16) is merely a case of (8), in view of D4.

There are related tautologous forms for the biconditional:

(17) $\ulcorner\phi \equiv \psi\ .\!\equiv\!.\ {\sim}\phi \equiv {\sim}\psi\urcorner$,
(18) $\ulcorner\phi \equiv {\sim}\psi\ .\!\equiv\!.\ {\sim}\phi \equiv \psi\urcorner$.

From (18) and (13) we see that the position of a denial sign in a biconditional is immaterial; it may apply to either component or to the whole with no difference of effect.

The next four tautologous forms show ways of transforming a conjunction, alternation, or conditional into a biconditional.

(19) $\ulcorner \phi \,.\, \psi \,.\!\equiv\!: \phi \equiv. \phi \supset \psi \urcorner$.

(20) $\ulcorner \phi \vee \psi \,.\!\equiv\!: \psi \equiv. \phi \supset \psi \urcorner$.

(21) $\ulcorner \phi \supset \psi \,.\!\equiv\!: \phi \equiv. \phi \,.\, \psi \urcorner$.

(22) $\ulcorner \phi \supset \psi \,.\!\equiv\!: \psi \equiv. \phi \vee \psi \urcorner$.

The *laws of development*, whereby a statement ϕ can be elaborated into an equivalent which incorporates any desired second statement ψ, are depicted in the tautologous forms

(23) $\ulcorner \phi \equiv. \phi \vee \psi \,.\, \phi \vee \sim \psi \urcorner$,

(24) $\ulcorner \phi \equiv: \phi \,.\, \psi \,.\!\vee\!.\, \phi \,.\, \sim \psi \urcorner$.

A mode of statement composition is said to be *transitive* if

$$\ulcorner \phi \,\kappa\, \psi \,.\, \psi \,\kappa\, \chi \,.\!\supset\!.\, \phi \,\kappa\, \chi \urcorner$$

is true for all statements ϕ, ψ, and χ. Conjunction, the conditional, and the biconditional are transitive; for, where ϕ, ψ, and χ are any statements, construction of tables shows that

(25) $\ulcorner \phi \,.\, \psi : \psi \,.\, \chi :\!\supset\!.\, \phi \,.\, \chi \urcorner$,

(26) $\ulcorner \phi \supset \psi \,.\, \psi \supset \chi \,.\!\supset\!.\, \phi \supset \chi \urcorner$,

(27) $\ulcorner \phi \equiv \psi \,.\, \psi \equiv \chi \,.\!\supset\!.\, \phi \equiv \chi \urcorner$

are tautologous. The tables wanted here will have eight rows apiece, in the manner of Table 1 of § 10. Note, however, that the labor of an eight-row table can often be reduced by a judicious approach. Consider, e.g., (26). Any chance of falsifying this conditional would require falsifying the consequent $\ulcorner \phi \supset \chi \urcorner$, and hence assigning 'T' to ϕ and 'F' to χ; consequently our construction can be limited to two rows, corresponding to the assignment of 'T' and 'F' to the remaining component ψ:

$\ulcorner\phi$	$\supset$	ψ	.	ψ	$\supset$	χ	.⊃.	ϕ	$\supset$	$\chi\urcorner$
T	T	T	F	T	F	F	T	T	F	F
T	F	F	F	F	T	F	T	T	F	F

The tautologous forms

(28) $$\ulcorner\phi \supset \psi \,.\!\supset\!:\, \phi \,.\, \chi \,.\!\supset\!.\, \psi \,.\, \chi\urcorner,$$

(29) $$\ulcorner\phi \supset \psi \,.\!\supset\!:\, \phi \vee \chi \,.\!\supset\!.\, \psi \vee \chi\urcorner$$

show that a true conditional remains true when antecedent and consequent are elaborated uniformly by conjunction or alternation. The tautologous forms

(30) $$\ulcorner\phi \supset\!.\, \psi \supset \chi \,:\!\equiv\!:\, \psi \supset\!.\, \phi \supset \chi\urcorner,$$

(31) $$\ulcorner\phi \supset\!.\, \psi \supset \chi \,:\!\equiv\!:\, \phi \,.\, \psi \,.\!\supset \chi\urcorner$$

show two useful ways of transforming an iterated conditional. A third way appears in (44), below.

A mode of statement composition is said to be *associative* if

$$\ulcorner(\phi \,\kappa\, \psi)\,\kappa\, \chi \,.\!\equiv\!.\, \phi \,\kappa\, (\psi \,\kappa\, \chi)\urcorner$$

is true for all statements ϕ, ψ, and χ; in short, if grouping is immaterial. Conjunction, alternation, and the biconditional are associative; for, where ϕ, ψ, and χ are any statements, the statements

(32) $$\ulcorner\phi \,.\, \psi \,:\, \chi \,:\!\equiv\!:\, \phi \,:\, \psi \,.\, \chi\urcorner,$$

(33) $$\ulcorner\phi \vee \psi \,.\!\vee \chi \,:\!\equiv\!:\, \phi \vee .\, \psi \vee \chi\urcorner,$$

(34) $$\ulcorner\phi \equiv \psi \,.\!\equiv \chi \,:\!\equiv\!:\, \phi \equiv\!.\, \psi \equiv \chi\urcorner$$

are tautologous. The associativity of conjunction and alternation was observed earlier (§ 7); but the associativity of the biconditional is perhaps surprising enough to move the reader to work out the table for (34) himself.

In the case of conjunction and alternation, associativity prompted a special notational convention (D6–7, § 9); but a corresponding convention for the biconditional is inadvisable. The notation $\ulcorner\phi \equiv \psi \equiv \chi\urcorner$ is likely to suggest, not $\ulcorner\phi \equiv \psi \,.\!\equiv \chi\urcorner$ nor $\ulcorner\phi \equiv\!.\, \psi \equiv \chi\urcorner$, but rather $\ulcorner\phi \equiv \psi \,.\, \psi \equiv \chi\urcorner$ — on the analogy of the continued equation '$x = y = z$'. The conditions under which $\ulcorner\phi \equiv \psi \,.\!\equiv \chi\urcorner$ (or $\ulcorner\phi \equiv\!.\, \psi \equiv \chi\urcorner$) is true are not those under which $\ulcorner\phi \equiv \psi \,.\, \psi \equiv \chi\urcorner$ is true; the latter is true when ϕ, ψ, and χ are alike in truth value, but the former proves to be true rather when just one or all of ϕ, ψ, and χ are true.

In an iterated conjunction, alternation, or biconditional the order and grouping of components have been seen to be immaterial (commutativity and associativity). In the case of conjunction and alternation, repetition of components has likewise been seen to be immaterial (idempotence). Any two iterated conjunctions, formed from the same components with or without repetitions, are equivalent — tautologously equivalent; and similarly for alternation.

But the biconditional, which lacks idempotence, happens to behave yet more simply under repetition of components: the repetitions cancel out in pairs, in view of the fact that

(35) $\ulcorner \phi \equiv : \phi \equiv \psi . \equiv \psi \urcorner$

is tautologous. Two iterated biconditionals are tautologously equivalent whenever the components appearing an odd number of times in each are the same. This principle is a consequence of another equally curious one: any iterated biconditional is tautologous if each component occurs an even number of times.[1] Further, any compound built up by the biconditional and denial is tautologous if each component occurs an even number of times and there are an even number of denial signs.[2] Note, e.g., that (1), (2), (9), (13), (17), (18), (34), and (35) involve ϕ, ψ, χ, and '$\sim$' twice each if at all.

Given any binary modes of statement composition, rendered say by the respective connectives κ and μ, the first is said to be *distributive* into the second if

$$\ulcorner \phi \; \kappa \; (\psi \; \mu \; \chi) \; . \equiv . \; (\phi \; \kappa \; \psi) \; \mu \; (\phi \; \kappa \; \chi) \urcorner$$

is true for all statements ϕ, ψ, and χ.[3] Distributivity always justifies an operation analogous to that of "multiplying through" in arithmetic: $x(y + z) = xy + xz$. Now conjunction turns out to be distributive into itself and alternation; and alternation and the conditional turn out to be distributive into conjunction, alternation, the conditional, and the biconditional. Where ϕ, ψ, and χ are any statements;

(36) $\ulcorner \phi : \psi . \chi : \equiv : \phi . \psi : \phi . \chi \urcorner$,

(37) $\ulcorner \phi . \psi \vee \chi . \equiv : \phi . \psi . \vee . \phi . \chi \urcorner$,

[1] Leśniewski, "Grundzüge," pp. 26, 29.

[2] This was pointed out to me by Dr. J. C. C. McKinsey.

[3] The first mode of composition may likewise be said to be distributive *out of* the second if $\ulcorner (\phi \; \mu \; \psi) \; \kappa \; \chi \; . \equiv . \; (\phi \; \kappa \; \chi) \; \mu \; (\psi \; \kappa \; \chi) \urcorner$ is true for all statements ϕ, ψ, and χ.

(38) $\ulcorner\phi \vee . \psi . \chi :\equiv . \phi \vee \psi . \phi \vee \chi\urcorner$,
(39) $\ulcorner\phi \vee . \psi \vee \chi :\equiv : \phi \vee \psi . \vee . \phi \vee \chi\urcorner$,
(40) $\ulcorner\phi \vee . \psi \supset \chi :\equiv : \phi \vee \psi . \supset . \phi \vee \chi\urcorner$,
(41) $\ulcorner\phi \vee . \psi \equiv \chi :\equiv : \phi \vee \psi . \equiv . \phi \vee \chi\urcorner$,
(42) $\ulcorner\phi \supset . \psi . \chi :\equiv . \phi \supset \psi . \phi \supset \chi\urcorner$,
(43) $\ulcorner\phi \supset . \psi \vee \chi :\equiv : \phi \supset \psi . \vee . \phi \supset \chi\urcorner$,
(44) $\ulcorner\phi \supset . \psi \supset \chi :\equiv : \phi \supset \psi . \supset . \phi \supset \chi\urcorner$,
(45) $\ulcorner\phi \supset . \psi \equiv \chi :\equiv : \phi \supset \psi . \equiv . \phi \supset \chi\urcorner$

are tautologous.

Along with (42) and (43), note the further tautologous forms

(46) $\ulcorner\phi . \psi . \supset \chi :\equiv : \phi \supset \chi . \vee . \psi \supset \chi\urcorner$,
(47) $\ulcorner\phi \vee \psi . \supset \chi :\equiv . \phi \supset \chi . \psi \supset \chi\urcorner$.

In view of (42), a conditional with conjunctive consequent amounts to a conjunction of conditionals; in view of (46), however, a conditional with conjunctive antecedent amounts to an alternation of conditionals. (47) contrasts with (43) in analogous fashion.

The form (45) shows one useful way of transforming $\ulcorner\phi \supset . \psi \equiv \chi\urcorner$; the following shows another:

(48) $\ulcorner\phi \supset . \psi \equiv \chi :\equiv : \phi . \psi . \equiv . \phi . \chi\urcorner$.

Continually, in subsequent developments (§§ 17 ff.), we shall find ourselves called upon to recognize simple forms of tautology. The method of truth tables is always available for the purpose; but since in many cases the form to be recognized will be one or another of the foregoing forms (1)–(48), or a minor variant thereof, the reader will save time by familiarizing himself with these.

Most of the tautologous forms (1)–(48) are cited, along with numerous others, by Whitehead and Russell (*1–*5). The bulk of them were known to earlier authors — Frege, Peano, Schröder, Peirce, Boole. Many were indeed anticipated, as Lukasiewicz points out ("Zur Geschichte"), by the Stoics in ancient times and by Petrus Hispanus, Duns Scotus, and others in the Middle Ages; what has come to be known as De Morgan's law, e.g., goes back to William of Ockham. One or more of (19), (20), (40), and (41) may be new to the literature.

The terms 'idempotent', 'commutative', 'associative', and 'distributive' are borrowed from abstract algebra, where their use is closely analogous to that described above (cf. § 33). The terms 'reflexive' and 'transitive' are from relation theory (cf. § 42; also § 25).

CHAPTER TWO

QUANTIFICATION

§ 12. *The Quantifier*

THE TRIVIAL statement:

(1) 9 is less than, equal to, or greater than 0

is translatable into arithmetical and logical symbols thus:

(2) $9 < 0 \;.\mathrm{v}.\; 9 = 0 \;.\mathrm{v}.\; 9 > 0$.

But the statement:

(3) Every number is less than, equal to, or greater than 0,

despite its superficial resemblance to (1), is not correspondingly translatable as:

(4) Every number < 0 .v. every number $= 0$ .v. every number > 0;

for (3) is true, if we suppose imaginary numbers excluded, while (4) is false. The alternation (4) says that some one of the three categories exhausts all numbers, while (3) says only that the three categories together exhaust all numbers. (3) calls for expansion rather into some such form as this:

> Whatever number you may select, it will be less than or equal to or greater than 0;

or, incorporating the arithmetical and logical notation:

(5) Whatever number you may select, it < 0 .v. it $= 0$ .v. it > 0.

As a step toward further logical notation we might compress (5) thus:

(6) whatever number (it < 0 .v. it $= 0$ .v. it > 0).

Note, by way of contrast, that this rudimentary notation endows the falsehood (4) rather with the following form:

> whatever number (it < 0) .v. whatever number (it $= 0$)
> .v. whatever number (it > 0).

What (2) says about the number 9 is, as we have noted, true of all numbers. This fact is expressed in (3), and again in (5) and (6).

It is expressed in (6) by putting 'it' for '9' in (2) and applying the prefix 'whatever number'. But note now that what (3) says about 0 is in turn true of every number. A statement to this effect can be formed from (3) just as (6) was formed from (2):

(7) whatever number (every number is less than, equal to, or greater than it).

Now the parenthesized part of (7), viz.:

(8) every number is less than, equal to, or greater than it,

still wants translation into our new notation. Since (8) differs from (3) only in containing 'it' instead of '0', the translation of (3) into (6) suggests translation of (8) into:

(9) whatever number (it $<$ it .v. it $=$ it .v. it $>$ it).

But this translation is inacceptable; it turns (7) into:

(10) whatever number (whatever number (it $<$ it .v. it $=$ it .v. it $>$ it)),

wherein the distinction is lost between the 'it's which correspond to the outer occurrence of 'whatever' and the 'it's which correspond to the inner occurrence of 'whatever'.

This difficulty can be overcome by returning to (6) and (7) and revising our basic notation, to the extent of tagging 'whatever' and 'it' with one or another arbitrary subscript to show that they belong together. (7) and (6) might be rendered respectively:

(11) whatever_1 number (every number is less than, equal to, or greater than it_1),

(12) whatever_2 number ($\text{it}_2 < 0$ .v. $\text{it}_2 = 0$ .v. $\text{it}_2 > 0$).

Now the parenthesized part of (11), viz.:

(13) every number is less than, equal to, or greater than it_1,

differs from (3) only in containing 'it_1' instead of '0'. Parallel to the translation of (3) into (12), we can translate (13) into:

(14) whatever_2 number ($\text{it}_2 < \text{it}_1$ .v. $\text{it}_2 = \text{it}_1$ .v. $\text{it}_2 > \text{it}_1$),

which differs from (12) only in containing 'it_1' instead of '0' Thus (11) in its entirety becomes:

(15) whatever$_1$ number (whatever$_2$ number (it$_2 <$ it$_1$ .v. it$_2 =$ it$_1$ .v. it$_2 >$ it$_1$)).

Such use of two sets of 'it's with distinguishing subscripts corresponds to the common use of 'former' and 'latter'. (15) might be rendered thus in ordinary language:

> Whatever number you may select, it will turn out, whatever number you may next select, that the latter is less than, equal to, or greater than the former.

In more complex cases there may be more than two distinct subscripts; such cases are commonly handled in ordinary language, not by the 'former-latter' idiom, but by the device of speaking of the first, second, third, etc. in an arbitrary order. It is after this device that the present use of numerical subscripts is patterned.

But a more compact and convenient notation consists in using simply different *letters*, instead of 'it' with different subscripts, and then using the same letter in parentheses instead of 'whatever' with the corresponding subscript. Thus revised, (12) and (15) appear as follows:

(16) (y) number $(y < 0 \text{ .v. } y = 0 \text{ .v. } y > 0)$,
(17) (x) number $((y)$ number $(y < x \text{ .v. } y = x \text{ .v. } y > x))$.

Similarly 'All men are mortal', or 'Every man is mortal', becomes:

(18) (x) man (x is mortal).

Again, 'No man has seen every city', or in other words 'Every man is in the position of not having seen all cities', becomes:

(x) man $\sim$ (x has seen all cities).

But 'x has seen all cities', or 'x has seen every city', gives way similarly to '(y) city (x has seen y)'; so the whole becomes:

(19) (x) man $\sim$ (y) city (x has seen y).

Such prefixes as '(x) number', '(x) man', '(y) city' may be read 'no matter what number x may be', 'no matter what man x may be', 'no matter what city y may be'. The tentative symbolism at which we have arrived is thus a regimentation of the familiar practice, in algebra and elsewhere, of using letters to achieve generality. An algebraic discussion which contains 'x', and is supposed to hold for every number x, can now be explained as tacitly prefaced in the fashion '(x) number'. The algebraic use of letters was characterized earlier (§ 6) in a rough, tentative way; the letters were spoken of as names of unspecified numbers, ambiguous names of numbers, signs replaceable by names of any numbers. Now the status of such letters is clearer; the use of 'x', 'y', etc. in connection with prefixes '(x) number', '(y) number', etc. involves nothing beyond what was just now explained in terms of 'whatever' and pronouns. There are indeed uses of 'x' in algebra which do not lend themselves immediately to formulation in terms of the idiom '(x) number', or 'no matter what number x may be'; often the appropriate prefix is rather 'there is a number x such that', or 'the number x such that', or 'the class of all numbers x such that'. It will be found, however (§§ 19, 26, 27) that all such apparently divergent uses admit finally of translation into forms involving just the idiom now under consideration.

What has been said concerning algebra applies also to the analogous use of letters in non-numerical fields; e.g., to the use of Greek letters in discussions of expressions. Use of 'μ' in saying something about all expressions, or of 'ϕ' in saying something about all statements, would now be explained in terms of such prefixes as '(μ) expression' and '(ϕ) statement' — or better '(x) expression', '(x) statement'. To insist on explicit reformulations of this kind in all future discussions would, however, be as unwarranted as to insist on elimination of the words 'or', 'and', and 'if-then' from future discussions in favor of '$\vee$', '$.$', and '$\supset$'. Most things can be explained clearly enough by the common-sense sort of statement; it is rather when the statement itself becomes an object of analysis that its full translation into the logical notation becomes particularly advantageous.

The notation exemplified in (16)–(19) is only tentative, and will now undergo further evolution. Consider the statement 'Everything is identical with itself'. Paraphrased in the manner of (5), this becomes:

(20) Whatever you may select, it = it.

But the limitation imposed by the word 'number', in (5), has no analogue in (20). We may appropriately put (20) into symbols by following the pattern of (16) and (18) but omitting any range-indicator such as 'number', 'man'. We have:

(21) $(x)(x = x)$.

Just as the prefix '(x) number' may be read 'no matter what number x may be', so the simple prefix '(x)', as of (21), may be read simply 'no matter what x may be'. Now a prefix such as '(x) number', '(x) man', etc. can be eliminated, always, in favor of the simple kind of prefix used in (21). E.g., the statement:

(22) No matter what man x may be, x is mortal

can be paraphrased as:

(23) No matter what x may be, if x is a man then x is mortal.

Instead of limiting our assertion to men at the start, we admit all entities at the start and then employ the conditional form to spare the non-men. Thus (18) gives way to:

(24) $(x)(x$ is a man $.\supset.$ x is mortal$)$.

In general, the form '(x) M (——)' gives way to '$(x)(x$ is an M $.\supset.$ ——$)$'. (16), (17), and (19) become:

(25) $(y)(y$ is a number $.\supset{:}$ $y < 0 .\vee. y = 0 .\vee. y > 0)$,

(26) $(x)(x$ is a number $.\supset$ $(y)(y$ is a number $.\supset{:}$ $y < x .\vee. y = x .\vee. y > x))$,

(27) $(x)(x$ is a man $.\supset \sim (y)(y$ is a city $.\supset.$ x has seen $y))$.

The notation (16)–(19) involving range-indicators will never be used hereafter; the notation exemplified by (21) and (24)–(27) is adopted as standard.

The prefixes '(x)', '(y)', etc. are called *quantifiers*, and their use in forming statements is known as *quantification*. The letters 'x', 'y', 'z', etc. themselves are called *variables*. In order that the letters available as variables may not be limited by the bounds of the alphabet, further ones are formed by means of accents; thus the variables will be understood as comprising 'w', 'x', 'y', 'z', 'w'', 'x'', 'y'', 'z'', 'w''', ...

Though the word 'variable' has been retained here out of consideration for established usage, the "variation" connoted belongs to a vague metaphor which is best forgotten. The variables have no meaning beyond the pronominal sort of meaning which is reflected in translations such as (20); they serve merely to indicate cross-references to various positions of quantification. Such cross-references could be made instead by curved lines or *bonds*; e.g., we might render (27) thus:

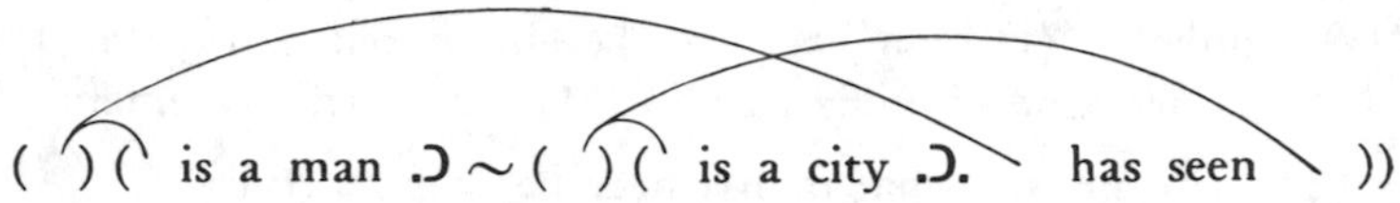
( ) (is a man .⊃ ∼ ( ) (is a city .⊃. has seen))

and (26) thus:

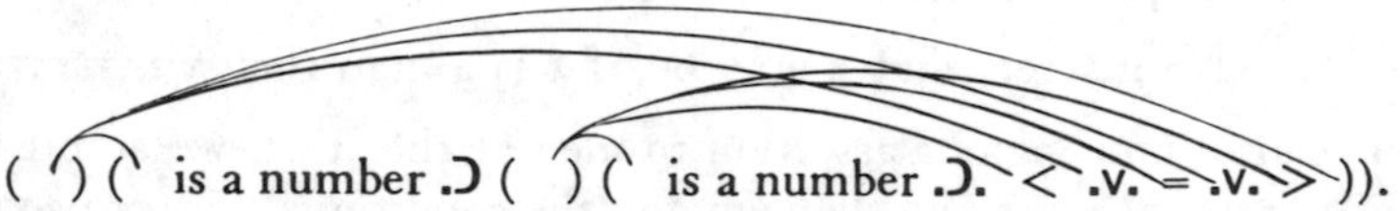
( ) (is a number .⊃ ( ) (is a number .⊃. < .v. = .v. >)).

But these "quantificational diagrams" are too cumbersome to recommend themselves as a practical notation; hence the use of variables.

Quantification cuts across the vernacular use of 'all', 'every', 'any', and also 'some', 'a certain', etc. (cf. § 19), in such fashion as to clear away a baffling tangle of ambiguities and obscurities. One of the anomalies of the common idiom was noted at the beginning of the section, and many further instances suggest themselves. Consider e.g. the statements:

(28) Smith can outrun every man on the team,
(29) Smith can outrun any man on the team,
(30) Smith cannot outrun every man on the team,
(31) Smith cannot outrun any man on the team.

Clearly (28) and (29) are equivalent, and (30) is the denial of (28); but (31) is *not* correspondingly the denial of (29). Whereas (28) and (29) are:

(32) $(x)(x$ is a man on the team $.\supset.$ Smith can outrun $x)$

and (30) is:

(33) $\sim(x)(x$ is a man on the team $.\supset.$ Smith can outrun $x)$,

(31) is rather:

(34) (*x*)(*x* is a man on the team .⊃. ∼ Smith can outrun *x*).

The faultiness of the vernacular is illustrated again in the tendency to blur the distinction which is dealt with below in *139 (§ 19). The device of quantification subjects this level of discourse, for the first time, to a clear and general algorithm. It reveals the precise connection, hitherto obscure, between general statements and truth-functional composition; in (25) and (32)–(34), e.g., unlike their idiomatic equivalents (3) and (28)–(31), the relevant truth-functional structure is clearly exhibited.

Frege (1879) was the first to devise a general notation of quantification, using auxiliary variables in the modern fashion. So important was this step that we might indeed look upon Frege, rather than Boole, as the founder of modern logic. The present notation, easier to print than Frege's, is from Whitehead and Russell. The pronominal character of the variable was clear to Peano (*Formulaire*, 1897, p. 26; 1901, p. 2); but it is only with the advent of combinatory logic, founded by Schönfinkel and developed by Curry, that the rôle of the variable as an index of cross-reference has received full analysis. The analysis consists in showing how variables can be eliminated in favor of a few constant terms designating functions of functions (or relations of relations). See Schönfinkel; also Curry's "Grundlagen," "Apparent Variables," and "Functionality," Rosser's "Mathematical Logic", and my "Reinterpretation" (which cites further papers by Curry).

§ 13. *Formulæ*

WHEREAS '(*x*)(*x* is blue)' is a false statement to the effect that everything is blue, and '(*x*)(*x* = *x*)' is a true statement to the effect that everything is itself, the components '*x* is blue' and '*x* = *x*' are neither true nor false; they are not statements at all, but statement *matrices* — expressions which would be statements if they contained names instead of variables. Between '*x* = *x*' and '*x* is blue' there is, indeed, this difference: all statements formed from '*x* = *x*' by putting a name in place of '*x*' are true, while some statements so formed from '*x* is blue' are false. But this difference does not make '*x* = *x*' itself true, nor '*x* is blue' false. Matrices are not statements, and only statements are true or false.

A matrix is not, however, meaningless, any more than an adverb or preposition is meaningless. Matrices are fragmentary, but meaningful in that they can serve as ingredients of genuine statements like '$(x)(x$ is blue)' and '$(x)(x = x)$'. Since '$(x)(x$ is blue)' and '$(x)(x = x)$' are translatable as 'Whatever you may select, it is blue' and 'Whatever you may select, it = it', the matrices 'x is blue' and '$x = x$' may be regarded as corresponding to the expressions 'it is blue' and 'it = it' — expressions resembling statements in form but incapable of standing alone because of their dangling pronouns.

Since matrices differ from statements only in exhibiting variables in place of names, matrices are wanted in all forms of which statements are capable. Quantification and the truth-functional modes of composition thus cease to be methods merely of constructing statements; they are to be regarded henceforth as methods of constructing statements and matrices indifferently. Furthermore the notion of tautology admits of immediate extension to matrices: a matrix is tautologous if it has the form of a tautologous statement, i.e., if it becomes a tautologous statement when variables are supplanted by names. A tautologous matrix is not, like a tautologous statement, true; still the tabular test of tautology can be applied directly to the matrix, for the pattern on which the test turns is unaffected by the presence of variables. The notions of tautologous implication and equivalence, and logical implication and equivalence, admit similarly of immediate extension to matrices. Because of these affinities between statements and matrices, it will be convenient henceforth to treat statements and matrices together under a single head; they will be called *formulæ*.

In Chapter I no attempt was made to mark out the category of expressions known as statements. We now proceed to impose bounds on that category; indirectly, however, through imposing bounds on the broader category of *formulæ*. This category will retain a degree of relativity, for the space of the present chapter, but is subjected to the following restriction: no formula can occur within another except through quantification or truth-functional composition. Instead of truth-functional composition, indeed, we might cite joint denial alone; for conjunction, denial, and the rest may be regarded as unofficial abbreviations after the manner of D1–5 (§ 9).

Thus we are to think of the totality of formulæ as comprising some manner of *atomic formulæ* — formulæ containing no shorter formulæ as parts — and in addition just the expressions thence constructible by continued application of joint denial and quantification. In the next chapter, expressions which are to be taken as atomic formulæ will be specified; but for present developments we may leave the class of atomic formulæ still undetermined. Relative to the atomic formulæ, however, the totality of formulæ is fixed; it comprises the class of atomic formulæ, together with all results of putting formulæ of that class in the blanks of '(↓)' or after a quantifier, together with all results of putting formulæ of the thus increased class in the blanks of '(↓)' or after a quantifier, and so on. If μ and ν are among the atomic formulæ, and α and β are variables ('w', 'x', 'y', or 'z' with or without accents), then

$$\mu,\ \nu,\ \ulcorner(\alpha)\mu\urcorner,\ \ulcorner(\beta)\mu\urcorner,\ \ulcorner(\alpha)\nu\urcorner,\ \ulcorner(\mu \downarrow \mu)\urcorner,\ \ulcorner(\mu \downarrow \nu)\urcorner,\ \ulcorner((\alpha)\mu \downarrow (\alpha)\mu)\urcorner,$$
$$\ulcorner((\alpha)\mu \downarrow \mu)\urcorner,\quad \ulcorner(\mu \downarrow (\beta)\nu)\urcorner,\quad \ulcorner((\alpha)\mu \downarrow (\alpha)\nu)\urcorner,\quad \ulcorner((\alpha)\mu \downarrow (\beta)\nu)\urcorner,$$
$$\ulcorner(\alpha)(\mu \downarrow \nu)\urcorner,\quad \ulcorner((\alpha)(\mu \downarrow \nu) \downarrow \mu)\urcorner,\quad \ulcorner((\alpha)(\mu \downarrow (\beta)\nu) \downarrow (\beta)\nu)\urcorner,$$

etc. are among the formulæ.

So long as the atomic formulæ remain unspecified, it is impossible to present any specific formulæ by way of examples; we can only give partially determinate examples with help of Greek letters, as above. When specific examples seem wanted, however, we may conveniently avail ourselves of random components '$x = x$', '$x < 5$', '$3 < 5$', 'Socrates is mortal', etc., under the fiction that these are among our atomic formulæ.

In the requirement that no formula contain another except in a context of joint denial or quantification, an extreme restriction would seem to be imposed. Actually, however, nothing is thereby barred which is needed anywhere in mathematics. For all mathematical purposes it is sufficient that formulæ occur in three ways: alone, in joint denial, in quantification. Such other manners of occurrence as may turn up in mathematical practice are reducible to these by definitional translation. There is reason to believe, indeed, that this is true not only of mathematics but of discourse generally; this thesis is a controversial one, however, being an extension of the thesis that all statement composition can be limited to truth-functional composition (cf. §§ 1, 5).

There are at the same time some respects in which the above characterization of formulæ is more liberal than § 12 would have led one to expect. It allows the attachment of any quantifier $\ulcorner(\alpha)\urcorner$ to any formula μ, whereas the explanation in § 12 covers only the case where μ is a matrix involving α. Toleration of $\ulcorner(\alpha)\mu\urcorner$ where μ lacks α is actually a matter only of convenience; $\ulcorner(\alpha)\mu\urcorner$ is to be construed in such cases as equivalent simply to μ, i.e., the quantifier is to be construed as vacuous. Thus the statement:

(x)(Socrates is mortal)

is explained as equivalent simply to 'Socrates is mortal'. The verbal analogue 'Whatever you may select, Socrates is mortal' (cf. (20), § 12) reflects the vacuousness of the quantifier.

In allowing attachment of every quantifier to every formula we let down the bars also to superimposition of identical quantifiers, as here:

(1) $(x)(x$ is a man . $(x)(x$ is a man .⊃. x is mortal) .⊃. x is mortal).

The question then arises: which of the two occurrences of '(x)' is supposed to govern the occurrences of 'x' in 'x is a man .⊃. x is mortal'? But the usage is unambiguous if we agree that occurrences of a variable α overlaid thus by duplicate occurrences of $\ulcorner(\alpha)\urcorner$ are to be governed always by the innermost occurrence of $\ulcorner(\alpha)\urcorner$. The interior segment:

(2) $(x)(x$ is a man .⊃. x is mortal)

of (1) is to be read, without regard to its context, in the usual fashion 'all men are mortal'. (1) as a whole thus becomes:

$(x)(x$ is a man . all men are mortal .⊃. x is mortal),

or, read entirely in words, 'Whatever you may select, if it is a man and all men are mortal then it is mortal'. The letter 'x' in (2) could be rewritten as 'y' without regard to the outer parts of (1). (1) is equivalent to:

(3) $(x)(x$ is a man . $(y)(y$ is a man .⊃. y is mortal) .⊃. x is mortal).

It is clearest in practice to choose different letters, as in (3), but quantification theory is greatly simplified by tolerating also the usage exemplified by (1).[1]

Strictly speaking the examples (1)–(3) are not formulæ, but abbreviations of formulæ; to get the corresponding formulæ we must expand the abbreviations '.' and '⊃' into terms of '↓' according to D1–5. In practice, however, this distinction can conveniently be slurred over by keeping to the abbreviations and imagining meanwhile that the full primitive notation is written out instead. Except in the few cases where the abbreviations themselves are the topic of discussion, remarks ostensibly applied to abbreviated expressions are to be thought of as really applying to the corresponding expressions in unabbreviated notation. Thus it is that (1)–(3) and their like are treated as formulæ.

Hereafter 'ϕ', 'ψ', 'χ', and their accented and subscripted variants will be used not merely to refer to statements but to refer to formulæ generally — statements or matrices. This extension is to be understood as applying also retroactively; thus the Greek letters in D1–7 (§ 9), also in the explanation of the dot notation (§ 7), and also in the formulation of tautology (§ 10, italicized passages), are to be reread as referring to any formulæ. Our formulation of tautology now comes to characterize tautologous statements and tautologous matrices simultaneously; and (1)–(48) of § 11 now present forms of tautologous formulæ — statements and matrices indifferently. The formulæ described in (1)–(48) are tautologous statements when ϕ, ψ, and χ happen to be chosen as statements, tautologous matrices otherwise.

This modification of our Greek-letter usage will be accompanied by another: 'α', 'β', 'γ', 'δ', and their accented and subscripted variants will not be used to refer to expressions indiscriminately, but will be limited to cases where the expressions referred to are the specific one-letter expressions '*w*', '*x*', '*y*', '*z*', '*w'*', etc. — the so-called variables.

[1] The usage in question is proscribed by Hilbert and Ackermann, tacitly avoided by Whitehead and Russell, and explicitly admitted by Frege (*Grundgesetze*, vol. 1, p. 13), von Neumann ("Beweistheorie," p. 7), and Carnap (*Syntax*, p. 88).

§ 14. *Bondage, Freedom, Closure*

THE NOTIONS of bondage, freedom, and closure, which are to be explained in this section, are used constantly in subsequent discussion. A given occurrence of α is said to be *bound to* a given occurrence of $\ulcorner(\alpha)\urcorner$ if *it stands in a formula beginning with the given occurrence of* $\ulcorner(\alpha)\urcorner$ *and stands in no formula beginning with a later occurrence of* $\ulcorner(\alpha)\urcorner$. In (1) of § 13, e.g., the first, second, and last of the six occurrences of 'x' are bound to the initial occurrence of '(x)', while the third, fourth, and fifth occurrences of 'x' are bound to the other occurrence of '(x)'.

It was explained (§ 13) that an occurrence of α, if overlaid by duplicate occurrences of $\ulcorner(\alpha)\urcorner$, is to be regarded as governed by the innermost of those occurrences and as having only an alphabetical accident in common with the other occurrences. It is to this innermost occurrence of $\ulcorner(\alpha)\urcorner$ that the given occurrence of α is said to be bound, under our newly introduced terminology. The occurrence of $\ulcorner(\alpha)\urcorner$ to which a given occurrence of α is bound is the occurrence of $\ulcorner(\alpha)\urcorner$ which governs the given occurrence of α. The terminology 'bound to' receives graphic significance if we recall the notation of quantificational diagrams (§ 12); that notation presents actual bonds, in the form of curved lines, connecting each quantifier-position with such subsequent variable-positions as are "bound thereto" in the above sense. (The little bond which always connects an occurrence of $\ulcorner(\alpha)\urcorner$ with the occurrence of α inside it is of course not shown.)

In applying the criterion of bondage italicized above, we must think of our binary statement compounds as fitted with the outside parentheses which belong to them under strict notation (cf. § 7). Otherwise one is tempted to say e.g. that the final occurrence of 'x' in (1) of § 13 is bound to the second occurrence of '(x)', on the ground that the final occurrence of 'x' stands in a formula:

$$(x)(x \text{ is a man } .\supset. \; x \text{ is mortal}) \; .\supset. \; x \text{ is mortal}$$

which begins with the occurrence of '(x)' in question.. Actually, however, (1) of § 13 becomes:

(1) $(x)((x \text{ is a man} \,.\, (x)(x \text{ is a man} \;.\supset.\; x \text{ is mortal})) \supset.\; x \text{ is mortal})$

on restoration of parentheses; and obviously its terminal fragment:

$$(x)(x \text{ is a man} \;.\supset.\; x \text{ is mortal})) \supset.\; x \text{ is mortal}),$$

with unmated parentheses, is no formula. Again, one might be tempted to say that the last occurrence of 'x' in:

(2) $(x)(x = y + x) \supset.\; y \leqslant x$

is bound to the occurrence of '(x)', on the ground that (2) is a formula beginning with the given occurrence of '(x)'; actually, however, (2) becomes:

(3) $((x)(x = y + x) \supset.\; y \leqslant x)$

on restoration of parentheses, and (3) does *not* begin with '(x)'.

Discernment of bondage relations does not in practice require that we restore absent parentheses, so long as we keep in mind the intended cleavages of our formulæ. We must keep in mind, e.g., that (2) is not formed by applying '(x)' to:

$$(x = y + x) \supset.\; y \leqslant x.$$

An occurrence of ψ in ϕ will be said to be *bound in ϕ with respect to α* if within that occurrence of ψ there is an occurrence of α which is bound to an occurrence of $\ulcorner(\alpha)\urcorner$ lying in ϕ but outside ψ. In (2), e.g., the occurrence of '$x = y + x$' is bound with respect to 'x'; for the occurrences of 'x' in '$x = y + x$' are bound to an occurrence of '(x)' in (2) which is outside '$x = y + x$'. In (1), each occurrence of 'x is a man' and 'x is mortal' is bound with respect to 'x'.

An occurrence of ψ will be said simply to be *bound in ϕ* if it is bound in ϕ with respect to at least one variable. Thus the occurrence of '$x = y + x$' is bound in (2), whereas the occurrence of '$y \leqslant x$' is not. In terms of quantificational diagrams, bondage of ψ in ϕ has this significance: there is at least one bond reaching into ψ from somewhere in ϕ anterior to ψ.

Obviously an occurrence of ψ may be bound in a broad context ϕ and yet not bound in a narrower context ϕ'; for, the bondage of ψ in ϕ may depend on occurrences of $\ulcorner(\alpha)\urcorner$, $\ulcorner(\beta)\urcorner$, etc. which, though within ϕ, are outside ϕ'. Diagrammatically: the bonds reaching into ψ from anterior parts of ϕ may begin anterior also to ϕ'. The first occurrence of 'x is a man', e.g., is bound in the broad context (1) but not in the narrower context:

(4) x is a man . $(x)(x$ is a man .⊃. x is mortal$)$.

The terminology 'bound in' will be applied not only to an occurrence of a formula ψ, as above, but also to an occurrence of a variable — in the following sense: an occurrence of α will be said to be *bound in* ϕ if it is bound to an occurrence of $\ulcorner(\alpha)\urcorner$ in ϕ. Thus the first three occurrences of 'x' are bound in (2); the fourth is not. An occurrence of α which is bound in a broad context ϕ need not be bound in a narrower context ϕ', since the occurrence of $\ulcorner(\alpha)\urcorner$ to which the given occurrence of α is bound may lie inside ϕ but outside ϕ'; the second and third occurrences of 'x' in (2), e.g., are bound in (2) and in '$(x)(x = y + x)$' but not in '$x = y + x$'.

The terminology 'bound in' will be applied not only to an occurrence of a formula or variable, as described, but also to the formula or variable itself: a formula or variable will be said to be *bound in* ϕ if an occurrence of it is bound in ϕ. Thus '$x = y + x$' is bound in (2); so also is 'x'. Variables bound in ϕ will be spoken of also as *bound variables* of ϕ.

An occurrence of a formula or variable in ϕ is said to be *free in* ϕ if it is not bound in ϕ; and a formula or variable is said to be free in ϕ if an occurrence of it is free in ϕ. Thus '$y \leqslant x$' is free in (2); so also are 'x' and 'y'. Variables free in ϕ will be spoken of also as *free variables* of ϕ.

To describe one and the same occurrence of μ as both bound and free in ϕ would be, of course, a flat contradiction. On the other hand μ itself may be both bound and free in ϕ; i.e., it may have both a bound and a free occurrence in ϕ. E.g., 'x is a man' is bound and free in (4). So is 'x'.

It is obvious from the foregoing formulations that the following hold in general. (i) An occurrence of ψ is bound in ϕ with respect to α if and only if it contains an occurrence of α which is bound in ϕ and free in ψ. (ii) An occurrence of α in ϕ is bound to the initial occurrence of $\ulcorner(\alpha)\urcorner$ in $\ulcorner(\alpha)\phi\urcorner$ if and only if it is free in ϕ.

The terminology which we have been developing enables us, among other things, to render explicit the distinction between statements and matrices. The distinction does not turn merely on the absence or presence of variables, for '$(x)(x$ is blue)' and '$(x)(x = x)$' are statements despite their variables; it turns rather on the fact that in a statement every occurrence of a variable is governed by a quantifier, while a matrix contains occurrences which are not thus governed. Thus, given the totality of *formulæ*, we may describe *matrices* and *statements* respectively as those formulæ which have free variables and those which have not.

Every matrix can be turned into a statement, true or false, simply by prefixing quantifiers corresponding to all the free variables. The matrix '$x < y .\equiv. y > x$', e.g., is turned in this way into the true statement:

$$(x)(y)(x < y .\equiv. y > x); \tag{5}$$

and the matrix '$x < y$' into the falsehood '$(x)(y)(x < y)$'. Such successive quantifiers apply, of course, one at a time to the matrix which immediately follows. Thus '$(x)(y)(x < y)$' is the result of applying the quantifier '(x)' to the matrix '$(y)(x < y)$', which is in turn the result of applying '(y)' to '$(x < y)$'; and similarly for (5). The strict reading of (5) is: 'No matter what x may be, it will be the case, no matter what y may be, that $x < y$ if and only if $y > x$.' It is more natural, however, to read such successive quantifiers together: 'No matter what x and y may be, $x < y$ if and only if $y > x$.' The order of the successive quantifiers is obviously immaterial; (5) could have been rendered equivalently as:

$$(y)(x)(x < y .\equiv. y > x). \tag{6}$$

(5) and (6) are logically equivalent statements; so are '$(x)(y)(x < y)$' and '$(y)(x)(x < y)$'.

If ϕ has n free variables, there are $n!$ statements which can be formed by prefixing distinct quantifiers in one or another order corresponding to all the free variables; though all $n!$ statements are, as just remarked, equivalent. An arbitrary one of these statements will be referred to as the *closure* of ϕ; the one, say, in which the added quantifiers are applied in alphabetic order. Accordingly, where $\alpha_1, \alpha_2, \ldots, \alpha_n$ ($n \geqslant 0$) are in alphabetic order all the free variables of ϕ, the closure of ϕ is $\ulcorner(\alpha_n)\ldots(\alpha_2)(\alpha_1)\phi\urcorner$; it is formed by applying $\ulcorner(\alpha_1)\urcorner$ to ϕ, then $\ulcorner(\alpha_2)\urcorner$ to the result, and so on. The *alphabetic* order of variables may for this and subsequent purposes be fixed arbitrarily thus:

$$w \quad x \quad y \quad z \quad w' \quad x' \quad y' \quad z' \quad w'' \quad \ldots$$

Thus (6), but not (5), is the closure of '$x < y \,.\!\equiv\!.\, y > x$'. Again, since '$w$', '$x$', and '$z$' are in alphabetic order all the free variables of:

$$(y)(w \text{ loves } y \,.\!\supset\!.\, x \text{ hates } z),$$

the closure of this matrix is the falsehood:

(7) $$(z)(x)(w)(y)(w \text{ loves } y \,.\!\supset\!.\, x \text{ hates } z).$$

At the same time (7) is the closure of:

$$(w)(y)(w \text{ loves } y \,.\!\supset\!.\, x \text{ hates } z)$$

and of: $$(x)(w)(y)(w \text{ loves } y \,.\!\supset\!.\, x \text{ hates } z).$$

Also it is the closure of itself; every statement is its own closure. On the other hand (7) is not the closure of:

$$w \text{ loves } y \,.\!\supset\!.\, x \text{ hates } z;$$

the closure of this is rather:

$$(z)(y)(x)(w)(w \text{ loves } y \,.\!\supset\!.\, x \text{ hates } z),$$

which is equivalent to (7) but is not the same statement.

Peano's nomenclature for bound and free variables, carried over by Whitehead and Russell, was 'apparent' and 'real'. The term 'free' was applied to variables by Hilbert (1922); but the full terminology of bondage and freedom elaborated above is, with the exception of 'bound with respect to', substantially von Neumann's ("Beweistheorie", p. 9). The notion of closure is an adaptation from Carnap (*Syntax*, p. 94).

§ 15. *Axioms of Quantification*

OUT OF THE totality of logically true statements, one special kind — the tautologous — has thus far been considered. Statements of this kind are true by virtue solely of their structure in terms of truth-functional composition; i.e., in terms of joint denial. Now the advent of quantification opens up a wider class of logical truths: statements which are true by virtue of their structure in terms of joint denial and quantification. These may be called *quantificationally* true. Church has shown [1] that it is impossible to devise a general test whereby, given any statement, we can mechanically decide whether or not it is quantificationally true. Quantificational truth is thus more elusive than tautology; for we have a mechanical test of tautology in the truth table.

It remains possible to mark out the class of quantificational truths by appeal to a generative principle, as follows. We first single out certain fundamental sets of quantificational truths which *are* satisfactorily recognizable, either on sight or by mechanical tests; then we describe a simple rule (*modus ponens*, § 16) whereby further quantificational truths can be progressively generated. The demarcation of quantificational truths which is thus achieved is *complete*, in that every quantificational truth can in fact be generated from the fundamental ones by some chain of applications of the rule; [2] but the appropriate chain of applications may elude us in particular cases, and herein consists the lack of a mechanical test.

The quantificational truths are but a species in turn of logical truths in a still more inclusive sense (§ 28). We shall eventually find (§ 60) that this latter category resists even the generative type of demarcation available for quantificational truth.

In this section the fundamental sets of quantificational truths alluded to above will be specified. One such set is typified by:

(1) $(x)(x \text{ is red} \;.\supset.\; x \text{ is red})$,

which is the closure of a tautologous matrix 'x is red $.\supset.$ x is red'. The quantifier is essential to the truth of this statement, for without it we have only a matrix, neither true nor false. Tautologous formulæ are not always true, for they are not always statements; their closures are statements, however, and true. Here we have a new set of logical truths, more inclusive than the tautologous state-

[1] Church, "A Note." See also Hilbert and Bernays, vol. 2, pp. 416 ff, and Rosser, "Informal Exposition."

[2] Gödel proved this in 1930. See also Henkin. These arguments relate to other versions of quantification theory, but can be transferred to the present one.

ments; namely, the closures of tautologous formulæ. Let us call this set A. It includes the tautologous statements, since a tautologous statement is a tautologous formula which is its own closure.

Next consider the statement:

$$(x)(x \text{ is colored } .\supset.\ x \text{ is extended}) \supset. \quad (x)(x \text{ is colored}) \supset (x)(x \text{ is extended});$$

in words, 'If everything colored is extended then if everything is colored everything is extended.' Any statement of this sort — any statement of the form

(2) $\ulcorner(\alpha)(\phi \supset \psi) \supset. (\alpha)\phi \supset (\alpha)\psi\urcorner$

— will of course be true. If ϕ or ψ has a free variable other than α, the whole will be a matrix rather than a statement; but still its closure will be true. The closures of formulæ of the form indicated in (2) thus constitute another set of logical truths, which we may call B.

Logical truths of a trivial kind, exemplified by:

Socrates is mortal $.\supset (x)$(Socrates is mortal),

issue from our decision (§ 13) to construe $\ulcorner(\alpha)\phi\urcorner$ as equivalent to ϕ in cases where ϕ lacks α. In general, where ϕ lacks α, the closure of $\ulcorner\phi \supset (\alpha)\phi\urcorner$ (and, indeed, of $\ulcorner\phi \equiv (\alpha)\phi\urcorner$) will be true. Furthermore, the requirement that ϕ "lack" α need not be taken to forbid occurrences of α within a shorter quantification $\ulcorner(\alpha)\psi\urcorner$ inside ϕ; for such occurrences are bound to that inner occurrence of $\ulcorner(\alpha)\urcorner$, and are irrelevant to the initial $\ulcorner(\alpha)\urcorner$ of $\ulcorner(\alpha)\phi\urcorner$ (cf. §§ 13, 14). The only occurrences of α in ϕ which are bound to the initial $\ulcorner(\alpha)\urcorner$ of $\ulcorner(\alpha)\phi\urcorner$ are the occurrences which are free in ϕ (cf. (ii), § 14); and it is only such occurrences that need be forbidden in order that $\ulcorner(\alpha)\phi\urcorner$ be a case of vacuous quantification, equivalent to ϕ. Here, then, is another set of logical truths: closures of formulæ $\ulcorner\phi \supset (\alpha)\phi\urcorner$ such that α is not free in ϕ. Let us call this set C.

Now consider the statements:

(3) $(x)(x$ is a man $.\supset.\ x$ is mortal$) \supset:$ Socrates is a man $.\supset.$ Socrates is mortal,

(4) (x)(God created x) $\supset.$ God created Socrates,

(5) (x)(God created x) $\supset.$ God created God;

in words, 'If all men are mortal then if Socrates is a man Socrates is mortal'; 'If God created everything then God created Socrates'; 'If God created everything then God created himself.' Each of these statements is a conditional of the form $\ulcorner(\alpha)\phi \supset \phi'\urcorner$ where ϕ' affirms of some specific object (Socrates, God) what $\ulcorner(\alpha)\phi\urcorner$ affirms of all objects. In each case ϕ' is like ϕ except for containing some name ζ ('Socrates', 'God') where ϕ contains α. More accurately, in general, ϕ' is like ϕ except for containing ζ wherever ϕ contains *free* occurrences of α; for it is only the free occurrences of α in ϕ that are relevant to the initial $\ulcorner(\alpha)\urcorner$ of $\ulcorner(\alpha)\phi\urcorner$. Any conditional statement of this kind will of course be true, since whatever can be affirmed of everything (as in $\ulcorner(\alpha)\phi\urcorner$) can be affirmed of any particular object we choose to name (as in ϕ').

Furthermore, the references to Socrates are obviously inessential to the truth of (3) and (4); corresponding conditionals are true of all men, indeed all entities, not just Socrates. Hence the general statement which is formed from (4) or from (3) by putting a variable for 'Socrates', and quantifying the whole, is likewise true:

(6) $(y)((x)(\text{God created } x) \supset. \text{ God created } y)$,

(7) $(y)((x)(x \text{ is a man} \,.\!\supset.\, x \text{ is mortal}) \supset: y \text{ is a man} \,.\!\supset.\, y \text{ is mortal})$.

Similarly the references to God are inessential to the truth of (5) and (6); the general statements formed from (5) and (6) by putting a variable for 'God', and quantifying the wholes, are likewise true:

(8) $(w)((x)(w \text{ created } x) \supset. w \text{ created } w)$,

(9) $(w)(y)((x)(w \text{ created } x) \supset. w \text{ created } y)$.

Whereas (3)–(5) are statements $\ulcorner(\alpha)\phi \supset \phi'\urcorner$ such that ϕ' is like ϕ except for containing a name ζ wherever ϕ contains free occurrences of α, on the other hand (6)–(9) are closures of formulæ $\ulcorner(\alpha)\phi \supset \phi'\urcorner$ such that ϕ' is like ϕ except for containing a variable α' wherever ϕ contains free occurrences of α. Let this latter set of statements, exemplified by (6)–(9), be called D′. (D will come later.) This set does not include (3)–(5). Indeed, the set exemplified by (3)–(5) is not satisfactorily specifiable at the present stage; for its specification depends on the notion of name, and nothing has been said thus far to indicate what sorts of expressions are to be classed as names. But the set exemplified by (3)–(5) can be passed over; we shall find later (§ 27) that names are eliminable — that statements involving names are translatable in a way which avoids recognition of any ultimate name-category.[1]

Along with (6)–(9), the set D′ as described includes the statement:

$$(10) \qquad (x){\sim}(y)(y = x) \supset {\sim}(y)(y = y);$$

for, (10) is the closure of $\ulcorner(\alpha)\phi \supset \phi'\urcorner$ where α is 'x', ϕ is '${\sim}(y)(y = x)$', and ϕ' is like ϕ except for containing 'y' wherever ϕ contains free occurrences of 'x'. But this reveals a defect in the formulation of D′, for (10) is in fact false. The antecedent of the conditional (10) says that, whatever entity may be selected, not every entity will be identical with it. This much is true; if e.g. the number 5 is selected then the number 1 is not identical with it, while if any other entity is selected then the number 5 is not identical with it. The consequent of the conditional (10), on the other hand, is obviously false; it denies that everything is self-identical. Thus (10) as a whole, having a true antecedent and false consequent, is false.

The formulæ whose closures were intended to belong to D′ were, roughly speaking, of the sort:

$$(x)(\ldots x \ldots) \supset \ldots y \ldots$$

where the consequent '$\ldots y \ldots$' imposes upon y the condition which '$\ldots x \ldots$' imposes upon x. Any such closure will indeed be true. But (10) is of a different sort; whereas its antecedent says:

$(x) \sim$ (everything is identical with x),

its consequent does *not* say correspondingly:

$\sim$ (everything is identical with y),

but rather:

$\sim$ (everything is self-identical).

[1] Names and other terms will subsequently make their formal appearance through certain conventions of notational abbreviation (§§ 24, 26, 27); and a principle covering the logical truths exemplified in (3)–(5) will then be derived (*231, § 31).

The source of this deviation on the part of (10) is apparent: when the free occurrence of α (i.e. 'x') in ϕ (i.e. '$\sim(y)(y = x)$') is supplanted by α' (i.e. 'y'), the resulting occurrence of α' comes inadvertently to be appropriated to the purposes of a quantifier $\ulcorner(\alpha')\urcorner$ which was lurking inside ϕ. Our characterization of D$'$ can be adjusted, therefore, by stipulating that the occurrences of α' in ϕ', to which the free occurrences of α in ϕ give way, remain *likewise free*. The revised set, which may be called D, comprises the closures of formulæ $\ulcorner(\alpha)\phi \supset \phi'\urcorner$ such that ϕ' is like ϕ except for containing *free* occurrences of α' wherever ϕ contains free occurrences of α.

Statements of the four kinds A, B, C, and D will play a basic rôle in subsequent developments, and a general term is needed for referring to them. For lack of a better term, they will be called *axioms of quantification*. Thus the axioms of quantification comprise the closures of all tautologous formulæ; also the closures of all formulæ of the form

$$\ulcorner(\alpha)(\phi \supset \psi) \supset . (\alpha)\phi \supset (\alpha)\psi\urcorner;$$

also the closures of all formulæ $\ulcorner\phi \supset (\alpha)\phi\urcorner$ such that α is not free in ϕ; also the closures of all formulæ $\ulcorner(\alpha)\phi \supset \phi'\urcorner$ such that ϕ' is like ϕ except for containing free occurrences of α' wherever ϕ contains free occurrences of α.

§ 16. *Theorems*

IT IS APPARENT, equally from the ordinary usage of 'if-then' and from the truth table of '$\supset$', that ψ will be true if $\ulcorner\phi \supset \psi\urcorner$ and ϕ are true. The form of inference which carries us thus from premisses $\ulcorner\phi \supset \psi\urcorner$ and ϕ to the conclusion ψ is known traditionally as the *modus ponens;* hence let us speak of ψ as the *ponential* of $\ulcorner\phi \supset \psi\urcorner$ and ϕ. In general, let us call ψ a ponential of $\phi_1, \phi_2, \phi_3, \ldots$ if one of $\phi_1, \phi_2, \phi_3, \ldots$ is a conditional having another of $\phi_1, \phi_2, \phi_3, \ldots$ as antecedent and ψ as consequent. Ponentials of true statements are true; and, in particular, ponentials of logically true statements are logically true.

Though the axioms of quantification comprise infinitely many logical truths, they leave further infinities of logical truths untouched. A broader class of logical truths is obtainable by throwing in the ponentials of the axioms of quantification; a still broader one is obtainable by throwing in the ponentials of the thus supplemented class; and so on.

The axioms of quantification, together with the ponentials of those axioms, together with the ponentials of this further totality, and so on, will be referred to collectively as *theorems*. To say that ϕ is a theorem is to say that there is a sequence of statements ψ_1, ψ_2, . . . , ψ_n, ϕ each of which is either an axiom of quantification or a ponential of earlier statements of the sequence. The theorems can be characterized recursively as follows: (i) axioms of quantification are theorems, and (ii) if $\ulcorner \phi \supset \psi \urcorner$ and ϕ are theorems then so is ψ.[1]

The following three statements, e.g., are theorems:

(1) $(y)((x)(y \text{ affects } x) \supset . \, y \text{ affects } y)$,

(2) $(y)((x)(y \text{ affects } x) \supset . \, y \text{ affects } y) \supset .$
$(y)(x)(y \text{ affects } x) \supset (y)(y \text{ affects } y)$,

(3) $(y)(x)(y \text{ affects } x) \supset (y)(y \text{ affects } y)$.

(1) and (2) are axioms of quantification, of the respective kinds D and B (cf. § 15). (3) is not an axiom of quantification, but it is a theorem because it is the ponential of the theorems (1) and (2). The sequence (1), (2), (3) is of the kind ψ_1, ψ_2, . . . , ψ_n, ϕ described above.

The tautologous statements form one class of logical truths; the axioms of quantification form a broader class; and the theorems form a still broader class. This third class differs from the preceding two in this important respect: we have no general procedure whereby we can test whether a given statement belongs to the class or not. With respect to tautologous statements, such a test is provided by the truth table; given any statement, framed

[1] Any general notion which is resoluble into an infinite sequence of special cases is said to be recursively characterized when we have explained the first case and added a general rule describing the $(i + 1)$st case, for each i, in terms of the first i cases. Here the axioms of quantification constitute the first case, and the ponentials of theorems falling under the first i cases constitute the $(i + 1)$st case.

in our logical notation, we can appraise it systematically by the tabular method and discover whether or not it is tautologous.[1] With respect to the broader class of quantificational axioms, such a test is again forthcoming. Given a statement, we may first determine whether it is an axiom of quantification of the kind A; this is accomplished by dropping off any initial quantifiers, to get the net formula ϕ whereof the original statement was the closure, and then testing ϕ for tautology by table. If this investigation yields a negative answer, we next determine whether the statement is of the kind B; i.e., we observe whether ϕ is a conditional whose antecedent and consequent are alike except for a distributed quantifier as in (2) of § 15. In case of a negative answer, we check the statement in similar fashion against the type C; then against D.

But with respect to the class of theorems we have no such complete test. A statement ϕ may be a theorem without our being able to discover the fact; there may be a sequence $\psi_1, \psi_2, \ldots, \psi_n, \phi$ of the required kind without our being lucky enough to find it. Failure to discover such a sequence is no evidence that ϕ is not a theorem, though the chance discovery of such a sequence does show that ϕ is a theorem.

With respect to the class of theorems we do have a partial test, but only a partial one; namely, given any sequence $\psi_1, \psi_2, \ldots, \psi_n, \phi$, we can decide whether *it* is of the kind which makes ϕ a theorem. We can check each statement of the sequence, from ψ_3 onward, against each pair of its predecessors to see if it is a ponential thereof; then, reserving the ponentials, we can check each of the remaining statements to see whether it is an axiom of quantification. This reflects the characteristic mathematical situation; the mathematician hits upon his proof by unregimented insight and good fortune, but afterwards other mathematicians can check his proof.[2]

[1] For this purpose it is ordinarily essential that the statement be given explicitly and in full, of course, rather than merely described by quasi-quotation. Forms exhibited in quasi-quotation are often capable of being shared by statements which are tautologous and statements which are not; cf. § 10.

[2] Cf. Introduction; also §§ 55, 60.

Let us write '$\vdash \phi$' to mean that the closure of ϕ is a theorem.[1] Then the characterization of theorem embodied in (i) and (ii) above takes on the following form, once the notion of axiom of quantification is analyzed out.

***100.** *If* ϕ *is tautologous,* $\vdash \phi$.
***101.** $\vdash \ulcorner(\alpha)(\phi \supset \psi) \supset. (\alpha)\phi \supset (\alpha)\psi\urcorner$.
***102.** *If* α *is not free in* ϕ, $\vdash \ulcorner\phi \supset (\alpha)\phi\urcorner$.
***103.** *If* ϕ' *is like* ϕ *except for containing free occurrences of* α' *wherever* ϕ *contains free occurrences of* α*, then* $\vdash \ulcorner(\alpha)\phi \supset \phi'\urcorner$.
***104.** *If* $\ulcorner\phi \supset \psi\urcorner$ *and* ϕ *are theorems, so is* ψ.

The principles *100–*104 are to be understood, of course, as applying to all choices of ϕ, ψ, α, and α'. The three-digit numerals and asterisks belong to a scheme of cross-reference which will become clear in the sequel.[2]

Note that *104 cannot be abbreviated to read:

(4) If $\vdash \ulcorner\phi \supset \psi\urcorner$ and $\vdash \phi$ then $\vdash \psi$;

this means something more complicated than *104, namely that if the *closures of* $\ulcorner\phi \supset \psi\urcorner$ and ϕ are theorems then so is that of ψ. It will be found later (*111, § 17) that (4) does in fact also hold; it will be found that whenever $\ulcorner\phi \supset \psi\urcorner$ and ϕ are formulæ whose closures are theorems according to *100–*104, the closure of ψ is also a theorem according to *100–*104. But this result is no mere translation of *104; its substantiation requires a considerable train of reasoning.

Frege was perhaps the first to distinguish clearly between axioms and the rules of inference whereby theorems are generated from the axioms. Once this distinction is drawn, a recursive characterization of the class of theorems is virtually at hand. But the highly explicit way of presenting formal deductive systems which is customary nowadays dates back only to Hilbert (1922) or Post (1921). Under this plan a class of expressions to be known as formulæ is specified by reference to purely

---

[1] The sign '$\vdash$' was borrowed from Frege by Whitehead and Russell. The sense which they attached to it is somewhat obscure, but comes near enough to this sense to justify my retention of the notation.

[2] A special sense will subsequently (§ 17) be attached to suppression of the asterisk; also an important distinction will be recorded by using daggers sometimes instead of asterisks (§ 25). Concerning the choice of numerals, see footnotes to § 17.

notational features — ordinarily through a recursive characterization (cf. §§ 13, 23); then a subclass of formulæ to be known as theorems is singled out by another recursive characterization, referring again to none but notational features. The so-called axioms, i.e. those initial theorems which are introduced by the first of the two clauses in a recursive characterization of theorem, used to be taken as finite in number and hence presented by list; the practice of describing an infinity of axioms dates only from von Neumann (1927). It is this innovation that has enabled us to keep to *modus ponens* as sole rule of inference, following Tarski (1935). The combination of principles *100–*104 was anticipated in its main outlines by Fitch in 1938.

In the first edition an additional kind of axioms of quantification was recognized, beyond the present kinds A, B, C, and D; viz., the closures of all formulæ of the form $\ulcorner(\alpha)(\beta)\phi \supset (\beta)(\alpha)\phi\urcorner$. Also the concept of closure differed slightly from the present one, in that the quantifiers were arranged in the opposite order. But Berry showed that by switching to the present version of closure we could dispense with those additional axioms and still get them as theorems. The proofs of *115 and *119 below are due in principle to Berry.

Most of the "metatheorems" running through the rest of the present chapter were recorded by one or another of Frege, Peirce, and Schröder, and comprehensively systematized by Whitehead and Russell (*9–*11).

§ 17. *Metatheorems*

STUDY of *100–*104 reveals an endless variety of general conditions under which statements will be theorems. By recording conditions of this kind once and for all, we avoid the labor of writing out specific sequences of quantificational axioms and ponentials to establish individual theorems falling under those conditions. We establish theorems wholesale, by arguments which show that the appropriate sequences *could* be found for each particular case. Such principles, describing general circumstances under which statements are theorems, will be called *metatheorems*. *100–*104 themselves are our initial metatheorems.

So long as the atomic formulæ remain unspecified, indeed, it is impossible to cite any specific formulæ (cf. § 13) and hence impossible to cite any specific theorems. Apart from fictitious examples in which makeshift atomic formulæ are borrowed from ordinary language (e.g. (3) of § 16), we can only cite the general forms of theorems — using Greek letters as in *100–*104. At a later stage (§§ 25 ff.), after the specification of atomic formulæ, individual theorems will be forthcoming.

The following metatheorem is merely a special case of *103.

***110.**[1] $\vdash \ulcorner (\alpha)\phi \supset \phi \urcorner$. *Proof:* *103 (taking α' as α).

A less trivial metatheorem, anticipated in earlier discussion (§ 16; see also Appendix), is this:

***111.** *If* $\vdash \ulcorner \phi \supset \psi \urcorner$ *and* $\vdash \phi$ *then* $\vdash \psi$.

Proof. Let a number n be called *favorable* if *111 holds for all conditionals $\ulcorner \phi \supset \psi \urcorner$ having n or fewer free variables. What we want to show is that every number is favorable. Suppose that k is favorable, and consider any conditional $\ulcorner \phi \supset \psi \urcorner$ which has just $k + 1$ free variables and is such further that

$$\vdash \ulcorner \phi \supset \psi \urcorner, \tag{1}$$

$$\vdash \phi. \tag{2}$$

Where α is alphabetically the first of those free variables, the closure of $\ulcorner \phi \supset \psi \urcorner$ is also the closure of $\ulcorner (\alpha)(\phi \supset \psi) \urcorner$ (cf. § 14); hence, by (1),

$$\vdash \ulcorner (\alpha)(\phi \supset \psi) \urcorner. \tag{3}$$

Since the conditional $\ulcorner (\alpha)(\phi \supset \psi) \supset . (\alpha)\phi \supset (\alpha)\psi \urcorner$ has just k free variables, and k is favorable, we see from *101 and (3) that

$$\vdash \ulcorner (\alpha)\phi \supset (\alpha)\psi \urcorner. \tag{4}$$

Now if α is free in ϕ, it is alphabetically the first free variable of ϕ; and in this case the closure of ϕ is also the closure of $\ulcorner (\alpha)\phi \urcorner$, so that

$$\vdash \ulcorner (\alpha)\phi \urcorner \tag{5}$$

by (2). If on the other hand α is not free in ϕ, the conditional $\ulcorner \phi \supset (\alpha)\phi \urcorner$ has at most k free variables; and in this case, since k is favorable, (5) is forthcoming in view of *102 and (2). In either case, thus, (5) holds. But the conditional $\ulcorner (\alpha)\phi \supset (\alpha)\psi \urcorner$ has just k free variables; by (4) and (5), then, since k is favorable,

[1] The numeration of metatheorems will not be altogether consecutive. Commonly, as in the present case, the numeral of the first metatheorem of a section will open a fresh decade. Numerals having '0' in the decade position are reserved to the initial metatheorems *100–*104 of quantification and the initial metatheorems *200–*202 of membership (§ 29).

$$\vdash \ulcorner(\alpha)\psi\urcorner. \qquad (6)$$

Moreover, by *110, $\vdash \ulcorner(\alpha)\psi \supset \psi\urcorner$. (7)

Now if α is free in ψ, it is alphabetically the first free variable of ψ; and in this case the closure of ψ is also the closure of $\ulcorner(\alpha)\psi\urcorner$, so that

$$\vdash \psi \qquad (8)$$

by (6) alone. If on the other hand α is not free in ψ, $\ulcorner(\alpha)\psi \supset \psi\urcorner$ has at most k free variables; and in this case, since k is favorable, (8) is forthcoming in view of (7) and (6). We thus conclude that $k + 1$ is a favorable number — having assumed that k was. But 0 is favorable, by *104; hence, taking k as 0 in the foregoing argument, we conclude that 1 is favorable; hence, taking k as 1, we conclude that 2 is favorable; and so on. *111 thus holds regardless of the number of free variables.

In future proofs, *111 will be made the basis of a condensed notation which may be illustrated thus:

	$\vdash \ulcorner(\beta)(\psi \supset \phi)\urcorner$.	(1)
*101	$\vdash \ulcorner[1 \supset .]\ (\beta)\psi \supset (\beta)\phi\urcorner$.	(2)

Suppose line (1) given or previously justified. Now the numeral '1' in line (2) stands as an abbreviation for the whole expression '$(\beta)(\psi \supset \phi)$' which appears within corners in (1). Thus, if for the moment we ignore the pair of square brackets, line (2) amounts to:

$$\vdash \ulcorner(\beta)(\psi \supset \phi) \supset . (\beta)\psi \supset (\beta)\phi\urcorner;$$

and it is this line that *101 is cited to justify. Now from this line and (1) we can conclude by *111 that

$$\vdash \ulcorner(\beta)\psi \supset (\beta)\phi\urcorner;$$

and it is this step that the square brackets express. Any subsequent citation of (2) refers to line (2) minus its bracketed part.

The conventions involved may be summed up as follows. Square brackets indicate deletion, on the basis of *111, of an antecedent whose closure has been shown to be a theorem. A reference at the left of a line justifies the line inclusive of any bracketed matter. A parenthesized numeral at the right of a line L labels L exclusive of any bracketed matter. Divorced of its parentheses, this numeral serves in the sequel as an abbreviation of the whole expression (exclusive of any bracketed matter) which appeared within corners in *L*.

Sometimes, as in the line:

$$\vdash \ulcorner [1 \supset : 2 \supset .]\ \phi_1 \supset \phi_3 \urcorner$$

which appears in the proof of the next metatheorem, square brackets indicate repeated use of *111. This line is merely a condensation of two lines:

$$\vdash \ulcorner [1 \supset :]\ 2 \supset .\ \phi_1 \supset \phi_3 \urcorner, \qquad (2')$$

$$(2') \qquad \vdash \ulcorner [2 \supset .]\ \phi_1 \supset \phi_3 \urcorner.$$

***112.** *If* $\vdash \ulcorner \phi_1 \supset \phi_2 \urcorner, \vdash \ulcorner \phi_2 \supset \phi_3 \urcorner, \ldots,$ *and* $\vdash \ulcorner \phi_{n-1} \supset \phi_n \urcorner$ *then* $\vdash \ulcorner \phi_1 \supset \phi_n \urcorner$.

Proof. hp 1 $\vdash \ulcorner \phi_1 \supset \phi_2 \urcorner$. (1)
hp 2 $\vdash \ulcorner \phi_2 \supset \phi_3 \urcorner$. (2)
*100 $\vdash \ulcorner [1 \supset : 2 \supset .]\ \phi_1 \supset \phi_3 \urcorner$. (3)
hp 3 $\vdash \ulcorner \phi_3 \supset \phi_4 \urcorner$. (4)
*100 $\vdash \ulcorner [3 \supset : 4 \supset .]\ \phi_1 \supset \phi_4 \urcorner$; and so on.

The notation 'hp 1' refers to the first of the hypotheses in the metatheorem which is being proved; 'hp 2' refers to the second hypothesis, and so on.

The citation of *100 calls upon the reader to observe that the formula depicted is tautologous, and hence that its closure is a theorem. The reader can always verify the claim of tautology, by truth table if not by inspection. In order that brief inspection may ordinarily suffice, the use of *100 will be limited to fairly simple forms of tautology: forms which, like (1)–(48) of § 11, are capable of depiction by means of not more than three distinct Greek letters and not more than seven Greek-letter occurrences.

Use of *112 will ordinarily be tacit, through the medium of a notation of *stacked conditionals*. If metatheorems $*m_1$, $*m_2$, ..., $*m_{n-1}$ show respectively that $\vdash \ulcorner \phi_1 \supset \phi_2 \urcorner$, $\vdash \ulcorner \phi_2 \supset \phi_3 \urcorner, \ldots,$ and $\vdash \ulcorner \phi_{n-1} \supset \phi_n \urcorner$, and we want then to draw the conclusion (*k*) that $\vdash \ulcorner \phi_1 \supset \phi_n \urcorner$, we depict the whole argument as follows:

$$\begin{array}{ll} *m_1 & \vdash \ulcorner \phi_1 \supset \phi_2 \urcorner \\ *m_2 & \quad\ \supset \phi_3 \urcorner \\ \cdot & \quad\ \cdot \\ \cdot & \quad\ \cdot \\ \cdot & \quad\ \cdot \\ *m_{n-1} & \quad\ \supset \phi_n \urcorner. \quad (k) \end{array}$$

In the fourth line of the proof of *117 below, thus, the entry '(2)' at the left is intended to show that $\vdash \ulcorner(\alpha_n)\phi \supset (\alpha_n)\psi\urcorner$; whereas what is labelled by the '(3)' at the right of that line is rather '$\vdash \ulcorner\phi \supset (\alpha_n)\psi\urcorner$'. Similarly, in the third line of the proof of *116 below, the entry '*101' is intended to show that

$$\vdash \ulcorner(\alpha)(\phi \supset \psi) \supset. (\alpha)\phi \supset (\alpha)\psi\urcorner;$$

whereas what is labelled by the '(2)' at the right of that line is rather:

$$\vdash \ulcorner(\alpha)(\phi \equiv \psi) \supset. (\alpha)\phi \supset (\alpha)\psi\urcorner$$

(bracketed matter being excluded as usual).

A similar procedure of stacking *biconditionals* is justified by the following metatheorem.

***113.** *If* $\vdash \ulcorner\phi_1 \equiv \phi_2\urcorner$, $\vdash \ulcorner\phi_2 \equiv \phi_3\urcorner$, . . . , *and* $\vdash \ulcorner\phi_{n-1} \equiv \phi_n\urcorner$ *then* $\vdash \ulcorner\phi_1 \equiv \phi_n\urcorner$.

Proof similar to that of *112.

***114.** $\vdash \ulcorner(\alpha_1)(\alpha_2)\ldots(\alpha_n)\phi \supset \phi\urcorner$.

Proof. Case 1: $n = 0$. *100.

Case 2: $n > 0$.

*110	$\vdash \ulcorner(\alpha_1)\ldots(\alpha_n)\phi$	$\supset (\alpha_2)\ldots(\alpha_n)\phi\urcorner$
*110		$\supset (\alpha_3)\ldots(\alpha_n)\phi\urcorner$
.		.
.		.
.		.
*110		$\supset \phi\urcorner$.

***115.** *If* $\vdash \phi$ *then* $\vdash \ulcorner(\alpha)\phi\urcorner$.

Proof. Case 1: α is not free in ϕ. By *102, hypothesis, and *111, $\vdash \ulcorner(\alpha)\phi\urcorner$.

Case 2: α is free in ϕ. Let $\beta_1, \ldots, \beta_m, \alpha, \gamma_1, \ldots, \gamma_n$ be the free variables of ϕ in alphabetic order, and let $\delta_1, \ldots, \delta_m$ be a segment of the alphabet beyond all variables of ϕ. For each i from 1 to m let us form ϕ_i from ϕ by changing $\beta_1, \ldots, \beta_i$ to $\delta_1, \ldots, \delta_i$. Now the closure of ϕ, by hypothesis a theorem, is simultaneously the closure of $\ulcorner(\beta_1)\phi\urcorner$; so

	$\vdash \ulcorner(\beta_1)\phi\urcorner$.	(1)
*103	$\vdash \ulcorner[1 \supset]\phi_1\urcorner$.	

But the closure of ϕ_1 is simultaneously that of $\ulcorner(\beta_2)\phi_1\urcorner$; so

$$\vdash \ulcorner(\beta_2)\phi_1\urcorner. \qquad (2)$$

$$*103 \qquad \vdash \ulcorner[2 \supset]\phi_2\urcorner.$$

Continuing thus, we conclude that $\vdash \phi_m$. But the closure of ϕ_m is $\ulcorner(\delta_m)\ldots(\delta_1)(\gamma_n)\ldots(\gamma_1)(\alpha)\phi_m\urcorner$; so

$$\vdash \ulcorner(\delta_m)\ldots(\delta_1)(\gamma_n)\ldots(\gamma_1)(\alpha)\phi_m\urcorner. \qquad (3)$$

$$*103 \qquad \vdash \ulcorner[3 \supset]\,(\delta_{m-1})\ldots(\delta_1)(\gamma_n)\ldots(\gamma_1)(\alpha)\phi_{m-1}\urcorner. \qquad (4)$$

$$*103 \qquad \vdash \ulcorner[4 \supset]\,(\delta_{m-2})\ldots(\delta_1)(\gamma_n)\ldots(\gamma_1)(\alpha)\phi_{m-2}\urcorner.$$

Continuing thus, we conclude that

$$\vdash \ulcorner(\gamma_n)\ldots(\gamma_1)(\alpha)\phi\urcorner. \qquad (5)$$

$$*114 \qquad \vdash \ulcorner[5 \supset]\,(\alpha)\phi\urcorner.$$

***116.** $\vdash \ulcorner(\alpha)(\phi \equiv \psi) \supset. (\alpha)\phi \equiv (\alpha)\psi\urcorner$.

Proof.

*100 $\vdash \ulcorner\phi \equiv \psi .\supset. \phi \supset \psi\urcorner$. (1)

*101 $\vdash \ulcorner[(\alpha)1 \supset.]\ (\alpha)(\phi \equiv \psi) \supset (\alpha)(\phi \supset \psi)\urcorner$

*101 $\supset. (\alpha)\phi \supset (\alpha)\psi\urcorner$. (2)

Similarly $\vdash \ulcorner(\alpha)(\phi \equiv \psi) \supset. (\alpha)\psi \supset (\alpha)\phi\urcorner$. (3)

*100 $\vdash \ulcorner[2 \supset:. 3 \supset:]\ (\alpha)(\phi \equiv \psi) \supset. (\alpha)\phi \supset (\alpha)\psi \,.\, (\alpha)\psi \supset (\alpha)\phi\urcorner$,

q.e.d. (cf. D5).

Attachment of a quantifier to a numeral within square brackets indicates that the application of *111 connoted by the brackets is to be preceded by use of *115. Thus the second line of the above proof embodies the following steps:

(1), *115 $\vdash \ulcorner(\alpha)1\urcorner$. (1′)

*101 $\vdash \ulcorner[1' \supset.]\ (\alpha)(\phi \equiv \psi) \supset (\alpha)(\phi \supset \psi)\urcorner$.

An expression following a citation of *100 may exhibit Greek letters in excess of the prescribed limit of three distinct letters and seven occurrences, so long as the whole exemplifies some general tautologous form which falls within the limit. At the end of the above proof, *100 is cited to introduce an expression which (when the abbreviations '2' and '3' are supplanted) contains fourteen occurrences of Greek letters; the whole is an instance, however, of the simple tautologous form

$$\ulcorner \chi_1 \supset \chi_2 \,.\supset:.\, \chi_1 \supset \chi_3 \,.\supset:\, \chi_1 \supset.\, \chi_2 \,.\, \chi_3 \urcorner$$

(a variant of (42), § 11).

***117.** *If* $\vdash \ulcorner \phi \supset \psi \urcorner$, *and none of* $\alpha_1, \ldots, \alpha_n$ *is free in* ϕ, *then* $\vdash \ulcorner \phi \supset (\alpha_1) \ldots (\alpha_n)\psi \urcorner$.

Proof. hp 1 $\vdash \ulcorner \phi \supset \psi \urcorner$. (1)
*101 $\vdash \ulcorner [(\alpha_n)1 \supset.] \; (\alpha_n)\phi \supset (\alpha_n)\psi \urcorner$. (2)
*102 (& hp 2) $\vdash \ulcorner \phi \supset (\alpha_n)\phi \urcorner$
(2) $\supset (\alpha_n)\psi \urcorner$. (3)

The argument leading from (1) to (3) leads in turn from (3) to:

$$\vdash \ulcorner \phi \supset (\alpha_{n-1})(\alpha_n)\psi \urcorner,$$

thence to: $\vdash \ulcorner \phi \supset (\alpha_{n-2})(\alpha_{n-1})(\alpha_n)\psi \urcorner$, and so on.

Just as '1', '2', etc. are used as abbreviations of the expressions appearing within corners in the lines (1), (2), etc. of a given proof, so '102', '110', etc. will be used as abbreviations of the expressions appearing within corners in the metatheorems *102, *110, etc.[1] Thus '102', '110', and '118', used in the next proof, are short for '$\phi \supset (\alpha)\phi$', '$(\alpha)\phi \supset \phi$', and '$\phi \equiv (\alpha)\phi$'.

***118.** *If* α *is not free in* ϕ, $\vdash \ulcorner \phi \equiv (\alpha)\phi \urcorner$.

Proof. *100 $\vdash \ulcorner [102 \supset.\, 110 \supset] \; 118 \urcorner$.

***119.** $\vdash \ulcorner (\alpha)(\beta)\phi \equiv (\beta)(\alpha)\phi \urcorner$.

Proof. *114 $\vdash \ulcorner (\alpha)(\beta)\phi \supset \phi \urcorner$. (1)
(1), *117 $\vdash \ulcorner (\alpha)(\beta)\phi \supset (\beta)(\alpha)\phi \urcorner$. (2)
Similarly $\vdash \ulcorner (\beta)(\alpha)\phi \supset (\alpha)(\beta)\phi \urcorner$. (3)
*100 $\vdash \ulcorner [2 \supset.\, 3 \supset] \; 119 \urcorner$.

[1] This convention was suggested by my student Miss Leigh D. Steinhardt. It would not work if the numeration of metatheorems had begun with '*1' rather than '*100', since *1, *2, etc. would cease to be distinguished from (1), (2), etc.

§18. *Substitutivity of the Biconditional*

IN § 13 a restriction was imposed according to which one formula can occur in another only in a context of quantification or truth-functional composition. This restriction gives rise, it will be found, to the following convenient *substitutivity principle:* if ϕ and ϕ' are statements agreeing in truth value, then ϕ' can be substituted for any occurrences of ϕ in any statement ψ without affecting the truth value of ψ. In other words, if the statements ψ and ψ' are alike except that ψ' contains the statement ϕ' in places where ψ contains the statement ϕ, then ψ and ψ' are alike in truth value if ϕ and ϕ' are. In other words, (I) *any statement of the form* $\ulcorner \phi \equiv \phi' . \supset . \psi \equiv \psi' \urcorner$ *is true, where* ψ' *is like* ψ *except for containing* ϕ' *in places where* ψ *contains* ϕ. E.g., the conditional:

(1) Smith met Jones .≡. Jones was in Omaha :⊃.
(x)(Smith met Jones .∨ ∼ (Smith sold x to Jones)) ≡
(x)(Jones was in Omaha .∨ ∼ (Smith sold x to Jones))

is true.

Moreover, any such conditional is true independently of the particular names which happen to occur in it; replacement of names by free variables leaves a matrix which is (roughly speaking) true for all values of the variables — a matrix whose closure is true. E.g., the matrix:

(2) y met z .≡. z was in w :⊃.
$(x)(y$ met z .∨ ∼ $(y$ sold x to $z)) \equiv$
$(x)(z$ was in w .∨ ∼ $(y$ sold x to $z))$

has a true closure:

(3) $(z)(y)(w)(y$ met z .≡. z was in w :⊃.
$(x)(y$ met z .∨ ∼ $(y$ sold x to $z)) \equiv$
$(x)(z$ was in w .∨ ∼ $(y$ sold x to $z)))$.

This does not, however, entitle us to generalize (I) in the following fashion: (II) *the closure of* $\ulcorner \phi \equiv \phi' . \supset . \psi \equiv \psi' \urcorner$ *is true where* ψ' *is like* ψ *except for containing* ϕ' *in places where* ψ *contains* ϕ. For, many conditional matrices of the kind described in (II) do not correspond in the intended fashion to the conditional statements described in (I). One is this:

(4) $x = x .\equiv. x < y :\supset. (x)(x = x) \equiv (x)(x < y).$

When names are put for the free variables of (2), we have a conditional statement — e.g. (1) — of the kind described in (I); but when names are put for the free variables of (4), we get conditional statements such as the following, altogether alien to (I):

(5) $0 = 0 .\equiv. 0 < 1 :\supset. (x)(x = x) \equiv (x)(x < 1).$

Far from being a conditional of the kind described in (I), (5) is not even true — as is readily seen by observing the truth of '$0 = 0 .\equiv. 0 < 1$' and '$(x)(x = x)$' and the falsehood of '$(x)(x < 1)$' and then calculating the truth value of the whole by truth-tables. (II) is too strong; the matrix (4) is not true for all values of its variables, i.e. does not have a true closure, as the counter-instance (5) demonstrates.

The essential difference between (2) and (4) is apparent. Roughly speaking, the trouble with (4) is that when the '$x = x$' and '$x < y$' of the antecedent recur in the consequent they fall under new quantifiers which capture free variables of the antecedent and divert them to their own purposes. The occurrences of '$x = x$' and '$x < y$' in the consequent of (4) resemble those in the antecedent by virtue only of an alphabetic accident; for the occurrences of 'x' in '$(x)(x = x)$' and '$(x)(x < y)$' are governed by the adjacent occurrences of the quantifier '(x)', and have nothing but an alphabetic coincidence in common with the occurrences of 'x' in '$x = x .\equiv. x < y$'. (4) could be rendered equivalently as:

(6) $x = x .\equiv. x < y :\supset. (z)(z = z) \equiv (w)(w < y),$ to which (II) is obviously irrelevant.

The corrected form of (II), which avoids such unintended cases as (4), is this: (III) *the closure of* $\ulcorner\phi \equiv \phi' .\supset. \psi \equiv \psi'\urcorner$ *is true where* ψ' *is like* ψ *except for containing* free *occurrences of* ϕ' *at some places where* ψ *contains* free *occurrences of* ϕ. (I) is that special case of (III) which arises when $\ulcorner\phi \equiv \phi' .\supset. \psi \equiv \psi'\urcorner$ is its own closure, i.e., a statement.

Nothing has been said thus far by way of establishing even (I), much less (III); all that has been shown is that (II) fails. But in the metatheorem *122 below it will be shown that the broad substitutivity principle (III) holds, and more: the closures of the conditionals described in (III) are not merely true, but theorems.

It turns out, moreover, that even such recalcitrant forms as (4) will hold if we modify them to the extent of attaching one or more appropriate quantifiers to the antecedent. In particular, whereas the closure of (4) as it stands is false, the closure of:

$$(x)(x = x \mathbin{.\equiv.} x < y) \mathbin{\supset.} (x)(x = x) \equiv (x)(x < y)$$

is true. A generalization of (III) or *122 which covers such cases appears below as *121. This is proved ahead of *122, which then follows as a special case.

Preparatory to proving *121 as a whole, it is convenient to prove the case where ψ' differs from ψ in point of just one occurrence of ϕ'. This case takes the form of the following metatheorem.

***120**. *If ψ' is like ψ except for containing ϕ' at a place where ψ contains ϕ, and $\alpha_1, \ldots, \alpha_n$ $(n \geqslant 0)$ exhaust the variables with respect to which these occurrences of ϕ and ϕ' are bound in ψ and ψ', then*

$$\vdash \ulcorner(\alpha_1)\ldots(\alpha_n)(\phi \equiv \phi') \mathbin{\supset.} \psi \equiv \psi'\urcorner.$$

Proof. Let $\phi_0, \phi_1, \ldots, \phi_k$ $(k \geqslant 0)$ be, in order of increasing length, all the formulæ which occur in ψ and contain the occurrence of ϕ in question; and let $\phi'_0, \phi'_1, \ldots, \phi'_k$ be the corresponding parts of ψ'. From the general method of constructing formulæ (§ 13) it is then apparent, for each i from 1 to k, that ϕ_i and ϕ'_i are either $\ulcorner\phi_{i-1} \downarrow \chi\urcorner$ and $\ulcorner\phi'_{i-1} \downarrow \chi\urcorner$ for some formula χ (*Case 1*) or else $\ulcorner\chi \downarrow \phi_{i-1}\urcorner$ and $\ulcorner\chi \downarrow \phi'_{i-1}\urcorner$ (*Case 2*) or else $\ulcorner(\beta)\ \phi_{i-1}\urcorner$ and $\ulcorner(\beta)\ \phi'_{i-1}\urcorner$ for some variable β (*Case 3*). Now if it happens that

$$\vdash \ulcorner(\alpha_1)\ldots(\alpha_n)(\phi \equiv \phi') \mathbin{\supset.} \phi_{i-1} \equiv \phi'_{i-1}\urcorner \tag{1}$$

it will follow that

$$\vdash \ulcorner(\alpha_1)\ldots(\alpha_n)(\phi \equiv \phi') \mathbin{\supset.} \phi_i \equiv \phi'_i\urcorner, \tag{2}$$

as will now be shown in each of the three cases.

Cases 1 & 2: (1) $\vdash \ulcorner(\alpha_1)\ldots(\alpha_n)(\phi \equiv \phi') \mathbin{\supset.} \phi_{i-1} \equiv \phi'_{i-1}\urcorner$
*100 $\mathbin{\supset.} \phi_i \equiv \phi'_i\urcorner.$

Case 3. Any occurrence of β, within the given occurrence of ϕ in ψ, will be bound to the occurrence of $\ulcorner(\beta)\urcorner$ which begins $\ulcorner(\beta)\ \phi_{i-1}\urcorner$ if not to a later occurrence of $\ulcorner(\beta)\urcorner$. Thus any occurrence of β which is free in ϕ will be bound to the occurrence of $\ulcorner(\beta)\urcorner$ which begins $\ulcorner(\beta)\ \phi_{i-1}\urcorner$ or to some later occurrence of $\ulcorner(\beta)\urcorner$ which is likewise outside ϕ. Hence if β is free in ϕ then the occurrence of ϕ is bound in ψ with respect to β. Then, by hp 2, β is among $\alpha_1, \ldots, \alpha_n$. By similar reasoning, if β is free in ϕ' then β is among $\alpha_1, \ldots, \alpha_n$. Thus, regardless of whether β is free in $\ulcorner\phi \equiv \phi'\urcorner$, β will not be free in $\ulcorner(\alpha_1)\ldots(\alpha_n)(\phi \equiv \phi')\urcorner$. Hence

$$\begin{array}{ll} (1), *117 & \vdash \ulcorner(\alpha_1)\ldots(\alpha_n)(\phi \equiv \phi') \supset (\beta)(\phi_{i-1} \equiv \phi'_{i-1})\urcorner \\ *116 & \qquad\qquad\qquad\qquad \supset .\ \phi_i \equiv \phi'_i\urcorner. \end{array}$$

In *Cases 1–3* alike, therefore, and hence for each i from 1 to k, (2) holds if (1) does. But, since ϕ_0 is ϕ and ϕ'_0 is ϕ',

$$*114 \qquad \vdash \ulcorner(\alpha_1)\ldots(\alpha_n)(\phi \equiv \phi') \supset .\ \phi_0 \equiv \phi'_0\urcorner.$$

Hence, taking i as 1, it follows that

$$\vdash \ulcorner(\alpha_1)\ldots(\alpha_n)(\phi \equiv \phi') \supset .\ \phi_1 \equiv \phi'_1\urcorner.$$

Hence, taking i as 2, it follows that

$$\vdash \ulcorner(\alpha_1)\ldots(\alpha_n)(\phi \equiv \phi') \supset .\ \phi_2 \equiv \phi'_2\urcorner.$$

By k such steps we conclude that

$$\vdash \ulcorner(\alpha_1)\ldots(\alpha_n)(\phi \equiv \phi') \supset .\ \phi_k \equiv \phi'_k\urcorner.$$

But ϕ_k is ψ and ϕ'_k is ψ'.

The more general metatheorem *121 is now easily established.

***121.** *If ψ' is like ψ except for containing ϕ' at some places where ψ contains ϕ, and $\alpha_1, \ldots, \alpha_n$ ($n \geqslant 0$) exhaust the variables with respect to which these occurrences of ϕ and ϕ' are bound in ψ and ψ', then*

$$\vdash \ulcorner(\alpha_1)\ldots(\alpha_n)(\phi \equiv \phi') \supset .\ \psi \equiv \psi'\urcorner.$$

Proof. Let those places be k in number; and, for each i from 0 to k, let ψ_i be like ψ except for containing ϕ' instead of ϕ at the first i of the k places. Then, for each i from 1 to k, ψ_i is like ψ_{i-1} except for containing an occurrence of ϕ' where ψ_{i-1} contains an occurrence of ϕ; wherefore, in view of hp 2,

$$*120 \quad \vdash \ulcorner(\alpha_1)\ldots(\alpha_n)(\phi \equiv \phi') \supset .\ \psi_{i-1} \equiv \psi_i\urcorner. \qquad (1)$$

$*100 \quad \vdash \ulcorner [1 \supset :] (\alpha_1) \ldots (\alpha_n)(\phi \equiv \phi') \,.\, \psi_{i-1} \,.\!\equiv .\, (\alpha_1) \ldots (\alpha_n)(\phi \equiv \phi') \,.\, \psi_i \urcorner.$

Thus $\vdash \ulcorner (\alpha_1) \ldots (\alpha_n)(\phi \equiv \phi') \,.\, \psi_0 \,.\!\equiv .\, (\alpha_1) \ldots (\alpha_n)(\phi \equiv \phi') \,.\, \psi_1 \urcorner$

$$\equiv .\, (\alpha_1) \ldots (\alpha_n)(\phi \equiv \phi') \,.\, \psi_2 \urcorner$$

$$\vdots$$

$$\equiv .\, (\alpha_1) \ldots (\alpha_n)(\phi \equiv \phi') \,.\, \psi_k \urcorner. \quad (2)$$

$*100 \quad \vdash \ulcorner [2 \supset :] (\alpha_1) \ldots (\alpha_n)(\phi \equiv \phi') \supset .\, \psi_0 \equiv \psi_k \urcorner.$

But ψ_0 is ψ and ψ_k is ψ'.

*120 differs from *121 in requiring that just one occurrence of ϕ in ψ give way to ϕ' in ψ'. This means, of course, just one occurrence from the point of view of unabbreviated notation. Even where only a single occurrence of ϕ appears in our definitionally abbreviated notations, expansion according to the definitions will invariably show that many occurrences are really present. In practice, therefore, *120 is not useful; the more general principle *121 is needed.

The principle *122 discussed earlier follows from *121 as a special case.

***122.** *If ψ' is like ψ except for containing free occurrences of ϕ' at some places where ψ contains free occurrences of ϕ, then*

$$\vdash \ulcorner \phi \equiv \phi' \,. \supset .\, \psi \equiv \psi' \urcorner.$$

Proof: *121 (taking n as 0).

*121 also has the following much used corollary, in which mention of freedom or bondage ceases to be needed: If $\vdash \ulcorner \phi \equiv \phi' \urcorner$ then ϕ and ϕ' are interchangeable anywhere. This takes the form of the following metatheorem.[1]

***123.** *If $\vdash \ulcorner \phi \equiv \phi' \urcorner$, and ψ' is formed from ψ by putting ϕ' for some occurrences of ϕ, then $\vdash \ulcorner \psi \equiv \psi' \urcorner$.*

Proof. Let $\alpha_1, \ldots, \alpha_n$ be all the variables of ϕ and ϕ'.

hp 1 $\quad \vdash \ulcorner \phi \equiv \phi' \urcorner.$ (1)

*121(& hp 2) $\quad \vdash \ulcorner [(\alpha_1) \ldots (\alpha_n) 1 \supset .] \, \psi \equiv \psi' \urcorner.$

The following is a corollary in turn of *123.

[1] A proof covering this metatheorem is given by Hilbert and Ackermann (Ch. 3, § 7); but they do not mention the stronger principles *121–*122.

***124.** *If* $\vdash \phi$ *and* $\vdash \chi$, *and* χ' *is formed from* χ *by putting* ψ *for some occurrences of* $\ulcorner \phi \supset \psi \urcorner$ *or* $\ulcorner \phi \equiv \psi \urcorner$ *or* $\ulcorner \psi \equiv \phi \urcorner$ *or* $\ulcorner \phi \,.\, \psi \urcorner$ *or* $\ulcorner \psi \,.\, \phi \urcorner$, *then* $\vdash \chi'$.

Proof of the case of $\ulcorner \phi \supset \psi \urcorner$:

hp 1	$\vdash \phi$.	(1)
hp 2	$\vdash \chi$.	(2)
*100	$\vdash \ulcorner [1 \supset :]\ \phi \supset \psi \,. \equiv \psi \urcorner$.	(3)
(3), *123	$\vdash \ulcorner \chi \equiv \chi' \urcorner$.	(4)
*100	$\vdash \ulcorner [4 \supset .\ 2 \supset]\ \chi' \urcorner$.	

The other cases are proved similarly.

*124 is made the basis of a powerful extension of the device of square brackets. Where $\vdash \phi$, *124 provides for the immediate deletion of $\ulcorner \phi \supset \urcorner$ or $\ulcorner \phi \equiv \urcorner$ or $\ulcorner \equiv \phi \urcorner$ or $\ulcorner \phi \,. \urcorner$ or $\ulcorner .\, \phi \urcorner$ from any position within any theorem. Square brackets, hitherto used only in the limited fashion justified by *111, can now be used to indicate deletions of all these kinds.[1]

§ 19. *Existential Quantification*

TO SAY that nothing fulfills a given condition is to say that everything fulfills the denial of that condition. To affirm e.g. that there are no objects x such that x is distinct from itself, we may write:

$$(x) \sim (x \text{ is distinct from } x);$$

i.e., 'No matter what x may be, $\sim$ (x is distinct from x).' To affirm that there are no black swans, we may write:

(1) $\qquad (x) \sim (x \text{ is black} \,.\, x \text{ is a swan}).$

In effect, thus, the composite prefix '$(x) \sim$' amounts to the words 'there are no objects x such that'.

To say that there *are* objects satisfying a given condition, we have only to deny that there are none. To say that there are black swans we have only to apply a denial sign to (1), obtaining:

[1] See the remarks on the proof of *136, § 19.

(2) $\sim (x) \sim (x \text{ is black} \,.\, x \text{ is a swan})$.

Just as '$(x) \sim$' amounts to 'there are no objects x such that', so '$\sim (x) \sim$' amounts to 'there is at least one object x such that', 'there is something x such that'. The parts of '$\sim (x) \sim$' do not, of course, hang together as a unit; in (2) the second '$\sim$' applies to:

$$(x \text{ is black} \,.\, x \text{ is a swan}),$$

the quantifier '(x)' applies to:

$$\sim (x \text{ is black} \,.\, x \text{ is a swan}),$$

and the first '$\sim$' applies to the whole quantification (1). But the configuration of prefixes '$\sim (x) \sim$' figures so prominently in subsequent developments that it is convenient to adopt a condensed notation for it; the customary one is '$(\exists x)$', which we may read 'there is something x such that'. Hence the following abbreviative convention:

D8. $\ulcorner(\exists\alpha)\urcorner$ *for* $\ulcorner\sim (\alpha) \sim\urcorner$.

Prefixes of the form $\ulcorner(\exists\alpha)\urcorner$ are known as *existential* (or *particular*) *quantifiers;* and, where distinction is necessary, the basic quantifiers $\ulcorner(\alpha)\urcorner$ are referred to as *universal* quantifiers.[1]

The first four of the ensuing metatheorems relate existential quantification with universal quantification in obvious fashion.

***130.** $\vdash \ulcorner\sim(\alpha)\phi \equiv (\exists\alpha)\sim\phi\urcorner$.

Proof. *100 $\vdash \ulcorner\phi \equiv \sim\sim\phi\urcorner$. (1)

(1), *123 $\vdash \ulcorner \sim(\alpha)\phi \equiv \sim(\alpha)\sim\sim\phi\urcorner$, q.e.d. (cf. D8).

Arguments of the above sort, wherein a tautology is adduced by *100 and then *123 is applied to it, will hereafter be rendered in a single line with the composite justification '*100, *123'; thus the above proof becomes:

*100, *123 $\vdash \ulcorner \sim (\alpha)\phi \equiv \sim (\alpha)\sim \sim \phi\urcorner$.

From such a condensed formulation we can always discern the two components of the relevant tautologous biconditional (in this case

[1] The version $\ulcorner\sim(\alpha)\sim\urcorner$ of $\ulcorner(\exists\alpha)\urcorner$ goes back to Frege. The notation $\ulcorner(\exists\alpha)\urcorner$ was adapted from Peano by Whitehead and Russell.

ϕ and $\ulcorner \sim \sim \phi \urcorner$), simply by comparing the two sides of the exhibited biconditional (in this case $\ulcorner \sim (\alpha)\phi \urcorner$ and $\ulcorner \sim (\alpha) \sim \sim \phi \urcorner$) and picking out their dissimilar ingredients.

The formulæ described in metatheorems and proofs are always to be thought of, theoretically, as written out in full without definitional abbreviations (cf. § 13). In practice, for the most part, we do not need to reflect on the details of the expanded form; but occasionally we do need to lift one or another abbreviation in order to recognize a particular step of proof. On such occasions the relevant definition is mentioned parenthetically, as in the above proof and the following one. Expansion of the abbreviation $\ulcorner (\exists \alpha) \urcorner$ into $\ulcorner \sim (\alpha) \sim \urcorner$, e.g., immediately reveals that the following metatheorem is covered by *100.

***131.** $\vdash \ulcorner \sim (\exists \alpha)\phi \equiv (\alpha) \sim \phi \urcorner$. *Proof:* *100 (& D8).

***132.** $\vdash \ulcorner \sim (\alpha_1) \ldots (\alpha_n)\phi \equiv (\exists \alpha_1) \ldots (\exists \alpha_n) \sim \phi \urcorner$.

Proof. *130 $\vdash \ulcorner \sim (\alpha_1) \ldots (\alpha_n)\phi \equiv (\exists \alpha_1) \sim (\alpha_2) \ldots (\alpha_n)\phi \urcorner$

*130, *123 $\equiv (\exists \alpha_1)(\exists \alpha_2) \sim (\alpha_3) \ldots (\alpha_n)\phi \urcorner$

.
.
.

*130, *123 $\equiv (\exists \alpha_1) \ldots (\exists \alpha_n) \sim \phi \urcorner$.

***133.** $\vdash \ulcorner \sim (\exists \alpha_1) \ldots (\exists \alpha_n)\phi \equiv (\alpha_1) \ldots (\alpha_n) \sim \phi \urcorner$.

Proof similar, using *131.

The next two metatheorems are the analogues, for existential quantification, of *103 and *110.

***134.** *If* ϕ' *is like* ϕ *except for containing free occurrences of* α' *wherever* ϕ *contains free occurrences of* α, *then* $\vdash \ulcorner \phi' \supset (\exists \alpha)\phi \urcorner$.

Proof. *103 (& hp) $\vdash \ulcorner (\alpha) \sim \phi \supset \sim \phi' \urcorner$. (1)

*100 (& D8) $\vdash \ulcorner [1 \supset] \, 134 \urcorner$.

***135.** $\vdash \ulcorner \phi \supset (\exists \alpha)\phi \urcorner$. *Proof:* *134 (taking α' as α).

A corollary follows.

***136.** $\vdash \ulcorner (\alpha)\phi \supset (\exists \alpha)\phi \urcorner$.

Proof. *100 $\vdash \ulcorner [110 \,.\, 135 \,.\!\supset] \, 136 \urcorner$.

The bracketing involved in the above proof, which gets rid of two numerals '110' and '135' at once, merely indicates repeated use of the simple bracketing device. It can be broken down thus:

$\vdash \ulcorner 110 \,.\, 135 \,.\!\supset 136 \urcorner$, $\quad \vdash \ulcorner [110 \,.] \; 135 \,.\!\supset 136 \urcorner$,
$\vdash \ulcorner 135 \supset 136 \urcorner$, $\quad \vdash \ulcorner [135 \supset] \; 136 \urcorner$,

or alternatively thus:

$\vdash \ulcorner 110 \,.\, 135 \,.\!\supset 136 \urcorner$, $\quad \vdash \ulcorner 110 \, [.\, 135] \,.\!\supset 136 \urcorner$,
$\vdash \ulcorner 110 \supset 136 \urcorner$, $\quad \vdash \ulcorner [110 \supset] \; 136 \urcorner$.

*124 is the justification of this bracketing out of '110 .' or '. 135'; and it justifies similar bracketing when '≡' occurs in place of the '.' of these examples. The bracketing of '135 ⊃' and '110 ⊃', above, is justified by *124 and *111 indifferently. Even in cases of this sort '[*n* ⊃]', however, *124 allows more freedom than *111 allowed; matter bracketed out according to *124 may stand in the interior of a line of proof, whereas *111 served only for bracketing out the main antecedent of the whole line.

The following metatheorem explains vacuous existential quantification. It is the analogue of *118.

***137.** *If* α *is not free in* ϕ, $\vdash \ulcorner \phi \equiv (\exists \alpha)\phi \urcorner$.

Proof. *118 (& hp) $\vdash \ulcorner {\sim} \phi \equiv (\alpha){\sim} \phi \urcorner$. (1)
*100 (& D8) $\vdash \ulcorner [1 \supset] \; 137 \urcorner$.

The next shows that consecutive existential quantifiers, like consecutive universal ones, are permutable. It is the analogue of *119.

***138.** $\vdash \ulcorner (\exists \alpha)(\exists \beta)\phi \equiv (\exists \beta)(\exists \alpha)\phi \urcorner$.

Proof. *100, *123 (& D8) $\vdash \ulcorner (\exists \alpha)(\exists \beta)\phi \equiv {\sim}\, (\alpha)(\beta){\sim}\, \phi \urcorner$
*119, *123 $\equiv {\sim}\, (\beta)(\alpha){\sim}\, \phi \urcorner$
*100, *123 (& D8) $\equiv (\exists \beta)(\exists \alpha)\phi \urcorner$.

The order in which consecutive quantifiers are written can make a difference, on the other hand, when one quantifier is existential and the other universal. '$(\exists x)(y)(x = y)$' is false, e.g., whereas '$(y)(\exists x)(x = y)$' is true. It does turn out that $\ulcorner (\exists \alpha)(\beta)\phi \urcorner$ implies $\ulcorner (\beta)(\exists \alpha)\phi \urcorner$, but the converse fails.

***139.** $\vdash \ulcorner (\exists \alpha)(\beta)\phi \supset (\beta)(\exists \alpha)\phi \urcorner$.

Proof. *101 $\vdash \ulcorner [(\beta) 135 \supset .] (\beta)\phi \supset (\beta)(\exists \alpha)\phi \urcorner$. (1)

*100 $\vdash \ulcorner[1 \supset.] \sim (\beta)(\exists\alpha)\phi \supset \sim (\beta)\phi\urcorner$. (2)
(2), *117 $\vdash \ulcorner\sim (\beta)(\exists\alpha)\phi \supset (\alpha)\sim (\beta)\phi\urcorner$. (3)
*100 (& D8) $\vdash \ulcorner[3 \supset]\ 139\urcorner$.

Now that the metatheorems have come to be scattered through a considerable number of pages, and an ever increasing number, the reader is advised to use the list at the end of the book in order to identify metatheorems cited in proofs. If a metatheorem is used only within the section in which it appears, it is not entered in the list; but in this case the reader needs only to glance back a page or two. The definitions are likewise assembled at the end of the book.

§ 20. *Distribution of Quantifiers*

WE TURN now to some principles dealing with the distribution of a quantifier through a binary compound.

***140.** $\vdash \ulcorner(\alpha)(\phi\,.\,\psi) \equiv.\ (\alpha)\phi\,.\,(\alpha)\psi\urcorner$.

Proof. *100 $\vdash \ulcorner\phi\,.\,\psi\,.\!\supset \phi\urcorner$. (1)
*101 $\vdash \ulcorner[(\alpha)1 \supset.](\alpha)(\phi\,.\,\psi) \supset (\alpha)\phi\urcorner$. (2)
Similarly $\vdash \ulcorner(\alpha)(\phi\,.\,\psi) \supset (\alpha)\psi\urcorner$. (3)
*100 $\vdash \ulcorner(\alpha)\phi\,.\,(\alpha)\psi\,.\!\supset (\alpha)\phi\urcorner$
*110 $\supset \phi\urcorner$. (4)
Similarly $\vdash \ulcorner(\alpha)\phi\,.\,(\alpha)\psi\,.\!\supset \psi\urcorner$. (5)
*100 $\vdash \ulcorner[2\,.\,3\,.\!\supset:](\alpha)(\phi\,.\,\psi) \supset.\ (\alpha)\phi\,.\,(\alpha)\psi\urcorner$. (6)
*100 $\vdash \ulcorner[4\,.\,5\,.\!\supset:](\alpha)\phi\,.\,(\alpha)\psi\,.\!\supset.\ \phi\,.\,\psi\urcorner$. (7)
(7), *117 $\vdash \ulcorner(\alpha)\phi\,.\,(\alpha)\psi\,.\!\supset (\alpha)(\phi\,.\,\psi)\urcorner$. (8)
*100 $\vdash \ulcorner[6\,.\,8\,.\!\supset]140\urcorner$.

***141.** $\vdash \ulcorner(\exists\alpha)(\phi \vee \psi) \equiv.\ (\exists\alpha)\phi \vee (\exists\alpha)\psi\urcorner$.

Proof. *100, *123 $\vdash \ulcorner(\alpha)\sim (\phi \vee \psi) \equiv (\alpha)(\sim \phi\,.\sim \psi)\urcorner$
*140 $\equiv.\ (\alpha)\sim \phi\,.\,(\alpha)\sim \psi\urcorner$. (1)
*100 (& D8) $\vdash \ulcorner[1 \supset]141\urcorner$.

*140 shows that a universal quantifier can be distributed through a conjunction, and *141 shows that an existential quantifier can be distributed through an alternation. But distribution of a universal quantifier through an alternation is not in general valid, nor is distribution of an existential quantifier through a conjunction. The biconditionals:

$(x)(x \text{ is red}. \vee \sim (x \text{ is red})) \equiv. (x)(x \text{ is red}) \vee (x) \sim (x \text{ is red}),$
$(\exists x)(x \text{ is red} \,.\, \sim (x \text{ is red})) \equiv. (\exists x)(x \text{ is red}) \,.\, (\exists x) \sim (x \text{ is red}),$

e.g., are false, as can be seen by observing the truth of:

$(x)(x \text{ is red} .\vee \sim (x \text{ is red})), \quad (\exists x)(x \text{ is red}), \quad (\exists x) \sim (x \text{ is red})$

and the falsehood of:

$(\exists x)(x \text{ is red} \,.\, \sim (x \text{ is red})), \quad (x)(x \text{ is red}), \quad (x) \sim (x \text{ is red})$

and then applying truth tables.

*141 has the following corollary.[1]

***142.** $\vdash \ulcorner (\exists \alpha)(\phi \supset \psi) \equiv. (\alpha)\phi \supset (\exists \alpha)\psi \urcorner.$

Proof. *141 (& D4, 8) $\vdash \ulcorner (\exists \alpha)(\phi \supset \psi) \equiv. (\alpha) \sim \sim \phi \supset (\exists \alpha)\psi \urcorner$
*100, *123 $\equiv. (\alpha)\phi \supset (\exists \alpha)\psi \urcorner.$

The citation of D4 and D8 in this proof calls attention to the fact that

$\ulcorner (\exists \alpha)(\sim \phi \vee \psi) \equiv. (\exists \alpha) \sim \phi \vee (\exists \alpha)\psi \urcorner$
and $\ulcorner (\exists \alpha)(\phi \supset \psi) \equiv. (\alpha) \sim \sim \phi \supset (\exists \alpha)\psi \urcorner$

are the same formula under different abbreviations.

*140 equates a quantified conjunction with a conjunction of quantifications; *141 does likewise with respect to alternation; and *142 does likewise with respect to the conditional. Various further principles of this kind also hold, but only when the biconditional is weakened to a conditional in one direction or the other. *101 is one such principle, and we now proceed to fourteen more.

***143.** $\vdash \ulcorner (\alpha)\phi \vee (\alpha)\psi . \supset (\alpha)(\phi \vee \psi) \urcorner.$

Proof. *100 $\vdash \ulcorner \phi \supset. \phi \vee \psi \urcorner.$ (1)
*101 $\vdash \ulcorner [(\alpha)1 \supset.](\alpha)\phi \supset (\alpha)(\phi \vee \psi) \urcorner.$ (2)
Similarly $\vdash \ulcorner (\alpha)\psi \supset (\alpha)(\phi \vee \psi) \urcorner.$ (3)
*100 $\vdash \ulcorner [2 \,.\, 3 \,.\supset] 143 \urcorner.$

[1] This principle, though not so well known as others in this section, goes back to Schröder (vol. 3, p. 30). But its analogue $\ulcorner (\alpha)(\phi \supset \psi) \equiv. (\exists \alpha)\phi \supset (\alpha)\psi \urcorner$, also given by Schröder (loc. cit.), is invalid. Half of it does hold; cf. *148.

***144.** $\vdash \ulcorner(\alpha)(\phi \vee \psi) \supset. (\exists\alpha)\phi \vee (\alpha)\psi\urcorner$.

Proof. *101 (& D4, 8) $\vdash \ulcorner(\alpha)(\sim \phi \supset \psi) \supset. (\exists\alpha)\phi \vee (\alpha)\psi\urcorner$. (1)
*100, *123 $\vdash \ulcorner[1 \equiv]144\urcorner$.

***145.** $\vdash \ulcorner(\alpha)(\phi \vee \psi) \supset. (\alpha)\phi \vee (\exists\alpha)\psi\urcorner$.

Proof. *144 $\vdash \ulcorner(\alpha)(\psi \vee \phi) \supset. (\exists\alpha)\psi \vee (\alpha)\phi\urcorner$
*100 $\supset. (\alpha)\phi \vee (\exists\alpha)\psi\urcorner$. (1)
*100, *123 $\vdash \ulcorner[1 \equiv]145\urcorner$.

***146.** $\vdash \ulcorner(\alpha)\phi \vee (\exists\alpha)\psi .\supset (\exists\alpha)(\phi \vee \psi)\urcorner$.

Proof. *100 $\vdash \ulcorner[136 \supset:](\alpha)\phi \vee (\exists\alpha)\psi .\supset. (\exists\alpha)\phi \vee (\exists\alpha)\psi\urcorner$. (1)
*141, *123 $\vdash \ulcorner 146 [\equiv 1]\urcorner$.

***147.** $\vdash \ulcorner(\exists\alpha)\phi \vee (\alpha)\psi .\supset (\exists\alpha)(\phi \vee \psi)\urcorner$. *Proof* similar.

***148.** $\vdash \ulcorner(\exists\alpha)\phi \supset (\alpha)\psi .\supset (\alpha)(\phi \supset \psi)\urcorner$.

Proof. *100 (& D8) $\vdash \ulcorner(\exists\alpha)\phi \supset (\alpha)\psi .\supset. (\alpha)\sim \phi \vee (\alpha)\psi\urcorner$
*143 (& D4) $\supset (\alpha)(\phi \supset \psi)\urcorner$.

***149.** $\vdash \ulcorner(\alpha)(\phi \supset \psi) \supset. (\exists\alpha)\phi \supset (\exists\alpha)\psi\urcorner$.

Proof. *145 (& D4) $\vdash \ulcorner(\alpha)(\phi \supset \psi) \supset. (\alpha)\sim \phi \vee (\exists\alpha)\psi\urcorner$
*100 (& D8) $\supset. (\exists\alpha)\phi \supset (\exists\alpha)\psi\urcorner$.

***150.** $\vdash \ulcorner(\exists\alpha)\phi \supset (\exists\alpha)\psi .\supset (\exists\alpha)(\phi \supset \psi)\urcorner$.

Proof. *100 (& D8) $\vdash \ulcorner(\exists\alpha)\phi \supset (\exists\alpha)\psi .\supset. (\alpha)\sim \phi \vee (\exists\alpha) \psi\urcorner$
*146 (& D4) $\supset (\exists\alpha)(\phi \supset \psi)\urcorner$.

***151.** $\vdash \ulcorner(\alpha)\phi \supset (\alpha)\psi .\supset (\exists\alpha)(\phi \supset \psi)\urcorner$.

Proof. *147 (& D4) $\vdash \ulcorner(\exists\alpha)\sim \phi \vee (\alpha)\psi .\supset (\exists\alpha)(\phi \supset \psi)\urcorner$. (1)
*130, *123 (& D4) $\vdash \ulcorner 151 [\equiv 1]\urcorner$.

***152.** $\vdash \ulcorner(\alpha)(\phi . \psi) \supset. (\exists\alpha)\phi . (\alpha)\psi\urcorner$.

Proof. *100 $\vdash \ulcorner[136 \supset:] (\alpha)\phi . (\alpha)\psi .\supset. (\exists\alpha)\phi .(\alpha)\psi\urcorner$. (1)
*140, *123 $\vdash \ulcorner 152 [\equiv 1]\urcorner$.

***153.** $\vdash \ulcorner(\alpha)(\phi . \psi) \supset. (\alpha)\phi . (\exists\alpha)\psi\urcorner$. *Proof* similar.

***154.** $\vdash \ulcorner(\alpha)\phi . (\exists\alpha)\psi .\supset (\exists\alpha)(\phi . \psi)\urcorner$.

Proof. *101 $\vdash \ulcorner(\alpha)(\phi \supset \sim \psi) \supset. (\alpha)\phi \supset (\alpha)\sim \psi\urcorner$. (1)
*100, *123 $\vdash \ulcorner[1 \equiv:](\alpha)\sim(\phi . \psi) \supset. (\alpha)\phi \supset (\alpha)\sim \psi\urcorner$. (2)
*100 (& D8) $\vdash \ulcorner[2 \supset]154\urcorner$.

***155.** $\vdash \ulcorner(\exists\alpha)\phi \,.\, (\alpha)\psi \,.\!\supset (\exists\alpha)(\phi \,.\, \psi)\urcorner$.

Proof. *100 $\vdash \ulcorner(\exists\alpha)\phi \,.\, (\alpha)\psi \,.\!\supset\!.\, (\alpha)\psi \,.\, (\exists\alpha)\phi\urcorner$
*154 $\supset (\exists\alpha)(\psi \,.\, \phi)\urcorner$. (1)
*100, *123 $\vdash \ulcorner[1 \equiv]155\urcorner$.

***156.** $\vdash \ulcorner(\exists\alpha)(\phi \,.\, \psi) \supset\!.\, (\exists\alpha)\phi \,.\, (\exists\alpha)\psi\urcorner$.

Proof. *100 $\vdash \ulcorner\phi \,.\, \psi \,.\!\supset \phi\urcorner$. (1)
*149 $\vdash \ulcorner[(\alpha)1 \supset\!.](\exists\alpha)(\phi \,.\, \psi) \supset (\exists\alpha)\phi\urcorner$. (2)
Similarly $\vdash \ulcorner(\exists\alpha)(\phi \,.\, \psi) \supset (\exists\alpha)\psi\urcorner$. (3)
*100 $\vdash \ulcorner[2 \,.\, 3 \,.\!\supset]156\urcorner$.

*143, *145, and *146 reveal a chain of four successively weaker formulæ, each of which implies its successor:

$\ulcorner(\alpha)\phi \vee (\alpha)\psi\urcorner$, $\ulcorner(\alpha)(\phi \vee \psi)\urcorner$, $\ulcorner(\alpha)\phi \vee (\exists\alpha)\psi\urcorner$, $\ulcorner(\exists\alpha)(\phi \vee \psi)\urcorner$.

*143–*156 and *101 reveal also five other such chains, thus:

*143, *144, *147:
$\ulcorner(\alpha)\phi \vee (\alpha)\psi\urcorner$, $\ulcorner(\alpha)(\phi \vee \psi)\urcorner$, $\ulcorner(\exists\alpha)\phi \vee (\alpha)\psi\urcorner$, $\ulcorner(\exists\alpha)(\phi \vee \psi)\urcorner$;

*148, *101, *151:
$\ulcorner(\exists\alpha)\phi \supset (\alpha)\psi\urcorner$, $\ulcorner(\alpha)(\phi \supset \psi)\urcorner$, $\ulcorner(\alpha)\phi \supset (\alpha)\psi\urcorner$, $\ulcorner(\exists\alpha)(\phi \supset \psi)\urcorner$;

*148, *149, *150:
$\ulcorner(\exists\alpha)\phi \supset (\alpha)\psi\urcorner$, $\ulcorner(\alpha)(\phi \supset \psi)\urcorner$, $\ulcorner(\exists\alpha)\phi \supset (\exists\alpha)\psi\urcorner$, $\ulcorner(\exists\alpha)(\phi \supset \psi)\urcorner$;

*153, *154, *156:
$\ulcorner(\alpha)(\phi \,.\, \psi)\urcorner$, $\ulcorner(\alpha)\phi \,.\, (\exists\alpha)\psi\urcorner$, $\ulcorner(\exists\alpha)(\phi \,.\, \psi)\urcorner$, $\ulcorner(\exists\alpha)\phi \,.\, (\exists\alpha)\psi\urcorner$;

*152, *155, *156:
$\ulcorner(\alpha)(\phi \,.\, \psi)\urcorner$, $\ulcorner(\exists\alpha)\phi \,.\, (\alpha)\psi\urcorner$, $\ulcorner(\exists\alpha)(\phi \,.\, \psi)\urcorner$, $\ulcorner(\exists\alpha)\phi \,.\, (\exists\alpha)\psi\urcorner$.

The next four metatheorems show that a quantifier covering a conjunction or alternation may equivalently be restricted to a single component in case the variable of quantification is not free in the other component.

***157.** *If α is not free in ϕ,* $\vdash \ulcorner(\alpha)(\phi \,.\, \psi) \equiv\!.\, \phi \,.\, (\alpha)\psi\urcorner$.

Proof. *118 (& hp), *123 $\vdash \ulcorner 157\,[\equiv 140]\urcorner$.

***158.** *If α is not free in ϕ,* $\vdash \ulcorner(\exists\alpha)(\phi \,.\, \psi) \equiv\!.\, \phi \,.\, (\exists\alpha)\psi\urcorner$.

Proof.
*137 (& hp), *123 $\vdash \ulcorner(\exists\alpha)(\phi \,.\, \psi) \supset\!.\, \phi \,.\, (\exists\alpha)\psi[:\equiv 156]\urcorner$. (1)
*118 (& hp), *123 $\vdash \ulcorner\phi \,.\, (\exists\alpha)\psi \,.\!\supset (\exists\alpha)(\phi \,.\, \psi)[:\equiv 154]\urcorner$. (2)
*100 $\vdash \ulcorner[1 \,.\, 2 \,.\!\supset]158\urcorner$.

***159.** *If* α *is not free in* ϕ, $\vdash \ulcorner(\alpha)(\phi \vee \psi) \equiv . \phi \vee (\alpha)\psi\urcorner$.

Proof.

*137 (& hp), *123	$\vdash \ulcorner(\alpha)(\phi \vee \psi) \supset . \phi \vee (\alpha)\psi[:\equiv 144]\urcorner$.	(1)
*118 (& hp), *123	$\vdash \ulcorner\phi \vee (\alpha)\psi . \supset (\alpha)(\phi \vee \psi)[:\equiv 143]\urcorner$.	(2)
*100	$\vdash \ulcorner[1 . 2 . \supset]159\urcorner$.	

***160.** *If* α *is not free in* ϕ, $\vdash \ulcorner(\exists\alpha)(\phi \vee \psi) \equiv . \phi \vee (\exists\alpha)\psi\urcorner$.

Proof. *137 (& hp), *123 $\vdash \ulcorner 160 [\equiv 141]\urcorner$.

In like fashion a quantifier covering a conditional may be restricted to the consequent if the variable of quantification is not free in the antecedent. This is already provided by *159 and *160, in view of D4. But a curious twist appears in the opposite case, where the variable of quantification is not free in the consequent: to confine the quantifier to the antecedent we must change it from universal to existential or vice versa.

***161.** *If* α *is not free in* ψ, $\vdash \ulcorner(\alpha)(\phi \supset \psi) \equiv . (\exists\alpha)\phi \supset \psi\urcorner$.

Proof.

*137 (& hp), *123	$\vdash \ulcorner(\alpha)(\phi \supset \psi) \supset . (\exists\alpha)\phi \supset \psi[:\equiv 149]\urcorner$.	(1)
*118 (& hp), *123	$\vdash \ulcorner(\exists\alpha)\phi \supset \psi . \supset (\alpha)(\phi \supset \psi)[:\equiv 148]\urcorner$.	(2)
*100	$\vdash \ulcorner[1 . 2 . \supset]161\urcorner$.	

***162.** *If* α *is not free in* ψ, $\vdash \ulcorner(\exists\alpha)(\phi \supset \psi) \equiv . (\alpha)\phi \supset \psi\urcorner$.

Proof. *137 (& hp), *123 $\vdash \ulcorner 162 [\equiv 142]\urcorner$.

The following corollary of *161 proves useful.

***163.** *If* $\vdash \ulcorner\phi \supset \psi\urcorner$, *and none of* $\alpha_1, \ldots, \alpha_n$ *is free in* ψ, *then*

$$\vdash \ulcorner(\exists\alpha_1)\ldots(\exists\alpha_n)\phi \supset \psi\urcorner.$$

Proof. hp 1	$\vdash \ulcorner\phi \supset \psi\urcorner$,	(1)
*161 (& hp 2)	$\vdash \ulcorner[(\alpha_n)\ 1 \equiv .](\exists\alpha_n)\phi \supset \psi\urcorner$,	(2)
*161 (& hp 2)	$\vdash \ulcorner[(\alpha_{n-1})\ 2 \equiv .](\exists\alpha_{n-1})(\exists\alpha_n)\phi \supset \psi\urcorner$,	

and so on.

§ 21. *Alphabetic Variance*

VARIABLES, as remarked (§ 12), serve merely for cross-reference to various positions of quantification. The particular choice of letters which happens to be made, in constructing a statement, is immaterial to the meaning so long as the system of cross-references remains the same. The statements:

(1) $(x)(y)(x < y . \supset (z)(z < x . \supset . z < y))$,
(2) $(x)(y)(x < y . \supset (w)(w < x . \supset . w < y))$,
(3) $(x)(z)(x < z . \supset (w)(w < x . \supset . w < z))$,
(4) $(y)(z)(y < z . \supset (w)(w < y . \supset . w < z))$,

e.g., are equivalent; they differ from one another only in an accidental detail of notation. If instead of using variables we were to indicate the cross-references by the method of curved lines (cf. § 12), differences of this sort would drop out altogether. The development of a technique for recognizing such purely alphabetic variants as (1)–(4), and interchanging them at will, is the price which we must pay for a system of notation which is in most respects more convenient than its known alternatives. The present section will be devoted to paying this price.

Each of the statements (1)–(4) is transformed into the next by relettering a constituent quantification $\ulcorner(\alpha)\psi\urcorner$; i.e., by replacing $\ulcorner(\alpha)\psi\urcorner$ by $\ulcorner(\alpha')\psi'\urcorner$[1] where ψ' is like ψ except for exhibiting α' in place of α. But such relettering cannot always be depended upon to preserve equivalence. The rewriting of 'z' in (1) as 'w' produces (2), which is indeed equivalent; the rewriting of 'z' as 'y', however, would have produced:

(5) $(x)(y)(x < y . \supset (y)(y < x . \supset . y < y))$,

which, far from being equivalent to the truths (1) and (2), is false. The relettering which leads from (1) to (5) alters the pattern of cross-references; the last occurrence of 'y' in (5) refers back to the last quantifier occurrence in (5), whereas the correspondingly

[1] Strictly speaking the conventions of quasi-quotation fall into ambiguity at this point, because of the fact that Greek and Latin letters are both eligible to accentuation. Is the quantifier in $\ulcorner(\alpha')\psi'\urcorner$ to be understood as the result of putting the variable α' (e.g. 'y') in the blank of '( )', or is it to be understood rather as the result of putting the variable α (e.g. 'x') in the blank of '(′)'? The former, obviously, is the intention. In general, let us understand accents in quasi-quotation as belonging thus always to the Greek letters rather than to the contextual framework.

placed occurrence of 'y' in (1) refers back rather to the second quantifier occurrence in (1).

In general, if the replacement of $\ulcorner(\alpha)\psi\urcorner$ by $\ulcorner(\alpha')\psi'\urcorner$ is not to affect the cross-references, the free occurrences of α' in ψ' must exactly match the free occurrences of α in ψ; for it is just those occurrences that refer back to the initial occurrence of the quantifier in $\ulcorner(\alpha)\psi\urcorner$ or $\ulcorner(\alpha')\psi'\urcorner$ (cf. (ii), § 14). Quantifications $\ulcorner(\alpha)\psi\urcorner$ and $\ulcorner(\alpha')\psi'\urcorner$ which are so related will be called *immediate alphabetic variants* of each other. In full: $\ulcorner(\alpha)\psi\urcorner$ and $\ulcorner(\alpha')\psi'\urcorner$ are immediate alphabetic variants if ψ and ψ' are alike except that ψ' contains free occurrences of α' at all and only those places where ψ contains free occurrences of α. Thus the quantification:

$$(w)(w < x \,.\supset.\, w < y),$$

which appears in (2), is an immediate alphabetic variant of the quantification:

$$(z)(z < x \,.\supset.\, z < y)$$

of (1); on the other hand this is not true of the quantification:

$$(y)(y < x \,.\supset.\, y < y)$$

which appears in (5).

Now the kind of relettering which leads from (1) to (2), from (2) to (3), and from (3) to (4) can be described as follows: it consists in replacing a constituent quantification by an immediate alphabetic variant thereof. Formulæ which are intertransformable by one or more such steps will be called, in general, *alphabetic variants*. Thus (1)–(4) are all alphabetic variants of one another; (5), on the other hand, is an alphabetic variant of none of them. The rigorous formulation is this: formulæ ϕ and ϕ' are alphabetic variants if there are formulæ $\phi_0, \phi_1, \ldots \phi_n$ such that ϕ_0 is ϕ, ϕ_n is ϕ', and, for each i from 1 to n, ϕ_i is formed from ϕ_{i-1} by replacing some constituent quantification $\ulcorner(\alpha_i)\psi_i\urcorner$ by an immediate alphabetic variant $\ulcorner(\alpha'_i)\psi'_i\urcorner$ thereof.

In the above explanation it was not required that ϕ and ϕ' be statements; the notion of alphabetic variant applies to statements and matrices indifferently. The relettering involved in passing to an alphabetic variant of ϕ is always, however, a relettering of occurrences which are bound in ϕ; we rewrite the variable within a quantifier and at such subsequent occurrences as refer back to that position of quantification. Thus matrices ϕ and ϕ' are not classed as alphabetic variants if they differ from each other in point of free occurrences of variables. This is as it should be; the particular choice of letters is indeed immaterial to a statement so long as a certain general structure is maintained, but this is not true of the free variables of a matrix. The matrices '$y = x$' and '$y = z$' actually differ in meaning — in this important sense: replacement of the one by the other can turn a true context into a falsehood. It turns the truth:

(6) $$(x)(y)(z)(x = y \,.\, y = z \,.\!\supset\!.\, x = z),$$

e.g., into the falsehood:

$$(x)(y)(z)(x = y \,.\, y = x \,.\!\supset\!.\, x = z).$$

A matrix ϕ (e.g. '$y = z$') is a fragment of one or another statement ψ (e.g. (6)), and any free variable α of ϕ is indeed a bound variable of ψ. But any rewriting of α which preserves the meaning of ψ must touch occurrences of α outside ϕ; the fragment ϕ itself and the relettered fragment ϕ' (e.g. '$y = x$') are not interchangeable.

Where ϕ is a statement and ϕ' is an alphabetic variant thereof, the biconditional $\ulcorner\phi \equiv \phi'\urcorner$ will of course be trivially true. Moreover, it will be a theorem. This will now be proved, and more: where ϕ is a matrix rather than a statement, the closure of the biconditional is still a theorem.

First we attend to the case of immediate alphabetic variants:

***170.** *If* $\ulcorner(\alpha)\psi\urcorner$ *and* $\ulcorner(\alpha')\psi'\urcorner$ *are immediate alphabetic variants,* $\vdash\ulcorner(\alpha)\psi \equiv (\alpha')\psi'\urcorner$.

Proof. By hp, ψ and ψ' are alike except that ψ' contains free occurrences of α' at all and only those places where ψ contains free occurrences of α. Hence, by *103 and *117,

$$\vdash\ulcorner(\alpha)\psi \supset (\alpha')\psi'\urcorner, \qquad (1)$$

$$\vdash\ulcorner(\alpha')\psi' \supset (\alpha)\psi\urcorner. \qquad (2)$$

*100 $\vdash\ulcorner[1 \,.\, 2 \,.\!\supset] \; 170\urcorner$.

The extension to the case of alphabetic variants in general is then obvious:

***171.** *If ϕ and ϕ' are alphabetic variants,* $\vdash \ulcorner \phi \equiv \phi' \urcorner$.

Proof. By hp, there are formulæ $\phi_0, \phi_1, \ldots \phi_n$ such that ϕ_0 is ϕ, ϕ_n is ϕ', and, for each i from 1 to n, ϕ_i is formed from ϕ_{i-1} by replacing a quantification $\ulcorner(\alpha_i)\, \psi_i \urcorner$ by an immediate alphabetic variant $\ulcorner(\alpha'_i)\, \psi'_i \urcorner$ thereof. For each i, by *170,

$$\vdash \ulcorner(\alpha_i)\, \psi_i \equiv (\alpha'_i)\, \psi'_i \urcorner,$$

and hence, by *123, $\vdash \ulcorner \phi_{i-1} \equiv \phi_i \urcorner$. Thus

$$\begin{aligned} \vdash \ulcorner \phi_0 &\equiv \phi_1 \urcorner \\ &\equiv \phi_2 \urcorner \\ &\;\;\vdots \\ &\equiv \phi_n \urcorner, \text{ q. e. d.} \end{aligned}$$

The transformation of a formula ϕ into an alphabetic variant ϕ' runs through a series of stages $\phi, \phi_1, \phi_2, \ldots \phi_{n-1}, \phi'$. Each component step of transformation consists in supplanting some quantification $\ulcorner(\alpha_i)\, \psi_i \urcorner$ by an immediate alphabetic variant $\ulcorner(\alpha'_i)\, \psi'_i \urcorner$ thereof. The end result ϕ' differs from ϕ with respect to many variables, in general, and $\phi_1, \phi_2, \ldots \phi_{n-1}$ are the intermediate results of making the alphabetic changes one at a time.

It might be supposed, then, given ϕ and ϕ', that we can always supply the intervening stages $\phi_1, \phi_2, \ldots, \phi_{n-1}$ by the simple process of putting the new letters for the old ones one at a time in an arbitrary order — say in the order of occurrence of the quantifiers concerned. Actually the matter is not so simple. Let us return to the examples (1)–(4) at the beginning of the section. (1) and (4) are known to be alphabetic variants; and (4) differs from (1) in exhibiting '*y*', '*z*', and '*w*' instead respectively of '*x*', '*y*', and '*z*'. Yet we cannot get from (1) to (4) by first putting '*y*' for '*x*', then '*z*' for '*y*', and then '*w*' for '*z*'. The first of these steps leads from (1) to:

$$(y)(y)(y < y \,.\!\supset (z)(z < y \,.\!\supset\!.\, z < y)),$$

which is in fact not an alphabetic variant of (1); and if we put '*z*' for '*y*' in this result and '*w*' for '*z*' in the new result we end up not with (4) but with the monotonous truism:

$$(w)(w)(w < w \,.\!\supset (w)(w < w \,.\!\supset\!.\, w < w)).$$

That (1) and (4) are alphabetic variants is shown rather by the sequence (1), (2), (3), (4) — the sequence obtained by rewriting '*z*' first as '*w*', then '*y*' as '*z*', then '*x*' as '*y*'.

Again, consider the following pair of alphabetic variants:

(7) $(x)(x \text{ is a number } .\supset \sim (y)(x = x + y))$,

(8) $(y)(y \text{ is a number } .\supset \sim (x)(y = y + x))$.

(8) differs from (7) in exhibiting '*y*' and '*x*' instead respectively of '*x*' and '*y*'; yet an appropriate sequence ϕ, ϕ_1, ... ϕ_{n-1}, ϕ' joining (7) and (8) can be found neither by first changing '*x*' to '*y*' and then changing '*y*' to '*x*', nor by following the reverse order. The desired connection between (7) and (8) can be made only through intervening stages such as the following, which involve temporary introduction of a new letter '*z*':

(9) $(z)(z \text{ is a number } .\supset \sim (y)(z = z + y))$,

(10) $(z)(z \text{ is a number } .\supset \sim (x)(z = z + x))$.

The statements (7), (9), (10), (8) do constitute a sequence ϕ, ϕ_1, ..., ϕ_{n-1}, ϕ' of the appropriate kind.

Returning to the general case, let ϕ' be an alphabetic variant of ϕ containing α'_1, α'_2, ... α'_k in place respectively of α_1, α_2, ... α_k. We see that the connecting links ϕ_1, ϕ_2, etc. which establish alphabetic variance cannot in general be found by the simple expedient of putting α'_1 for α_1 in ϕ, then α'_2, for α_2 in the result, and so on. This expedient has been seen to fail where ϕ is (1) and ϕ' is (4), and again where ϕ is (7) and ϕ' is (8). Any number of other examples to the same purpose could be constructed, sharing always this feature: various of α_1, α_2, ... α_k reappear among α'_1, α'_2, ... α'_k.

A method is apparent, however, which reveals suitable links ϕ_1, ϕ_2, etc. in the general case. It turns on temporary introduction of new letters as in (9) and (10). Where β_1, β_2, ... β_k are any variables which are distinct from one another and from all variables of ϕ and ϕ', construct the sequence ϕ_1, ϕ_2, ... ϕ_k by putting β_1 for α_1 in ϕ, β_2 for α_2 in the result, and so on; then form the continuation ϕ_{k+1}, ϕ_{k+2}, ... ϕ_{2k-1}, ϕ' by putting α'_1 for β_1 in ϕ_k, α'_2 for β_2 in the result, and so on. The total sequence ϕ, ϕ_1, ϕ_2, ..., ϕ_{2k-1}, ϕ' does establish alphabetic variance of ϕ and ϕ'. Where ϕ and ϕ' are (1) and (4), e.g , and β_1 is chosen as 'x'', β_2 as 'y'', and β_3 as 'z'', the described method supplies ϕ_1, ϕ_2, ϕ_3, ϕ_4, and ϕ_5 as follows:

(11) $(x')(y)(x' < y \;.\supset\; (z)(z < x' \;.\supset.\; z < y))$,
(12) $(x')(y')(x' < y' \;.\supset\; (z)(z < x' \;.\supset.\; z < y'))$,
(13) $(x')(y')(x' < y' \;.\supset\; (z')(z' < x' \;.\supset.\; z' < y'))$,
(14) $(y)(y')(y < y' \;.\supset\; (z')(z' < y \;.\supset.\; z' < y'))$,
(15) $(y)(z)(y < z \;.\supset\; (z')(z' < y \;.\supset.\; z' < z))$.

Each succeeding formula of the sequence (1), (11), (12), (13), (14), (15), (4) differs from its predecessor merely through replacement of a constituent quantification by an immediate alphabetic variant thereof; the alphabetic variance of (1) and (4) is thus verified.

This general method delivers somewhat longer sequences than necessary. We know, e.g., that the sequence (1), (2), (3), (4) would suffice in place of the sequence (1), (11)–(15), (4). But the method has the virtue of simplicity in general formulation. Its existence will excuse us, in future pages, from citing intervening formulæ ϕ_1, ϕ_2, etc. by way of establishing any alleged case of alphabetic variance. The reader can be left always to check the matter for himself, by the general method if not by inspection.

The discussion of alphabetic variance has been limited to variables governed by universal quantifiers, but it carries over automatically to variables governed by existential quantifiers; for an existential quantifier is merely shorthand for a universal quantifier with a couple of denial signs.

CHAPTER THREE

TERMS

§22. *Class and Member*

IN CHAPTER I connectives were studied which answer to the statement connectives 'neither-nor', 'if-then', 'and', etc. of ordinary discourse. In Chapter II the essential notions were formulated which underlie the common use of the words 'whatever', 'all', 'there is', 'some'. We now turn to a connective 'ϵ' which embodies the principal meaning of the ambiguous word 'is'.[1]

Sometimes 'is' has the sense of '=', or 'is the same as'; such is its sense in 'Paris is the capital of France', 'Tully is Cicero'. But in 'Paris is a city' or 'Tully is wise' or 'Socrates is wise' the word cannot be so construed; from 'Tully = wise' and 'Socrates = wise', indeed, we could infer that Tully = Socrates. In such contexts 'is' expresses rather possession of a property, or membership in a class: Paris belongs to the class of cities, and Socrates belongs to the class of wise beings. It is this sense of 'is' that is rendered symbolically by the connective 'ϵ': 'Paris ϵ city', 'Socrates ϵ wise'. The connective 'ϵ' is (like '=', 'hates', etc.) a *binary predicate* (cf. § 5); whereas '$\downarrow$', '$\supset$', '.', etc. yield statements when put between statements, 'ϵ' yields statements when put between names.

In the examples which have been cited for one purpose or another in the course of foregoing sections, the names which figure most prominently are 'Jones', 'Smith', 'Socrates', 'Boston', 'Paris'. But the names which 'ϵ' is used to connect are not exclusively of this sort. What is ordinarily wanted to the right of 'ϵ' is a general term, e.g. 'wise' or 'city', rather than the name of some specific man or place. It is convenient, however, to regard such general terms as names on the same footing as 'Socrates' and 'Paris': names each of a single specific entity, though a less tangible entity than the man Socrates or the town Boston.[2] The word 'wise' may be treated as name of the *class* of wise beings,

[1] The notation 'ϵ', short for 'ἐστί', is Peano's.

[2] For a divergent, more philosophical account, see my *Methods*, §§ 34, 38.

taken as a single object of an abstract kind; and 'city' may be treated as a name of the class of cities. The statement 'Socrates ϵ wise', now, says something about two objects, a man and a class; namely it says that the one is a member of the other. Correspondingly for 'Paris ϵ city'.

The reassuring phrase 'mere aggregates' must be received warily as a description of classes. Aggregates, perhaps; but not in the sense of composite concrete objects or heaps. Continental United States is an extensive physical body (of arbitrary depth) having the several states as parts; at the same time it is a physical body having the several counties as parts. It is the same concrete object, regardless of the conceptual dissections imposed; the heap of states and the heap of counties are identical. The class of states, however, cannot be identified with the class of counties; for there is much that we want to affirm of the one class and deny of the other. We want to say e.g. that the one class has exactly 48 members, while the other has 3075. We want to say that Delaware is a member of the first class and not of the second, and that Nantucket is a member of the second class and not of the first. These classes, unlike the single concrete heaps which their members compose, must be accepted as two entities of a non-spatial and abstract kind.

Once classes are freed thus of any deceptive hint of tangibility, there is little reason to distinguish them from *properties*. It matters little whether we read '$x \epsilon y$' as 'x is a member of the class y' or 'x has the property y'. If there is any difference between classes and properties, it is merely this: classes are the same when their members are the same, whereas it is not universally conceded that properties are the same when possessed by the same objects. The class of all marine mammals living in 1940 is the same as the class of all whales and porpoises living in 1940, whereas the property of being a marine mammal alive in 1940 might be regarded as differing from the property of being a whale or porpoise alive in 1940. But classes may be thought of as properties if the latter notion is so qualified that properties become identical when their instances are identical. Classes may be thought of as properties in abstraction from any differences which are not reflected in differences of instances. For mathematics certainly, and perhaps for discourse generally, there is no need of countenancing properties in any other sense.

Discourse in general, mathematical and otherwise, involves continual reference to abstract entities of this sort — classes or properties. One may prefer to regard abstractions as fictions or manners of speaking; one may hope to find a method whereby all ostensible reference to abstract entities can be explained as mere shorthand for a more basic idiom involving reference only to concrete objects (in some sense or other).[1] Such a nominalistic program presents extreme difficulty, if much of standard mathematics and natural science is to be really analyzed and reduced rather than merely repudiated; however, it is not known to be impossible. If a nominalistic theory of this sort should be achieved, we may gladly accept it as the theoretical underpinning of our present ostensible reference to so-called abstract entities; meanwhile, however, we have no choice but to admit those abstract entities as part of our ultimate subject matter.

We are thus to recognize as names of entities not only such expressions as 'Nantucket', 'the capital of France', 'the northernmost chimney of Craigie House', but also general terms, conceived as naming abstract classes. And when through the medium of quantification we affirm something as true of every object *x*, we mean it to apply not merely to every spatially extended concrete object *x*, but to every object, abstract or concrete, class or individual.

Our working ontology is thus pretty liberal. But in mitigation it may now be said that this is the end; no abstract objects other than classes are needed — no relations, functions, numbers, etc., except insofar as these are construed simply as classes. In addition to concrete objects we need recognize only classes having such

[1] See my "Designation and Existence," "Logistical Approach," and "Theory of Classes." Note that Russell's contextual definition of class names ("Mathematical Logic"; also *Principia*, vol. 1, pp. 71 ff, 187 ff) does not dispense with abstract entities, but only eliminates so-called classes in favor of properties. This point is obscured by ambiguity of the phrase 'propositional function', which sometimes means 'property' and sometimes 'statement matrix'.

objects as members, then classes whose members are drawn from the thus supplemented totality, and so on. This is presumably all the ontology that is needed for discourse in general; certainly it is all that is needed for mathematics.

It is by no means clear what objects are to be regarded as concrete. E.g., should we regard men as concrete objects, or should we regard events as concrete objects and then explain men as classes of events? For present purposes it does not matter, for there will be no occasion in the formal developments to speak specifically of concrete objects. It would even be possible, compatibly with the projected formal developments and indeed with the whole of mathematics, to repudiate concrete objects altogether — to recognize just classes, each of which has classes in turn as members or else no members whatever. Such an ontology would recognize an empty or null class, a class whose sole member is the null class, a class whose sole member is the latter, and so on; also, classes whose members are variously chosen from the latter series; also, classes whose members are variously chosen from the thus supplemented totality; and so on.[1] This exclusively abstract ontology has little naturalness to recommend it, but there is no need here to reject or accept it.

Where y is a class, '$x \epsilon y$' means that x is a member of it. But what does it mean where y is not a class? We are free to decide this question as we like, by arbitrary supplementary interpretation of the sign 'ϵ'. The decision which seems to offer the greatest simplicity, as measured by the general principles produced, is this: where y is not a class, '$x \epsilon y$' means '$x = y$'. Hence '$x \epsilon y$', briefly read 'x is a member of y', is to have as its full translation the following: 'x is a member of y or is the same as y according as y is or is not a class'. We may continue to use the brief reading if we agree that 'is a member of y' is to mean 'is the same as y' when y is not a class.

The meaning of 'ϵ' reduces to that of '$=$' just in the extreme

[1] Cf. Fraenkel, "Untersuchungen," pp. 255f. The essential principle of this ontology might, in view of the developments which are immediately to follow, be given the symbolic rendering '$(y) \sim (x)(x \epsilon y \,.\!\equiv.\, x = y)$'; though this version is not suited to Fraenkel's own system.

case where the second object involved is not a class. Thus it does not follow that 'ϵ' will in general serve the purposes of '='; the latter connective continues to be needed to carry the meaning 'is the same as' as applied to classes and non-classes generally. It does turn out, however, that '=' is dispensable as a primitive connective; it can be defined in terms of 'ϵ', '$\downarrow$', and quantification (cf. § 25).

Note further that the notion of class itself is not needed as a separate assumption; 'y is a class' can be paraphrased in terms exclusively of 'ϵ', '$\downarrow$', and quantification. For, under the supplementary interpretation of 'ϵ' just now adopted, the distinguishing feature of a non-class y is this: '$x \epsilon y$' amounts, for every object x, to '$x = y$'. Hence, once having defined '=' in terms of 'ϵ', '$\downarrow$', and quantification, we can paraphrase 'y is not a class' as:

$$(x)(x \epsilon y \,.\!\equiv\!.\, x = y)$$

and 'y is a class' as:

$$\sim (x)(x \epsilon y \,.\!\equiv\!.\, x = y).$$

But none of these considerations, having to do with the interpretation of '$x \epsilon y$' in cases where y is not a class, will figure explicitly in the formal developments which are to follow. Situations will never arise where an object y is known to be a non-class. As mentioned, the formal developments are compatible with the nonexistence of entities other than classes; and under this extreme alternative the adopted supplementary interpretation of 'ϵ' evaporates altogether.

§ 23. *Logical Formulæ*

THE CONNECTIVE 'ϵ' of membership is adopted as part of our primitive logical notation, along with the connective '$\downarrow$' of joint denial and the quantifiers and variables which constitute the notation of quantification. By putting 'ϵ' between variables, we obtain formulæ $\ulcorner \alpha \epsilon \beta \urcorner$ — e.g. '$x \epsilon x$', '$x \epsilon y$', '$y \epsilon z$' — which are *atomic* in the sense of having no other formulæ as parts. This is the first time atomic formulæ have come to hand. Hitherto, though formulæ in general were described as built up of atomic formulæ by joint denial and quantifiers, the atomic formulæ themselves were left unspecified (cf. § 13).

Strictly, these newly acquired atomic formulæ should be described not as $\ulcorner \alpha \epsilon \beta \urcorner$ but as $\ulcorner (\alpha \epsilon \beta) \urcorner$; for '$\epsilon$', like all binary connectives, is to be construed as having a pair of parentheses associated with it. In practice, under the working conventions explained in § 7, $\ulcorner (\alpha \epsilon \beta) \urcorner$ loses its parentheses when it stands alone or as a component of a conjunction, alternation, conditional, or biconditional.[1]

Hitherto no actual expressions could be cited as formulæ, because of the fact that the atomic formulæ were not given. Now, however, one important sort of formulæ is definitely fixed; it comprises atomic formulæ of the kind $\ulcorner (\alpha \epsilon \beta) \urcorner$ together with all other formulæ whose atomic parts are of this kind. Such formulæ will be spoken of henceforward as *logical formulæ*. Thus the logical formulæ comprise, first, all expressions obtainable by putting variables 'x', 'y', etc. in the blanks of '(ϵ)'; second, all expressions obtainable by putting expressions of the first totality in the blanks of '($\downarrow$)' or after a parenthesized variable; third, all expressions obtainable by putting expressions of the thus supplemented totality in the blanks of '($\downarrow$)' or after a parenthesized variable; and so on. Synoptically: $\ulcorner (\alpha \epsilon \beta) \urcorner$ is a logical formula, and if ϕ and ψ are logical formulæ then so are $\ulcorner (\phi \downarrow \psi) \urcorner$ and $\ulcorner (\gamma)\phi \urcorner$.

Thus the logical formulæ comprise such expressions as:

$(x \epsilon x)$, $(x \epsilon y)$, $(z \epsilon w)$, $(x)(x \epsilon x)$, $(x)(x \epsilon y)$,
$(y)(x \epsilon y)$, $(x)(y)(x \epsilon y)$, $((x \epsilon x) \downarrow (x \epsilon x))$, $((x \epsilon x) \downarrow (x \epsilon y))$,
$((x \epsilon y) \downarrow (y \epsilon x))$, $((x \epsilon y) \downarrow (z \epsilon w))$, $(x)((x \epsilon x) \downarrow (x \epsilon x))$,
$((x)(x \epsilon x) \downarrow (y)(x \epsilon y))$, $(w)(x)(((x \epsilon x) \downarrow (y)(x \epsilon y)) \downarrow (x \epsilon w))$,

[1] Actually the notation $\ulcorner \alpha \epsilon \beta \urcorner$ could wholly supplant $\ulcorner (\alpha \epsilon \beta) \urcorner$ at this stage — even in the absence of special conventions such as those of § 7 — without danger of ambiguity. Coupled with notations subsequently to be introduced by definition, however, this course would result in ambiguity, as my student K. R. Symon has noted. Parentheses are needed e.g. to distinguish $\ulcorner \hat{\alpha}(\hat{\beta}\hat{\gamma}\phi \epsilon \delta) \urcorner$ from $\ulcorner \hat{\alpha}\hat{\beta}(\hat{\gamma}\phi \epsilon \delta) \urcorner$ (cf. p. 202).

etc. Such of the logical formulæ as have no free variables — e.g. the fourth, seventh, twelfth, and last of the above list — constitute the *logical statements*, true or false. These four examples are, as it happens, all false.

There are no atomic logical statements, for the atomic logical formulæ $\ulcorner(\alpha \epsilon \beta)\urcorner$ have free variables. The shortest possible logical statements are those which are closures of the one-variable matrices $\ulcorner(\alpha \epsilon \alpha)\urcorner$; in other words, '$(x)(x \epsilon x)$' and its alphabetic variants '$(y)(y \epsilon y)$', '$(z)(z \epsilon z)$', etc. These statements happen, incidentally, to be false; for they say that every entity (and hence every class) is a member of itself. Many classes are indeed members of themselves, but many also are not; the class of cats, e.g., is not a cat.

The shortest logical statements which are true are:

(1) $$((x)(x \epsilon x) \downarrow (x)(x \epsilon x))$$

and its alphabetic variants, e.g.:

$$((z)(z \epsilon z) \downarrow (y)(y \epsilon y)).$$

The truth (1), i.e. '$\sim (x)(x \epsilon x)$', denies that every entity is a member of itself — though allowing that some may be. This truth will not appear as a theorem until a considerably later point (†240, § 32).

It will be noted that the logical formulæ contain no names — not even names of abstract entities, classes. In the atomic formulæ $\ulcorner(\alpha \epsilon \beta)\urcorner$ the α and β are never names, but merely 'x', 'y', 'z', etc.; in effect, pronouns (cf. § 12). The logical formulæ do indeed comprise statements as well as matrices; the statements have definite import not by virtue of exhibiting names in place of variables, however, but merely by virtue of containing enough quantifiers to render the variables bound at all occurrences. All this is apparent from the foregoing description and the examples. Subsequently, indeed, the logical formulæ will be abbreviated through definitions in such a way that names of classes will take form within them; but such names will have the status merely of shorthand, capable always of elimination in favor of the frugal notation described above.

We shall see in subsequent sections that the notions of identity, relation, number, function, sum, product, power, limit, derivative, etc., are all definable in terms of our three primitive notational devices: membership, joint denial, and quantification with its variables. Under those definitions every true and every false statement which is couched in purely mathematical terms becomes an abbreviation of a logical formula, in particular a logical statement, in the sense described above. The three primitives thus provide a complete mathematical language. The language is not, indeed, a convenient one for mathematical practice; the simple statement '$1 + 1 = 2$', if translated into terms of these three primitives, would run to a length of many pages. For mathematical practice — demonstration and application of theorems — definitional abbreviations are thus indispensable. For metamathematical practice, on the other hand — formulation and investigation of the general notions of formula, theorem, mathematical truth, etc. — the reduction of concepts to a few primitives is indispensable; the subject matter of our metamathematical investigations is thereby simplified to the point of manageability. (Cf. Introduction.) Reduction of the concepts of mathematics to the three primitives is of theoretical significance also as affording a measure of the net conceptual presuppositions involved in mathematical theory; and the definitions are significant as an analysis of the various derivative mathematical concepts.

Frege was the first to show (1884) that the notions of arithmetic could be defined in purely logical terms. A more refined and detailed construction of arithmetic and derivative disciplines from logic was carried out by Whitehead and Russell. In the reduction of logic in turn to the three present primitives, one essential step was Russell's discovery of how to define complex terms in context (cf. § 26); a second was Sheffer's reduction of conjunction, alternation, and denial to joint denial (§ 9); and a third was Wiener's definition of the ordered pair (§ 36). The adequacy of substantially the three present primitives was mentioned by both Tarski and Gödel in 1931.

The systems of Tarski and Gödel can in fact be translated, in turn, into terms of still fewer primitives: just *inclusion* (§ 34) and *abstraction* (§ 24). See my "Inclusion and Abstraction"; also "Theory of Types". But the systematization of logic developed in the present book turns out, curiously, to resist such translation. The reason is that there are differences in detail between the notion of membership developed here and the notion of membership which Tarski and Gödel took over from Whitehead and Russell. The divergence is such that the equivalence of

⌜$(\alpha)\phi$⌝ with ⌜$\hat{\alpha}\phi = V$⌝, on which the definition of quantification in terms of inclusion and abstraction depends, fails in the present system (cf. § 32).

The notion of logical formula does not supersede the original relative notion of formula. Logical formulæ are formulæ of a restricted kind, inadequate to the expression of anything beyond the bounds of mathematics. For other subject matters we would need additional formulæ, describable by stipulating new atomic formulæ appropriate to that subject matter. The atomic formulæ formed by flanking 'ϵ' with variables happen merely to be the ones appropriate to logico-mathematical matters. A statement may be a theorem under *100–*104 without being so limited in point of atomic parts as to be classifiable as a logical formula, logical statement; and all such statements are logically true, regardless of their constituent vocabulary. The theorems which we shall have occasion to consider in subsequent sections, however, will all be logical statements, logical formulæ.

Hitherto we were able to specify only the outward forms of formulæ, leaving the content — the atomic formulæ — unspecified; but now, so far as logical formulæ are concerned, form and content are both at hand. We need no longer limit ourselves to setting down metatheorems, to the effect that all formulæ of such and such forms are theorems; we can now set down actual theorems. From *103, e.g., we see that the closure:

$$(y)((x)(x \epsilon y) \supset .\ y \epsilon y)$$

of '$(x)(x \epsilon y) \supset .\ y \epsilon y$' is a theorem; from *157, again, we see that:

$$(z)(y)((x)(y \epsilon z \,.\, x \epsilon y) \equiv .\ y \epsilon z \,.\, (x)(x \epsilon y))$$

is a theorem.

One might indeed think of logic, in a narrow sense, as comprising just the theory of statement composition and quantification, and hence as dealing with just the outward form of formulæ through the medium of metatheorems. The level upon which we are now entering, then, might be regarded as forming a second level of mathematics — the theory of classes, otherwise known as the theory of sets, the theory of aggregates, *Mengenlehre*. At this level for the first time actual formulæ are forthcoming, as theorems about a mathematical subject matter. Relation theory, arithmetic, and other branches of mathematics then fall into place as specializations or subdivisions of the theory of classes. From this point of view, incidentally, the so-called "logical formulæ" would be more appropriately describable as "mathematical formulæ"; the former designation will be retained, however, and the narrow sense of 'logic' just now suggested will not be urged.

§ 24. *Abstraction*

THE USUAL way of specifying a class is by citing a necessary and sufficient condition for membership in it. Such is the method when one speaks of "the class of all entities x such that . . . ," appending one or another matrix. The class of all entities x such that x writes poems, e.g., is the class of poets; the class of all entities x such that

$$(\exists y)(y \in \text{integer} \,.\, x = y^2),$$

again, is the class of square numbers. Despite its sanction from the side of usage and common sense, however, this method of specifying classes leads to trouble. Applied to certain matrices, the prefix 'the class of all entities x such that' produces expressions which cannot consistently be regarded as designating any class whatever. One matrix of this kind, discovered by Russell,[1] is '$\sim(x \in x)$'; there is no such thing as the class of all entities x such that $\sim(x \in x)$. For, suppose w were such a class. For every entity x, then,

$$x \in w \,.\!\equiv \sim(x \in x).$$

Taking x in particular as w itself, we are led to the contradiction:

$$w \in w \,.\!\equiv \sim(w \in w).$$

The matrix '$\sim(x \in x)$' is the first and simplest of an infinite series of matrices, viz.:

$$\sim(x \in x), \qquad (y) \sim (x \in y \,.\, y \in x), \qquad (y)(z) \sim (x \in y \,.\, y \in z \,.\, z \in x),$$

. . . , all of which share the same peculiarity. In the case of each

[1] *Principles*, Ch. X. See also Frege, *Grundgesetze*, vol. 2, pp. 253 ff.

of these matrices, as seen in the case of '$\sim(x \in x)$', the assumption of a "class of all entities x such that . . ." leads to contradiction. There is no class w such that

$$(x)(x \in w \,.\!\equiv \sim (x \in x)),$$

nor any such that

$$(x)(x \in w \,.\!\equiv (y) \sim (x \in y \,.\, y \in x)),$$

nor any such that

$$(x)(x \in w \,.\!\equiv (y)(z) \sim (x \in y \,.\, y \in z \,.\, z \in x)),$$

and so on. This is apparent from the metatheorem *181, below, according to which the statements:

$$(w) \sim (x)(x \in w \,.\!\equiv \sim (x \in x)),$$
$$(w) \sim (x)(x \in w \,.\!\equiv (y) \sim (x \in y \,.\, y \in x)),$$
$$(w) \sim (x)(x \in w \,.\!\equiv (y)(z) \sim (x \in y \,.\, y \in z \,.\, z \in x)),$$

etc. are all theorems.

It is convenient for purposes of future reference to prove a somewhat more complicated metatheorem *180 first and then derive *181 from it. In the proof of *180 a new sort of abbreviation is used, involving the letters 'L' and 'R'. The notation 'L180' stands as an abbreviation for the matter:

$$(\alpha)(\alpha \in \beta \,.\!\equiv\!.\, \phi \,.\, (\gamma_1) \ldots (\gamma_n) \sim (\alpha \in \gamma_1 \,.\, \gamma_1 \in \gamma_2 \,.\, \ldots \,.\, \gamma_n \in \alpha))$$

which appears to the *left* of the main connective in the expression for which '180' stands. Similarly 'R4' stands as an abbreviation for the matter:

$$\sim(\beta \in \gamma_2 \,.\, \ldots \,.\, \gamma_n \in \alpha \,.\, \alpha \in \beta)$$

which appears to the *right* of the main connective in the expression for which '4' stands. 'L1' and 'R6' are analogous.

***180.** *If* α, β, $\gamma_1, \ldots, \gamma_n$ *are distinct, and* ϕ' *is like* ϕ *except for containing free occurrences of* β *wherever* ϕ *contains free occurrences of* α, *then*

$$\vdash \ulcorner(\alpha)(\alpha \in \beta \,.\!\equiv\!.\, \phi \,.\, (\gamma_1) \ldots (\gamma_n) \sim (\alpha \in \gamma_1 \,.\, \gamma_1 \in \gamma_2 \,.\, \ldots \,.\, \gamma_n \in \alpha)) \supset \sim \phi'\urcorner.$$

Proof. Case 1: $n = 0$. By *103 (& hp 2),

$$\vdash \ulcorner(\alpha)(\alpha \in \beta \,.\!\equiv\!.\, \phi \,.\sim (\alpha \in \alpha)) \supset\!:\, \beta \in \beta \,.\!\equiv\!.\, \phi' \,.\sim(\beta \in \beta)\urcorner$$

*100 $\qquad \supset \sim \phi'\urcorner.$

Case 2: $n > 0$.

*114 $\vdash \ulcorner(\gamma_1)\ldots(\gamma_n) \sim (\alpha \epsilon \gamma_1 \,.\ldots.\, \gamma_n \epsilon \alpha)$

$\supset \sim (\alpha \epsilon \gamma_1 \,.\ldots.\, \gamma_n \epsilon \alpha)\urcorner$. (1)

*103 (& hp 1) $\vdash \ulcorner[(\gamma_1)1 \supset .]\, \mathrm{L}1 \supset \sim (\alpha \epsilon \beta \,.\, \beta \epsilon \gamma_2 \,.\ldots.\, \gamma_n \epsilon \alpha)\urcorner$. (2)

*100 $\vdash \ulcorner \phi \,.\, \mathrm{L}1 \,.\supset \mathrm{L}1\urcorner$. (3)

*110 $\vdash \ulcorner(\alpha)(\alpha \epsilon \beta \,.\supset \mathrm{L}1) \supset : \alpha \epsilon \beta \,.\supset \mathrm{L}1\urcorner$

*100 $\supset .\, [2 \supset] \sim (\beta \epsilon \gamma_2 \,.\ldots.\, \gamma_n \epsilon \alpha \,.\, \alpha \epsilon \beta)\urcorner$. (4)

*121 $\vdash \ulcorner \mathrm{L}180 \supset .\, (\alpha)(\alpha \epsilon \beta \,.\supset \mathrm{L}1)\, [\equiv (\alpha)3]\urcorner$

(4), *117 (& hp 1) $\supset (\gamma_2)\ldots(\gamma_n)(\alpha)\, \mathrm{R}4\urcorner$. (5)

*171 (& hp 1)

$\vdash \ulcorner[5 \equiv .]\, \mathrm{L}180 \supset (\gamma_1)\ldots(\gamma_n) \sim (\beta \epsilon \gamma_1 \,.\, \gamma_1 \epsilon \gamma_2 \,.\ldots.\, \gamma_n \epsilon \beta)\urcorner$. (6)

*103 (& hp 1) $\vdash \ulcorner \mathrm{R}6 \supset (\gamma_2)\ldots(\gamma_n) \sim (\beta \epsilon \beta \,.\, \beta \epsilon \gamma_2 \,.\ldots.\, \gamma_n \epsilon \beta)\urcorner$

*103 (& hp 1) $\supset (\gamma_3)\ldots(\gamma_n) \sim (\beta \epsilon \beta \,.\, \beta \epsilon \beta \,.\ldots.\, \gamma_n \epsilon \beta)\urcorner$

$\vdots$ $\vdots$

*103 $\supset \sim (\beta \epsilon \beta \,.\, \beta \epsilon \beta \,.\ldots.\, \beta \epsilon \beta)\urcorner$

*100 $\supset \sim (\beta \epsilon \beta)\urcorner$. (7)

*103 (& hp) $\vdash \ulcorner \mathrm{L}180 \supset : \beta \epsilon \beta \,.\equiv .\, \phi' \,.\, \mathrm{R}6\urcorner$

*100 $\supset : [7 \supset .]\, \mathrm{R}6 \supset \sim \phi'\urcorner$. (8)

*100 $\vdash \ulcorner[8 \supset .\, 6 \supset]\, 180\urcorner$.

***181.** *If* $\alpha, \beta, \gamma_1, \ldots, \gamma_n$ *are distinct,*

$\vdash \ulcorner \sim (\alpha)(\alpha \epsilon \beta \,.\equiv (\gamma_1)\ldots(\gamma_n) \sim (\alpha \epsilon \gamma_1 \,.\, \gamma_1 \epsilon \gamma_2 \,.\ldots.\, \gamma_n \epsilon \alpha))\urcorner$.

Proof. *100 $\vdash \ulcorner \alpha \epsilon \alpha \,.\supset .\, \alpha \epsilon \alpha\urcorner$. (1)

*180 (& hp) $\vdash \ulcorner(\alpha)(\alpha \epsilon \beta \,.\equiv .\, [1 \,.]\, (\gamma_1)\ldots(\gamma_n) \sim (\alpha \epsilon \gamma_1 \,.\ldots$

$.\, \gamma_n \epsilon \alpha)) \supset \sim (\beta \epsilon \beta \,.\supset .\, \beta \epsilon \beta)\urcorner$. (2)

*100 $\vdash \ulcorner[2 \supset]\, 181\urcorner$.

We are thus forced to recognize that the idiom 'the class of all entities x such that ...' cannot in general be trusted to express a class, common sense to the contrary notwithstanding. We must restrict this idiom in one way or another. There are a number of ways; we may revise the phrase 'all entities x', or we may restrict the matrix which follows 'such that', or we may do both. The first of these courses, adopted by von Neumann,[1] seems to recommend itself above the others. Under this procedure the realm of

[1] "Eine Axiomatisierung." For historical remarks see below, § 29.

entities which can be formed into classes is narrowed somewhat, by deciding that certain classes are incapable of being members of classes. The idiom 'the class of all entities x such that' is then revised to read 'the class of all *membership-eligible* entities x such that . . .'.

For brevity let us refer to "membership-eligible" entities hereafter as *elements*. An *element* is thus to be understood simply as the sort of entity that belongs to classes; 'x is an element' can be rendered symbolically '$(\exists y)(x \in y)$'. Observe now how our revised idiom, 'the class of all elements x such that . . .', fares in connection with Russell's matrix '$\sim(x \in x)$'. Where w is the class of all elements x such that $\sim(x \in x)$, we are not called upon to equate '$x \in w$' with '$\sim(x \in x)$'. Rather, '$x \in w$' means that x *is an element and* $\sim(x \in x)$; symbolically,

$$(\exists y)(x \in y) \,.\, \sim(x \in x).$$

Instead of having '$(x)(x \in w \,.\!\equiv \sim (x \in x))$', and the self-contradictory consequence '$w \in w \,.\!\equiv \sim (w \in w)$', we have merely:

$$(x)(x \in w \,.\!\equiv\!.\, (\exists y)(x \in y) \,.\, \sim (x \in x)),$$

whose consequence:

$$w \in w \,.\!\equiv\!.\, (\exists y)(w \in y) \,.\, \sim (w \in w),$$

far from being self-contradictory, is equivalent to:

$$\sim (\exists y)(w \in y)$$

and thus tells us merely that w itself is not an element.

The status of element must be withheld not only from the class w of all elements x such that $\sim(x \in x)$, but also from the class of all elements x such that $(y) \sim (x \in y \,.\, y \in x)$, and from the class of all elements x such that $(y)(z) \sim (x \in y \,.\, y \in z \,.\, z \in x)$, and so on. The theorems to this effect, viz. the closures of:

$$(x)(x \in w \,.\!\equiv\!.\, (\exists y)(x \in y) \,.\, \sim(x \in x)) \supset \sim(\exists y)(w \in y),$$
$$(x)(x \in w \,.\!\equiv\!.\, (\exists y)(x \in y) \,.\, (y) \sim (x \in y \,.\, y \in x)) \supset \sim(\exists y)(w \in y),$$
$$(x)(x \in w \,.\!\equiv\!.\, (\exists y)(x \in y) \,.\, (y)(z) \sim (x \in y \,.\, y \in z \,.\, z \in x)) \supset \sim(\exists y)(w \in y),$$

etc., are all provided by *180 (ϕ and ϕ' being taken as $\ulcorner(\exists \delta)(\alpha \in \delta)\urcorner$ and $\ulcorner(\exists \delta)(\beta \in \delta)\urcorner$). The question of determining what entities *are* to be rated as elements will be dealt with later (§ 28); for the present, suffice it to say that the only entities which will not be elements are certain queer classes such as the ones just now noted.

The prefix 'the class of all elements x such that' will be rendered symbolically '$\hat{x}$'; thus the class w discussed above is $\hat{x} \sim (x \in x)$.[1] The forming of class names by such prefixes will be called *abstraction;* and the result, e.g. '$\hat{x} \sim (x \in x)$' or in general $\ulcorner \hat{\alpha}\phi \urcorner$, will be called an *abstract.* Seeking now to define the notation $\ulcorner \hat{\alpha}\phi \urcorner$ on the basis of our primitive notation, we encounter the following difficulty. Our primitive notation comprises machinery for the construction solely of statements and statement matrices — atomic matrices $\ulcorner(\alpha \in \beta)\urcorner$, quantifications $\ulcorner(\alpha)\phi\urcorner$, joint denials $\ulcorner(\phi \downarrow \psi)\urcorner$; and an abstract $\ulcorner \hat{\alpha}\phi \urcorner$ cannot be construed as an abbreviation of any expression of this sort, since it is supposed to behave rather as a noun. The way out of this difficulty is suggested, however, by reflecting on fragmentary expressions such as '$\equiv$', '$\supset$', $\ulcorner \supset \psi \urcorner$. By no means identifiable with statements or statement matrices, these expressions have none the less been introduced by definitions; they have been defined *in context.* They are fragments of expressions $\ulcorner(\phi \equiv \psi)\urcorner$ and $\ulcorner(\phi \supset \psi)\urcorner$ which have been defined entire as abbreviations of appropriate formulæ. Now the same sort of contextual definition is available for abstracts. Turning our attention to such contexts of $\ulcorner \hat{\alpha}\phi \urcorner$ as are appropriately identifiable with formulæ, we can frame suitable definitions of these contexts as wholes. By adopting such definitions in sufficient variety, we can account for the occurrence of $\ulcorner \hat{\alpha}\phi \urcorner$ in statement contexts of every desired kind.

One definition is wanted, e.g., explaining contexts of the form $\ulcorner \beta \in \hat{\alpha}\phi \urcorner$: contexts affirming *membership* in the class which $\ulcorner \hat{\alpha}\phi \urcorner$ describes. Other definitions are wanted explaining other contexts; but let us postpone them (to § 26) and deal with this first case. We want to define $\ulcorner \beta \in \hat{\alpha}\phi \urcorner$ in such way that, where '$\ldots x \ldots$' is thought of as any formula involving 'x', the whole:

(1) $$y \in \hat{x}(\ldots x \ldots)$$

[1] This notation was adapted from Frege by Russell. But my reference to elementhood, in interpreting the notation, is a departure. Cf. § 29.

will agree in meaning with the words:

(2) y is a member of the class of all elements x such that . . . x . . ..

This requirement proves to be met by the definition:

D9. $\ulcorner(\beta \in \hat{\alpha}\phi)\urcorner$ *for* $\ulcorner(\exists\gamma)(\beta \in \gamma \,.\, (\alpha)(\alpha \in \gamma \,.\!\supset \phi))\urcorner$

(γ being any new variable). D9 explains (1) as an abbreviation of:

(3) $(\exists z)(y \in z \,.\, (x)(x \in z \,.\!\supset\! . \ldots x \ldots))$,

whose equivalence with (2) is easily seen. Where z is the class of all elements x such that . . . x . . . , clearly

$$(x)(x \in z \,.\!\supset\! . \ldots x \ldots);$$

hence (3) is true when (2) is. Conversely, since (3) says that y is a member of a class all of whose members are entities x such that . . . x . . . , we can conclude from (3) both that y is an element (being a member of a class) and that y is one of the entities x such that . . . x . . . ; hence (2) is true when (3) is.

D9 has the peculiarity of imposing like abbreviations upon unlike formulæ; e.g., the formulæ:

$$(\exists z)(y \in z \,.\, (x)(x \in z \,.\!\supset\! .\, x \in x)),$$
$$(\exists w)(y \in w \,.\, (x)(x \in w \,.\!\supset\! .\, x \in x))$$

both become '$y \in \hat{x}(x \in x)$'. But such formulæ will always be alphabetic variants of one another, and hence interchangeable in view of *171 (and *123). Shift from one choice of γ to another, in expanding $\ulcorner\beta \in \hat{\alpha}\phi\urcorner$ by D9, amounts merely to relettering the expanded form by *171 — so long as γ is kept distinct from α, β, and the variables of ϕ.

The formulæ and in particular the axioms of quantification and the theorems are, properly speaking, composed of the primitive notation without definitional abbreviations (cf. § 13). When a definitionally abbreviated expression is said to be a theorem, what is properly meant is rather that it is an abbreviation of a theorem. In all discourse about theorems, defined notations are imagined expanded into primitive notation. Hence caution is needed in adopting definitions which impose like abbreviations upon primitively unlike formulæ. If a definition of that kind happened to cause some theorem ψ and some non-theorem χ to be abbreviated alike, fallacy would result; by merely abbreviating ψ according to the definition, and then expanding the result according to the definition, one would reason from the fact that ψ is a theorem to the false conclusion that χ is a theorem. However, *171 has shown that D9 (with γ understood as distinct from α, β, and the variables of ϕ) is not capable of thus blurring the distinction between theorems and non-theorems. Adoption of such a definition as D9 amounts to tacit use of *171.

Definitions which impose like abbreviations upon primitively unlike formulæ will never obtrude themselves hereafter except in cases similar to D9: cases where a bound variable appears in the unabbreviated rendering and disappears in the abbreviation. But definitions of just this sort will turn up so frequently that it is simplest not to stop each time for an arbitrary alphabetical stipulation of the odd variable; rather it will merely be understood, as in the case of D9, that the variable is not to be so chosen as to conflict with the others at hand. The puristic reader can still, if he prefers, reconstrue all such definitions arbitrarily as calling for the alphabetically earliest variable other than those at hand; for he can always rewrite his variable afterward at will by *171.

§ 25. *Identity*

WE TURN now to the problem of so defining '$x = y$', in terms of 'ϵ' and our other primitives, that it will carry the intended sense 'x and y are the same object'. In the trivial case where y is not a class, indeed, $x \epsilon y$ if and only if $x = y$ in this sense (cf. § 22); but our problem remains, since '$x \epsilon y$' diverges in meaning from '$x = y$' in case y is a class. We must find a formula, composed of 'x' and 'y' by means of 'ϵ' and our other primitives, which will be true just in case x and y are the same object — whether a class or a non-class.

The requirement is met by:

(1) $$(z)(z \epsilon x \mathrel{.{\equiv}.} z \epsilon y)$$

when x and y are classes, since classes are the same when their members are the same (cf. § 22). Moreover, (1) continues to meet the requirement when x and y are not classes. For, in this case '$z \epsilon x$' and '$z \epsilon y$' identify z with x and with y; and (1) as a whole then says that whatever is the same as x is the same as y, thus identifying x and y. Both where x and y are classes and where they are not, therefore, (1) meets our requirements; (1) is true if and only if x and y are the same. We are thus led to introduce '$x = y$' as an abbreviation of (1).

This definition of '$=$' has one rather curious but harmless result: where x is not a class, we do *not* have over and above x any class whose sole member is x. What would appear to be the class whose sole member is x reduces to the non-class x itself. For, suppose x is a non-class and y is the class whose sole member is x. Then $z \epsilon y$ if and only if z is x. But also, according to the interpretation of '$z \epsilon x$' adopted for the case where x is not a class, $z \epsilon x$ if and only if z is x. Hence (1) is true; i.e., according to our definition, $x = y$.

This result — suppression of the class of a single non-class in favor of the non-class itself — may just as well be conceived in reverse fashion as an assimilation of non-classes to classes. What was regarded as a non-class becomes reconstrued as a special sort of class, viz. a class having itself as sole member. If we think of the matter in this way, we must of course abandon the term 'non-class' in favor of a more neutral designation — say 'individual'. Everything comes now to be thought of as a class, but individuals are distinguished from other classes by the peculiar circumstance of being their own sole members. This way of phrasing the situation is more convenient than the other, for it enables us to give '$x \epsilon y$' always the uniform reading 'x is a member of the class y'. Note, in any case, that we are far from identifying every object x with the class whose sole member is x; this happens only if x is an individual. From a formal standpoint, actually, there is no need of assuming that there is any such x at all (cf. § 22).

Variables and abstracts will be spoken of collectively as *terms*. Now let us supplement our Greek-letter conventions to this extent: just as we use 'ϕ', 'ψ', and 'χ', to refer to any formulæ, and 'α', 'β', 'γ', and 'δ' to refer to any variables, so let us use 'ζ', 'η', and 'θ' (along with their accented and subscripted variants) to refer in general to any *terms*. With help of this convention we can express the general definition of identity as follows, for application to variables and abstracts indifferently:

D10. $\ulcorner(\zeta = \eta)\urcorner$ *for* $\ulcorner(\alpha)(\alpha \in \zeta \,.\!\equiv\!.\, \alpha \in \eta)\urcorner$.

The notation $\ulcorner\alpha \in \zeta\urcorner$ here is either primitive or explained by D9, according as ζ is a variable or an abstract; and similarly for $\ulcorner\alpha \in \eta\urcorner$. The choice of α follows the usual understanding (§ 24); it may be any variable other than the ones in ζ and η. The parentheses in $\ulcorner(\zeta = \eta)\urcorner$ record the fact that '=', like all binary connectives, is to be construed as carrying with it a pair of parentheses.[1]

It is more common to define $\ulcorner\zeta = \eta\urcorner$ as $\ulcorner(\alpha)(\zeta \in \alpha \,.\!\equiv\!.\, \eta \in \alpha)\urcorner$ or $\ulcorner(\alpha)(\zeta \in \alpha \,.\!\supset\!.\, \eta \in \alpha)\urcorner$. Both of these versions are given by Peirce (3.398, 400) and one or the other is rather hinted by Leibniz (Lewis, *Survey*, p. 373); but they cease to be available once non-elements are recognized (cf. § 32). D10, as a definition of class identity, was stated in effect by Peirce (3.47) and in its present form by Fraenkel (*Einleitung*, 3d ed., p. 272). Its availability as a general definition of identity turns on the special way in which I have dealt with individuals.

My treatment of individuals is reminiscent of the days before Frege and Peano urged the importance of distinguishing between an object and its *unit class*, i.e. the class which has that object as sole member (cf. § 35). Commonly the distinction in question is indeed vital. We must distinguish between the null class (§ 26) and its unit class, for the former has no members while the latter has one; and we must distinguish between the class of Apostles and its unit class, for Peter belongs to the one class and not to the other. But the arguments which show that the distinction must commonly be made do not militate against the identification, in particular, of individuals with their unit classes (and hence with the unit classes of their unit classes and so on). Within the domain of individuals, retention of the pre-Fregean attitude leads to no trouble.

Though I have been led to this course by considerations only of technical efficacy, a certain naturalness may be argued for it on the basis of such illustrations as the following. Let us think of points as individuals, and of lines and planes as classes of points. Then the intersection or logical product (§ 33) of two planes is a line; but, if we distinguish individuals from their unit classes, the intersection of two lines is only the unit class of a point and not a point. Again, whereas a line in a given plane is related to the plane by inclusion (§ 34), a point in a given line is not related to the line by inclusion if we distinguish individuals from their unit

[1] In the matter of parentheses, '=' is like '$\in$'; the parentheses are adopted with an eye to later definitions. See p. 124, footnote.

classes; the point is rather a member of the line, while the unit class of the point is included in the line. Under the proposed theory, on the other hand, intersections of lines are points, and points bear to lines the same inclusion relation which lines bear to planes. (This illustration is due to Miss L. D. Steinhardt.)

The usual compact way of denying identity is adopted in the following definition.

D11. $\ulcorner(\zeta \neq \eta)\urcorner$ *for* $\ulcorner\sim(\zeta = \eta)\urcorner$.

We proceed to three theorems.

†182. $(x) \quad x = x$ *Proof:* *100 (& D10).
†183. $(y)(x) \quad x = y \,.\!\equiv\!.\, y = x$ *Proof:* *100, *123 (& D10).
†184. $(z)(y)(x) \quad x = y \,.\!\supset\!:\, x = z \,.\!\equiv\!.\, y = z$ *Proof:* *121 (& D10).

In stating theorems it is convenient to leave a space between the initial group of universal quantifiers and the matrix which they govern, on the understanding that those quantifiers are to cover the matrix in its entirety. The theorem is thus the closure of this segregated matrix; †182–†184 are:

$$(x)(x = x),$$
$$(y)(x)(x = y \,.\!\equiv\!.\, y = x),$$
$$(z)(y)(x)(x = y \,.\!\supset\!:\, x = z \,.\!\equiv\!.\, y = z).$$

The fact that †182–†184 are theorems, rather than metatheorems, is given recognition by attaching a dagger '†' rather than a star to the numerals. Apart from the distinguishing stars and daggers, the numerical designations of metatheorems and theorems run along together in a single increasing sequence; first we have metatheorems (*100–*181), then some theorems (†182–†185), then more metatheorems (*186–*188), and so on.

The proofs of †182–†184, indicated above, call for little comment. †182 is an abbreviation, by D10, of:

$$(x)(w)(w \in x \,.\!\equiv\!.\, w \in x),$$

which is the closure of a tautologous matrix $\ulcorner\phi \equiv \phi\urcorner$; and *100 says that closures of such matrices are theorems. †183 is an abbreviation, by D10, of:

$$(y)(x)((w)(w \in x \,.\!\equiv\!.\, w \in y) \equiv (w)(w \in y \,.\!\equiv\!.\, w \in x)),$$

which is the closure of a matrix $\ulcorner(\alpha)\phi \equiv (\alpha)\psi\urcorner$ such that $\ulcorner\phi \equiv \psi\urcorner$

is tautologous; and *100 and *123 say that closures of such matrices are theorems. †184 is an abbreviation, by D10, of:

$$(z)(y)(x)$$
$$((w)(w \in x \,.\!\equiv.\, w \in y) \supset .\, (w)(w \in x \,.\!\equiv.\, w \in z) \equiv (w)(w \in y \,.\!\equiv.\, w \in z)),$$

which is the closure of a matrix $\ulcorner(\alpha)(\phi \equiv \phi') \supset .\, \psi \equiv \psi'\urcorner$ of the kind described in *121.

†182 and †183 are known respectively as the principles of the *reflexivity* and *commutativity* (or symmetry) of identity, by a natural extension of the notions of reflexivity and commutativity which were applied to statement composition (§ 11).[1] The following, by a similar extension, is called the principle of the *transitivity* of identity:

$$(z)(y)(x) \qquad x = y \,.\, y = z \,.\!\supset.\, x = z.$$

†184 is a stronger form of this principle.

The proof of the next theorem is more elaborate. It illustrates the notation which will be used hereafter in proofs of theorems, as distinct from metatheorems.

†185. $(y)(x) \qquad y \in x \,.\!\equiv (\exists z)(y \in z \,.\, z = x)$

Proof. *110 (& D10) $z = x \,.\!\supset: y \in z \,.\!\equiv.\, y \in x$ (1)

*100, *163 $(\exists z)([1\,.]\, y \in z \,.\, z = x) \supset .\, y \in x$ (2)

*134 $y \in x\,[.\, 182] \,.\!\supset (\exists z)(y \in z \,.\, z = x)$ (3)

*100 $[2\,.\,3\,.\!\supset]\, 185$

The expressions which occupy the several lines in the proof of a theorem are not in general theorems themselves but rather formulæ whose closures are theorems. The references at the left of the line substantiate the fact that the closure of the exhibited formula is a

[1] The above instance suggests this simple formulation of reflexivity as applied to predicates: a reflexive predicate is one whose insertion in the blank of '$(x)(x \quad x)$' yields a truth. Such reflexivity may be called *total reflexivity* (cf. Carnap, *Abriss*, p. 40); for the term 'reflexive' is also commonly applied, more broadly, to any predicate whose insertion in the three blanks of:

$$(x)(y)(x \quad y \,.\!\vee.\, y \quad x :\supset .\, x \quad x)$$

yields a truth. Note that the notion of reflexivity as applied to statement composition (§ 11) can be broadened in analogous fashion; conjunction then comes to be classed as reflexive, though not totally reflexive.

theorem. Thus, in the first line of the above proof we cite *110 (and D10) to show that

$$\vdash `z = x \,.\supset:\, y \in z \,.\equiv.\, y \in x',$$

i.e., to show that the closure:

$$(z)(y)(x) \qquad z = x \,.\supset:\, y \in z \,.\equiv.\, y \in x$$

is a theorem. The numeral at the right of a line serves, in its unparenthesized recurrences, as an abbreviation of the formula appearing in that line (minus any bracketed matter); thus '1' in the second line of the proof is an abbreviation of:

$$z = x \,.\supset:\, y \in z \,.\equiv.\, y \in x,$$

and '2' in the last line is an abbreviation of:

$$(\exists z)(y \in z \,.\, z = x) \supset.\, y \in x.$$

Brackets indicate use of *124, as hitherto; having shown e.g. that the closure of the matrix which '1' abbreviates is a theorem, we bracket '1 .' out of the second line of the proof on the strength of *124. Divorced of its dagger, the numeral of a theorem serves as an abbreviation of the theorem minus such initial quantifiers as are spaced off to the left; thus '182' and '185' are abbreviations of the respective matrices:

$$x = x, \qquad y \in x \,.\equiv (\exists z)(y \in z \,.\, z = x),$$

whose closures are the theorems †182 and †185.

The combination of *100 and *163, as in the second line of the above proof, is one which will appear frequently hereafter. In such a step of proof we observe first that the whole line minus the existential quantifier, in this case:

$$1 \,.\, y \in z \,.\, z = x \,.\supset.\, y \in x,$$

is tautologous; then we introduce the quantifier on the strength of *163, noting that the variable of quantification is not free in the consequent. The number of quantifiers thus introduced need not be limited to one; in general, *100 and *163 show that

$$\vdash \ulcorner(\exists \alpha_1) \ldots (\exists \alpha_n)\phi \supset \psi\urcorner$$

where $\ulcorner\phi \supset \psi\urcorner$ is tautologous and $\alpha_1, \ldots, \alpha_n$ are not free in ψ.

Combination of *100 with *117 will also be common; in such cases the added quantifiers are universal rather than existential, and apply to the consequent rather than the antecedent.

§ 26. *Abstraction Resumed*

LET US now return to the task of defining abstracts $\ulcorner\hat{\alpha}\phi\urcorner$ in context. D9 provides for all contexts in which the abstract follows an occurrence of 'ϵ' which is preceded by a variable. But now we must provide for contexts in which the abstract precedes an occurrence of 'ϵ' which is followed by a variable; and also we must provide for contexts in which abstracts stand on both sides of 'ϵ'. These two cases can be lumped together: our problem is to define expressions of the form $\ulcorner\hat{\alpha}\phi \epsilon \zeta\urcorner$, where ζ is a variable or an abstract. Now that D10 is at hand, explaining '=' in application to variables and abstracts alike, our problem is easily solved:

D12. $\ulcorner(\hat{\alpha}\phi \epsilon \zeta)\urcorner$ *for* $\ulcorner(\exists\beta)(\beta = \hat{\alpha}\phi \,.\, \beta \epsilon \zeta)\urcorner$.

The notation $\ulcorner\beta \epsilon \zeta\urcorner$ here is either primitive or explained by D9, according as ζ is a variable or an abstract. The choice of β follows the usual understanding; it may be any variable other than α and the ones in ϕ and ζ.

The notation $\ulcorner\zeta \epsilon \eta\urcorner$ is now generally accounted for, no matter what terms ζ and η may be. If ζ and η are variables, $\ulcorner\zeta \epsilon \eta\urcorner$ is primitive; if ζ is a variable and η an abstract, $\ulcorner\zeta \epsilon \eta\urcorner$ is explained by D9; and if ζ is an abstract, $\ulcorner\zeta \epsilon \eta\urcorner$ is explained by D12. Two minor conventions of abbreviation will now be added.

D13. $\ulcorner(\zeta \bar{\epsilon} \eta)\urcorner$ *for* $\ulcorner\sim (\zeta \epsilon \eta)\urcorner$.

D14. $\ulcorner(\zeta_1, \zeta_2, \ldots, \zeta_n \epsilon \eta)\urcorner$ *for* $\ulcorner(\zeta_1 \epsilon \eta \,.\, \zeta_2 \epsilon \eta \,.\, \ldots \,.\, \zeta_n \epsilon \eta)\urcorner$.

Just as $\ulcorner\alpha \epsilon \beta\urcorner$ and $\ulcorner\beta \epsilon \alpha\urcorner$ are atomic contexts of α, so we may speak of $\ulcorner\hat{\alpha}\phi \epsilon \eta\urcorner$ and $\ulcorner\eta \epsilon \hat{\alpha}\phi\urcorner$ as *quasi-atomic* contexts of $\ulcorner\hat{\alpha}\phi\urcorner$. More generally, a quasi-atomic context of a term ζ is any result of putting ζ for one variable and perhaps other terms for other variables in an atomic matrix. Now D9 and D12 explain all those quasi-atomic contexts of $\ulcorner\hat{\alpha}\phi\urcorner$ which correspond to logical atomic matrices $\ulcorner \alpha \epsilon \beta \urcorner$; for D9 explains $\ulcorner \gamma \epsilon \hat{\alpha}\phi \urcorner$ and D12 explains $\ulcorner \hat{\alpha}\phi \epsilon \eta \urcorner$ and $\ulcorner \hat{\gamma}\psi \epsilon \hat{\alpha}\phi \urcorner$. But extra-mathematical discourse would involve atomic matrices of other forms than $\ulcorner \alpha \epsilon \beta \urcorner$, and would thus call for definition of $\ulcorner \hat{\alpha}\phi \urcorner$ in supplementary quasi-atomic contexts. All such definitions can follow the pattern of D12; corresponding to an atomic matrix $\ulcorner \mu\, \beta_1\, \beta_2 \ldots \beta_n \urcorner$, in general, we can introduce quasi-atomic contexts by a sequence of n definitions as follows:

Da_1. $\ulcorner \mu\, \beta_1\, \beta_2 \ldots \beta_{n-2}\, \beta_{n-1}\, \hat{\alpha}\phi \urcorner$ *for* $\ulcorner (\exists \beta_n)(\beta_n = \hat{\alpha}\phi \,.\, \mu\, \beta_1\, \beta_2 \ldots \beta_{n-2}\, \beta_{n-1}\, \beta_n) \urcorner$,

Da_2. $\ulcorner \mu\, \beta_1\, \beta_2 \ldots \beta_{n-2}\, \hat{\alpha}\phi\, \zeta_n \urcorner$ *for* $\ulcorner (\exists \beta_{n-1})(\beta_{n-1} = \hat{\alpha}\phi \,.\, \mu\, \beta_1\, \beta_2 \ldots \beta_{n-2}\, \beta_{n-1}\, \zeta_n) \urcorner$,

. . . .

. . . .

. . . .

Da_n. $\ulcorner \mu \hat{\alpha}\phi\, \zeta_2 \ldots \zeta_{n-2}\, \zeta_{n-1}\, \zeta_n \urcorner$ *for* $\ulcorner (\exists \beta_1)(\beta_1 = \hat{\alpha}\phi \,.\, \mu\, \beta_1\, \zeta_2 \ldots \zeta_{n-2}\, \zeta_{n-1}\, \zeta_n) \urcorner$.

Since from the standpoint of primitive notation any free occurrence of a variable has an atomic matrix as its immediate context, our definitions of the quasi-atomic contexts of $\ulcorner \hat{\alpha}\phi \urcorner$ explain abstracts in all manners of occurrence which are available to free variables. Definitions which abbreviate contexts of free variables can then be so framed as to apply at the same time to abstracts, as already observed in the case of D10, D11, D13, and D14. All occurrences of an abstract will turn out to stand in quasi-atomic contexts when the overlying definitional abbreviations as of D10, D11, D13, D14, etc. are resolved.

An occurrence of $\ulcorner \hat{\alpha}\phi \urcorner$, in any quasi-atomic context, comes to have an immediate context of one or other of the forms $\ulcorner \gamma \epsilon \hat{\alpha}\phi \urcorner$ and $\ulcorner \beta = \hat{\alpha}\phi \urcorner$ when overlying abbreviations are resolved; this is clear from D12 and Da_1–a_n. Furthermore, $\ulcorner \beta = \hat{\alpha}\phi \urcorner$ in turn is short for

$$\ulcorner (\gamma)(\gamma \epsilon \beta \,.\!\equiv\!.\, \gamma \epsilon \hat{\alpha}\phi) \urcorner$$

where the immediate context of $\ulcorner \hat{\alpha}\phi \urcorner$ is again of the form $\ulcorner \gamma \epsilon \hat{\alpha}\phi \urcorner$. The latter is thus the fundamental sort of context for abstracts; every occurrence of $\ulcorner \hat{\alpha}\phi \urcorner$, in any quasi-atomic context and therefore in any context whatever, comes to have an immediate context $\ulcorner\gamma \in \hat{\alpha}\phi\urcorner$ on definitional expansion. Once $\ulcorner\gamma \in \hat{\alpha}\phi\urcorner$ is expanded in turn, by D9, the abstract $\ulcorner\hat{\alpha}\phi\urcorner$ disappears altogether.

Inspection of D9 shows that all free occurrences of variables in ϕ remain free in $\ulcorner\gamma \in \hat{\alpha}\phi\urcorner$, except for occurrences of α; but that all occurrences of α in ϕ are bound in $\ulcorner\gamma \in \hat{\alpha}\phi\urcorner$. Moreover, this is true not only in the case of $\ulcorner\gamma \in \hat{\alpha}\phi\urcorner$ but in the case of any other quasi-atomic context ψ of $\ulcorner\hat{\alpha}\phi\urcorner$ as well; for we have noted that the expansion of ψ by D12 or Da_1–a_n (and D10) will merely give us $\ulcorner\gamma \in \hat{\alpha}\phi\urcorner$ again within a broader context, and by the general convention governing the choice of extra variables in definitions we are assured that the quantifiers of this broader context will not involve variables of ϕ. In all contexts, therefore, an abstraction prefix $\ulcorner\hat{\alpha}\urcorner$ operates in the same way as a quantifier $\ulcorner(\alpha)\urcorner$ so far as bondage and freedom are concerned.

When we speak *about* a formula, we are ordinarily to think of it as written out in full primitive notation (cf. § 13). But it is sometimes convenient to speak of occurrences of an abstract $\ulcorner\hat{\alpha}\phi\urcorner$ in a formula ψ; and this usage does compel us to think of ψ as temporarily subject to definitional abbreviations, to the extent of accommodating the occurrences of $\ulcorner\hat{\alpha}\phi\urcorner$ in question. The abbreviations thus imagined in ψ are still conveniently limited to quasi-atomic contexts of $\ulcorner\hat{\alpha}\phi\urcorner$. Insofar as a formula ψ is discussed from this point of view, questions of bondage and freedom call for explicit consideration of abstraction prefixes along with quantifiers. As an adjunct of discourse about occurrences of abstracts, accordingly, the following practical elaboration of the terminology of of § 14 is needed. A given occurrence of α is *bound to* a given occurrence of $\ulcorner(\alpha)\urcorner$ *or* $\ulcorner\hat{\alpha}\urcorner$ if it stands in a formula or abstract beginning with the given occurrence of $\ulcorner(\alpha)\urcorner$ or $\ulcorner\hat{\alpha}\urcorner$ and stands in no formula or abstract beginning with a later occurrence of $\ulcorner(\alpha)\urcorner$ or $\ulcorner\hat{\alpha}\urcorner$. An occurrence of a formula or term ν is *bound in* a formula or term μ *with respect to* α if within that occurrence of ν there is an occurrence of α which is bound to an occurrence of $\ulcorner(\alpha)\urcorner$ or $\ulcorner\hat{\alpha}\urcorner$ lying in μ but outside ν. An occurrence of ν is *bound in* μ if it is bound in μ with respect to at least one variable; and ν is bound in μ if an occurrence of ν is bound in μ. An occurrence of ν in μ

is *free in* μ if it is not bound in μ; and ν is free in μ if an occurrence of ν is free in μ. Clearly an occurrence of χ will be bound in ψ with respect to α, in the sense of the present paragraph, if and only if the corresponding occurrence or occurrences of χ in the primitive expansion of ψ are bound with respect to α in the sense of § 14; and the rest of the terminology of the present paragraph meshes with that of § 14 in similar fashion.

The following analogues of *121–*123 are typical of the metatheorems which involve use of this extended terminology.

***186.** *If* η' *is like* η *except for containing* ϕ' *at some places where* η *contains* ϕ, *and* $\alpha_1, \ldots, \alpha_n$ *exhaust the variables with respect to which those occurrences of* ϕ *and* ϕ' *are bound in* η *and* η', *then*

$$\vdash \ulcorner(\alpha_1) \ldots (\alpha_n)(\phi \equiv \phi') \supset . \eta = \eta'\urcorner.$$

Proof: *121, *117 (& D10).

***187.** *If* η' *is like* η *except for containing free occurrences of* ϕ' *at some places where* η *contains free occurrences of* ϕ, *then*

$$\vdash \ulcorner\phi \equiv \phi' . \supset . \eta = \eta'\urcorner.$$

Proof: *186 (taking n as 0).

***188.** *If* $\vdash \ulcorner\phi \equiv \phi'\urcorner$, *and* η' *is formed from* η *by putting* ϕ' *for some occurrences of* ϕ, *then* $\vdash \ulcorner\eta = \eta'\urcorner$.

Proof from *186 like that of *123 from *121.

But it must be borne in mind that the theorems provided by such metatheorems as *186–*188 are still to be thought of finally as in full primitive notation. *187, e.g., amounts properly to the following: If η' is like η except for containing free occurrences of ϕ' at some places where η contains free occurrences of ϕ, then the formula (in primitive notation) which the definitions abbreviate as $\ulcorner\phi \equiv \phi' . \supset . \eta = \eta'\urcorner$ is one whose closure is a theorem.

The following theorem shows that '$\hat{y}$' cancels '$y \in$'; every entity is the class of all those elements which are its members.

†189. $(x) \quad x = \hat{y}(y \in x)$

Proof. *100	$y \in x . \supset . y \in x$	(1)
*134	$z \in x [. (y)1] . \supset (\exists w)(z \in w . (y)(y \in w . \supset . y \in x))$	(2)
*103	$(y)(y \in w . \supset . y \in x) \supset : z \in w . \supset . z \in x$	(3)

*100, *163 $(\exists w)([3\,.]\, z \in w \,.\, (y)(y \in w \,.\supset.\, y \in x)) \supset.\, z \in x$ (4)

*140 (& D5, 9, 10) 189 $[\equiv.\, (z)2 \,.\, (z)4]$

The following two definitions introduce abbreviations for two particularly important abstracts.

D15. 'V' *for* '$\hat{x}(x = x)$',

D16. 'Λ' *for* '$\hat{x}(x \neq x)$'.

V is, by definition, the class of all those elements which are self-identical; i.e., since everything is self-identical (†182), V is simply the class of all elements.

†189a. (x) $x \in V \,.\equiv (\exists y)(x \in y)$

Proof. *100 $[182\,.]\, x \in y \,.\supset.\, x = x$ (1)

*100 (& D9, 15) $x \in V \,.\equiv (\exists y)(x \in y\, [.\, (x)1])$

†190. $(y)(x)$ $x \in y \,.\supset.\, x \in V$

†191. $(y)(x)$ $x \in y \,.\equiv.\, x \in V \,.\, x \in y$

Proofs. *135 $x \in y \,.\supset (\exists y)(x \in y)$ (1)

*100 $[189a \,.\, 1 \,.\supset]$ 190, 191

Λ is defined as the class of all those elements which are not self-identical; hence it is the memberless class, the so-called *null class*.

†192. (x) $x \bar{\in} \Lambda$

Proof. *110 $(x)(x \in y \,.\supset.\, x \neq x) \supset: x \in y \,.\supset.\, x \neq x$ (1)

*100 $[1 \,.\, 182 \,.\supset] \sim (x \in y \,.\, (x)(x \in y \,.\supset.\, x \neq x))$ (2)

*131 (& D9, 16) 192 $[\equiv (y)2]$

Citation of the trivial definitions D11 and D13 will ordinarily be omitted, as in the above proof. Similarly for D14, D6, and D7.

†192a. (y) $y = \Lambda \,.\equiv (x)(x \bar{\in} y)$

Proof.[1] *100, *123 (& D10) $y = \Lambda \,.\equiv (x)(x \bar{\in} y\, [.\equiv 192])$

†192b. (y) $y \neq \Lambda \,.\equiv (\exists x)(x \in y)$

Proof. *100 (& D8) $[192a \supset]$ 192b

***193.** $\vdash \ulcorner(\alpha)\phi \supset.\, \hat{\alpha}\phi = V\urcorner$.

Proof. *103 $\vdash \ulcorner[(x)182 \supset.]\, \alpha = \alpha\urcorner$. (1)

*186 $\vdash \ulcorner(\alpha)(\phi\, [\equiv 1]) \supset.\, \hat{\alpha}\phi = \hat{\alpha}(\alpha = \alpha)\urcorner$

*171 (& D15) $\supset.\, \hat{\alpha}\phi = V\urcorner$.

[1] This short proof was suggested by I. Kaplansky and M. J. Norris.

***194.** $\vdash \ulcorner(\alpha)\sim\phi \supset .\ \hat{\alpha}\phi = \Lambda\urcorner$.

Proof. *103 $\vdash \ulcorner[(x)182 \supset .]\ \alpha = \alpha\urcorner$. (1)

*186 $\vdash \ulcorner(\alpha)(\sim\phi[\equiv 1])\supset .\ \hat{\alpha} \sim\sim\phi = \hat{\alpha}(\alpha \neq \alpha)\urcorner$

*171 (& D16) $\supset .\ \hat{\alpha} \sim\sim\phi = \Lambda\urcorner$

*100, *123 $\supset .\ \hat{\alpha}\phi = \Lambda\urcorner$.

In the above proofs, *171 is used as if the biconditional $\ulcorner\phi \equiv \phi'\urcorner$ thereof were a mere conditional $\ulcorner\phi \supset \phi'\urcorner$. In the last step of the proof of *194, again, *123 is used as if the $\ulcorner\psi \equiv \psi'\urcorner$ thereof were $\ulcorner\psi \supset \psi'\urcorner$. This procedure involves tacit application of *100, of such kind as to weaken the biconditional to a conditional. Ellipsis of this sort will be common hereafter; a biconditional $\ulcorner\phi \equiv \psi\urcorner$ whose closure is cited as a theorem will be subject to any of the variant readings $\ulcorner\psi \equiv \phi\urcorner$, $\ulcorner\phi \supset \psi\urcorner$, and $\ulcorner\psi \supset \phi\urcorner$ at convenience. But this innovation applies only to citations by starred and daggered numerals; divorced of its star or dagger and used as an abbreviation, the numeral continues to stand for just the one expression for which it has hitherto stood.

In precisely the sense in which D9 and D12 constitute contextual definitions of $\ulcorner\hat{\alpha}\phi\urcorner$, D1–5 may be said to constitute contextual definitions of '$\sim$', '.', 'v', '$\supset$' and '$\equiv$'; likewise D8 of '$\exists$', and D10 of '='. The special subtlety of D9 and D12 is merely this: they endow $\ulcorner\hat{\alpha}\phi\urcorner$ with contexts notationally similar to contexts primitively accessible to variables, thus enabling us to handle $\ulcorner\hat{\alpha}\phi\urcorner$ as a *term* (unlike '$\sim$', '.', '$\exists$', '=', etc.). Such contextual definition of terms was initiated by Russell (1905, 1908) in his introduction of *descriptions* (§ 27); I have preferred to apply the method to abstracts, however, and then explain descriptions as direct abbreviations of certain abstracts. (Russell's theory of classes also proceeds by contextual definition, but in a fashion irrelevant to present concerns; cf. footnote to § 22.) Russell's method of contextual definition differs in detail from that exhibited above, and is more complex; it depends, for the avoidance of certain ambiguities which would otherwise arise, upon use of an auxiliary scheme of prefixes (cf. *Principia*, *14). The fact that Russell is concerned with descriptions rather than abstracts is not responsible for those complications; in "New Foundations," D9–10, I have defined description along lines similar to the above contextual definitions of abstraction.

The so-called universal and null classes V and Λ played a prominent role in the class algebra developed by Boole and his successors (§ 33). The signs 'V' and 'Λ' are Peano's. But the present use of 'V' as a means of affirming elementhood is new.

§ 27. *Descriptions and Names*

WHEREAS the prefix '$\hat{x}$' corresponds to the words 'the class of all elements x such that', the prefix '$(\imath x)$' is to correspond rather to the words 'the one and only object (entity) x such that'. This idiom '$(\imath x)(\ldots x \ldots)$', i.e.:

(1) the object x such that $\ldots x \ldots$,

is known as *description*. It has now to be defined.

Actually, the meaning of a description is not always clear under ordinary usage. The meaning of the description '$(\imath x)(x$ is capital of France)', or 'the object x such that x is capital of France', or briefly 'the capital of France', is clear enough; this description is a complex name which designates the city of Paris. Again the meaning of '$(\imath x)(5 + x = 9)$' is clear; this description designates the number 4. But what of the descriptions '$(\imath x)(9 > x)$', '$(\imath x)(x \neq x)$', '$(\imath x)$(Jones loves x)'? We cannot say that '$(\imath x)(9 > x)$' designates *the* number less than 9, for there are many such numbers. We cannot say that '$(\imath x)(x \neq x)$' designates the object which is distinct from itself, for there is no such thing. We cannot say that '$(\imath x)$(Jones loves x)' designates *the* person loved by Jones, for we do not know whether there is such a person; Jones may love several, or none at all. Ordinary usage gives us no clue as to the meaning of such descriptions, since use of the idiom (1) is ordinarily limited to cases where one and only one object is believed to satisfy the condition '$\ldots x \ldots$'. Everyday use of descriptions is indeed often elliptical, essential parts of the condition '$\ldots x \ldots$' being left understood; thus we may say simply 'the yellow house' (i.e. '$(\imath x)(x \in$ yellow $. \; x \in$ house)') when what is to be understood is rather 'the yellow house in the third block of Lee Street, Tulsa.' But when in '$\ldots x \ldots$' there is nothing either expressed or tacitly understood which is adequate to determining the object x (to the best of the speaker's belief), the idiom (1) is not ordinarily used.

Does this mean that we should reject the notation $\ulcorner(\imath\alpha)\phi\urcorner$ as meaningless in those cases where ϕ is satisfied by no object or by many, rather than by exactly one? This would fit ordinary usage of the idiom (1), but from the standpoint of logical analysis such a formulation is awkward. It is awkward, in general, to let questions of meaningfulness or meaninglessness rest upon casual matters of fact which are not open to any systematic and conclusive method of decision. We may never know whether Jones loves none, one, or many; and it is best not to have to wait for that information in order to decide whether to accord the expression '$(\imath x)$(Jones loves x)' a place in our language. The truth or falsehood of statements must indeed wait, in general, upon inquiries which lack any systematic and conclusive technique; but the *meaningfulness* of an expression — the eligibility of an expression to occur in statements at all, true or false — is a matter over which we can profitably maintain control. Let us then so define $\ulcorner(\imath\alpha)\phi\urcorner$ as to provide for its significant use regardless of the number of objects happening to satisfy the condition ϕ. In accordance with (1) we want $(\imath x)(\ldots x \ldots)$ to be the sole object x such that $\ldots x \ldots$ so long as there is such an object; but in the remaining cases we may construe $(\imath x)(\ldots x \ldots)$ in any arbitrary fashion that proves convenient — let us say as Λ — since ordinary usage affords no preconceptions regarding such cases.

It will be found shortly (*196, *197) that such a version of $(\imath x)$ $(\ldots x \ldots)$ is achieved by the following definition.

D17. $\ulcorner(\imath\alpha)\phi\urcorner$ for $\ulcorner\hat{\beta}(\exists\gamma)(\beta \in \gamma \,.\, (\alpha)(\alpha = \gamma . \equiv \phi))\urcorner$.

The choice of β and γ in D17 is subject to the obvious convention: they are to be distinct from each other, from α, and from the variables of ϕ.

Just as the prefix '$\hat{y}$' cancels '$y \in$' (cf. †189), so the prefix '$(\imath y)$' cancels '$y =$'. This is shown in the following theorem.

†195. (x) $x = (\imath y)(y = x)$

Proof. †185, *123	$[189 \equiv .]\ x = \hat{y}(\exists z)(y \in z \,.\, z = x)$	(1)
*171	$[1 \equiv .]\ x = \hat{w}(\exists z)(w \in z \,.\, z = x)$	(2)
*103	$[(x)182 \supset .]\ z = z$	(3)
*103	$(y)(y = z . \equiv . y = x) \supset : [3 \equiv .]\ z = x$	(4)
*121 (& D10), *117	$z = x . \supset (y)(y = z . \equiv . y = x)$	(5)
*121 (& D5, 17)	$[(z)(5 \,.\, 4) \supset .\ 2 \equiv]\ 195$	

The bracketing in the last line, above, can be resolved into steps as follows:

$(z)(5\,[.\,4])\supset.\,2 \equiv 195$, $\quad [(z)5 \supset.]\; 2 \equiv 195$, $\quad [2 \equiv\,]195.$

The matrix '$(x)(x = y \,.\!\equiv.\, \ldots x \ldots)$', according to which '$\ldots x \ldots$' is true of anything x just in case x is y, constitutes a simple symbolic translation of the words 'y is the sole object x such that $\ldots x \ldots$'. Now we want $(\iota x)(\ldots x \ldots)$ to be this object y, if such there be; and otherwise we want $(\iota x)(\ldots x \ldots)$ to be Λ. I.e., the closures of:

$$(x)(x = y \,.\!\equiv.\, \ldots x \ldots) \supset.\; y = (\iota x)(\ldots x \ldots)$$

and: $$\sim(\exists y)(x)(x = y \,.\!\equiv.\, \ldots x \ldots) \supset.\; (\iota x)(\ldots x \ldots) = \Lambda$$

are wanted as theorems. That D17 achieves these objectives is established in the following two metatheorems.

***196.** *If* β *is distinct from* α,

$$\vdash \ulcorner(\alpha)(\alpha = \beta \,.\!\equiv \phi) \supset.\; \beta = (\iota\alpha)\phi\urcorner.$$

Proof. *171 (& hp) $\vdash \ulcorner[(x)195 \equiv]\,(\beta)(\beta = (\iota\alpha)(\alpha = \beta))\urcorner.$ (1)
*110 $\vdash \ulcorner[1 \supset.]\;\beta = (\iota\alpha)(\alpha = \beta)\urcorner.$ (2)
*121 $\vdash \ulcorner(\alpha)(\alpha = \beta \,.\!\equiv \phi)\supset:[2 \equiv.]\;\beta = (\iota\alpha)\phi\urcorner.$

***197.** *If* β *is not* α *nor free in* ϕ,

$$\vdash \ulcorner\sim(\exists\beta)(\alpha)(\alpha = \beta \,.\!\equiv \phi) \supset.\; (\iota\alpha)\phi = \Lambda\urcorner.$$

Proof. Let γ be new.
*156 $\vdash \ulcorner(\exists\beta)(\gamma \in \beta \,.\, (\alpha)(\alpha = \beta \,.\equiv \phi)) \supset.$
$(\exists\beta)(\gamma \in \beta) \,.\, (\exists\beta)(\alpha)(\alpha = \beta \,.\equiv \phi)\urcorner.$ (1)
*100, *117 $\vdash \ulcorner\text{L}197 \supset (\gamma)([1 \supset] \sim \text{L}1)\urcorner$
*194 (& D17 & hp) $\supset.\; (\iota\alpha)\phi = \Lambda\urcorner.$

Since $(\iota x)(\ldots x \ldots)$ as of D17 is patently a class, whereas the one and only object fulfilling '$\ldots x \ldots$' may be any sort of object, the identification of the two will perhaps seem strange; but only in case the reader has lost sight of the assimilation of non-classes to classes (§ 25). An individual is a class, for it is identical with the class whereof it is sole member. Or, if the reader prefers the equivalent but less convenient phrasing, according to which an individual is neither a class nor the sole member of a class, then he must recognize that an abstract may in particular designate an individual rather than a class; so that there is again no strangeness in the identification of $(\iota x)(\ldots x \ldots)$ with an individual. In any case the argument in *196 is completely general, and shows that $(\iota x)(\ldots x \ldots)$ is the one and only object x such that $\ldots x \ldots$ whenever there is such a unique object.

Frege (*Grundgesetze*, vol. 1, p. 19) was the first to offer an explicit formulation of description. His version is equivalent to that provided by D17 except for the treatment of those uninteresting cases where ϕ is fulfilled by more than one value of α; in such cases he arbitrarily gives $\ulcorner(\iota\alpha)\phi\urcorner$ the sense of $\ulcorner\hat{\alpha}\phi\urcorner$. Still a third version of $\ulcorner(\iota\alpha)\phi\urcorner$, equivalent both to Frege's and to D17 except for those waste cases, is provided by defining $\ulcorner(\iota\alpha)\phi\urcorner$ in the very simple fashion $\ulcorner\hat{\beta}(\exists\alpha)(\beta \in \alpha \,.\, \phi)\urcorner$; but I have preferred so to frame D17 as to identify the waste cases uniformly with Λ.

The notation $\ulcorner(\iota\alpha)\phi\urcorner$, somewhat different from Frege's, was adapted from Peano by Russell. The sign 'ι' is supposed to suggest a certain inverseness to the notion of unit class expressed by 'ι' (§ 35).

Concerning the philosophical background of description theory, the limited remarks of the present section may well be supplemented by reading Russell "On Denoting."

All the *names* we ever need, for pure logic and mathematics or for any other sort of discourse, are adequately provided by abstraction. For, suppose there is a name, say 'Europe', which is not defined as an abbreviation of an abstract but is instead simply a primitive term of geography. By a trivial revision of our geographical primitives and definitions we can reconstrue this name as an abstract, and in particular as a description (which is a special sort of abstract, by D17). This is accomplished as follows. Instead of adopting the name 'Europe' as primitive, we may adopt a primitive form of atomic matrix $\ulcorner$eur $\alpha\urcorner$ having the following sense: 'eur x', taken as a whole, amounts to what would have been expressed in ordinary language as 'x is Europe'. Now we can introduce the name 'Europe' by definition as an abbreviation of '(ιx) eur x'; for if there is one and only one object x such that eur x then by *196 that object is (ιx) eur x. The predicate 'eur' itself is not a name, but a fragmentary sign like 'ϵ', '$=$', '$\sim$', '('; it can occur only attached, in contexts of the form $\ulcorner$ eur α $\urcorner$.

There is little point, one may suppose, in avoiding primitive names by such artificial methods. Why is it better to adopt the predicate 'eur', and then define 'Europe', than to adopt 'Europe' outright? Actually the suggested course has, despite such artificiality, a considerable theoretical advantage; for the following difficulty otherwise arises in connection with the notion of existence.

To say that *something* does not *exist*, or that there *is* something which *is not*, is clearly a contradiction in terms; hence '$(x)(x$ exists)' must be true. Moreover, we should certainly expect leave to put any primitive name of our language for the 'x' of any matrix '$\ldots x \ldots$', and to infer the resulting singular statement from '$(x)(\ldots x \ldots)$'; it is difficult to contemplate any alternative logical rule for reasoning with names. But this rule of inference leads from the truth '$(x)(x$ exists)' not only to the true conclusion 'Europe exists' but also to the controversial conclusion 'God exists' and the false conclusion 'Pegasus exists', if we accept 'Europe', 'God', and 'Pegasus' as primitive names in our language. The atheist seems called upon to repudiate the very name 'God', thus depriving himself of vocabulary in which to affirm his atheism; and those of us who disbelieve in Pegasus would seem to be in a similar position.

But this difficulty is resolved by the suggested procedure of adopting matrices $\ulcorner$eur $\alpha\urcorner$, $\ulcorner$god $\alpha\urcorner$, $\ulcorner$peg $\alpha\urcorner$, and then introducing 'Europe', 'God', and 'Pegasus' as abbreviations of '$(\imath x)$ eur x', '$(\imath x)$ god x', and '$(\imath x)$ peg x'. There is no question of the existence of these three entities; there is question only as to their nature. If there is a unique object x such that god x, i.e. if

$$(1) \qquad (\exists y)(x)(x = y \,.\!\equiv \text{god } x),$$

then $(\imath x)$ god x is that object; and otherwise, by *197,

$$(2) \qquad (\imath x) \text{ god } x = \Lambda.$$

Monotheists and atheists now need disagree only on the truth values of statements such as (1) and (2), not on questions of meaningfulness. The artificial dodge of dispensing with primitive names in favor of descriptions or other abstracts is a way of maintaining control over questions of vocabulary independently of questions of fact.

Russell undertook to resolve the anomalies of existence by admitting the word 'exists' only in connection with descriptions, and explaining the whole context '$(\iota x)(\ldots x \ldots)$ exists' as short for '$(\exists y)(x)(x = y \,.\!\equiv \ldots x \ldots)$' (cf. his "Mathematical Logic," p. 253; *Principia*, pp. 66 ff, 174 f). This course supplies a strict technical meaning for Kant's vague declaration that 'exists' is not a predicate; namely, 'exists' is not grammatically combinable with a variable to form a matrix 'y exists'. Russell's version of descriptional existence statements perhaps recommends itself still as an adjunct to the procedure which I have proposed above; alternatively such existence statements forfeit their usual function and fall into uniform triviality, since $\ulcorner(\iota\alpha)\phi\urcorner$ designates something (viz. Λ) even when $\ulcorner(\exists\beta)(\alpha)(\alpha = \beta \,.\!\equiv \phi)\urcorner$ is false.

The difficulty which we noted earlier in connection with existence would have survived in another form even if the usage '$(x)(x$ exists$)$' had been proscribed at the start in conformity with Russell's dictum. Independently of any predicate 'exists', we presuppose that a noun designates something whenever we deduce a singular statement from a universal quantification by substituting the noun for the variable. The quantification makes an affirmation regarding all entities, and we assume that the substituted noun designates one of those entities. So long as there are primitive expressions whose possession of designata is undecided, the logic of quantification remains indeterminate.

Names are indeed describable in general as expressions which (unlike '(', '$\downarrow$', 'ϵ', 'eur', etc.) admit thus of substitution for variables in deriving singular statements from quantifications. Cf. my "Designation and Existence," "Logistical Approach." The sort of primitive notation which I urge is free from such expressions; they enter afterward as abstracts by contextual definition, and the principle allowing their substitution for variables emerges as a metatheorem (*231, § 31).

Such avoidance of primitive names, in the construction not only of the purely logical language but also of formalized languages of broader scope, has incidentally the further advantage of making the quantification theory of Chapter II adequate for all extra-logical as well as purely logical applications of quantification. If primitive names were allowed, the quantificational axioms of the kind D (cf. § 15, also *103) would have to be extended to include statements in which names occur in place of the variable α'. But we have seen that such names can be avoided; languages for any extra-logical purposes can presumably be formed merely by supplementing our primitive logical notations by indefinitely many extra-logical predicates such as 'eur', 'red', etc. Such signs are of the same category as 'ϵ'; each attaches to one or more variables to produce an atomic formula.

For us, in any case, all names will be abstracts. The numerals

'0', '1', etc., and indeed all expressions which are to behave in the manner of names at all, will turn out in the course of the definitions to be abbreviations of descriptions or other abstracts.

Altogether we have three sorts of *terms*, or substantives: there are variables, which are essentially pronouns (cf. § 12), and there are abstracts with and without free variables. Abstracts without free variables, e.g. '$\hat{x}(x \in x)$', '$\hat{x}(x \bar{\in} x)$', '$\hat{x}(x = x)$' (or 'V'), and '$\hat{x}(x \neq x)$' (or 'Λ'), are names of definite classes. But abstracts with free variables, e.g. '$\hat{x}(y \in x)$' and '$\hat{x}(y \in x \,.\, x \in z)$', are not. They amount to substantive phrases with unattached pronouns; e.g., '$\hat{x}(y \in x)$' corresponds to the phrase 'the class of all those elements whereof *it* is a member'. Such expressions might be classed as *name matrices*, for they are related to names as statement matrices are related to statements. Variables themselves may also conveniently be classed as name matrices.

Thus, just as *formulæ* comprise *statements* and *statement matrices*, so *terms* comprise *names* and *name matrices*. A formula without free variables is a statement, and a term without free variables is a name. All names are abstracts, but not conversely; and all variables are name matrices but not conversely. From the standpoint of primitive notation our only terms are variables, pronouns; all abstracts, and in particular all names, admit ultimately of elimination through D9.

CHAPTER FOUR

EXTENDED THEORY OF CLASSES

§ 28. *Stratification*

WE CAN decide whether or not a statement is an axiom of quantification by examining its structure with regard merely to quantifiers and statement connectives (ultimately, quantifiers and '↓'); the question is independent of the properties of 'ϵ'. Again, we can decide whether or not a statement is a ponential of two given statements by examining the structure of the statements with regard merely to '$\supset$', hence ultimately '↓'. The notion of theorem hitherto defined turns, therefore, merely upon structure in terms of quantifiers and joint denial. Indeed, the notion of theorem was explained before membership had been discussed at all. The theorems are true statements which involve only quantifiers and '↓' essentially; 'ϵ' occurs vacuously. Even the theorems of Chapter III relating to membership, or to the derivative notions of identity, abstraction, and description, are thus independent of any special properties of membership; they would remain true if the sign 'ϵ' (throughout the primitive expansions of those theorems) were reconstrued as meaning 'is not a member of', or 'is older than', or 'loves', or anything else we like.

We have yet to consider the logical truths which depend for their truth upon the meaning of 'ϵ', rather than merely upon the quantifiers and joint denial. In order to provide for such truths, a new infinite set of axioms will be introduced by way of supplement to the axioms of quantification. Since they are various of the logical truths which involve 'ϵ' essentially, these new axioms will be called *axioms of membership*. The notion of *theorem* will then be extended to cover all ponentials of ponentials of . . . ponentials of axioms in general, whether of quantification or of membership (rather than just those of quantification).

Axioms are wanted specifying circumstances under which a class is to be an element. What conditions shall we regard as sufficient for elementhood? Common sense is of no avail in this question; common sense would construe every class indifferently as an element, i.e. as capable of membership in another class, whereas we already know that this status must be withheld from the classes

(1) $\hat{x}(x \bar{\epsilon} x)$, $\hat{x}(y) \sim (x \epsilon y \,.\, y \epsilon x)$, $\hat{x}(y)(z) \sim (x \epsilon y \,.\, y \epsilon z \,.\, z \epsilon x)$,

etc. (cf. § 24). It remains natural, however, to construe the range of elements as liberally as possible. Wherever $\ulcorner \hat{\alpha}\phi \urcorner$ names a class which is not by necessity barred from elementhood as are the classes cited in (1), the statement $\ulcorner \hat{\alpha}\phi \epsilon \mathrm{V} \urcorner$ recommends itself as an axiom of membership.

It might appear reasonable, then, to specify as axioms of membership all those statements $\ulcorner \hat{\alpha}\phi \epsilon \mathrm{V} \urcorner$ whose denials are not theorems. But there are objections to this course. For one thing, we must decide whether the notion of theorem involved in this description is to be understood in the narrower or the broader sense: ponentials of . . . ponentials of axioms of quantification, or ponentials of . . . ponentials of axioms of quantification and membership. The broader sense would vitiate the suggested explanation of axioms of membership, through circularity; for the term 'theorem' in the explanation would presuppose that we already knew what the axioms of membership were to be. We must therefore choose the narrower sense. There remains, however, an unanswerable objection of another kind: the lack of a test. Even when 'theorem' is taken in the narrower sense, there is no process for deciding that a given formula is not a theorem (cf. § 16); yet this decision would have to be made before we could be sure that $\ulcorner \hat{\alpha}\phi \epsilon \mathrm{V} \urcorner$ is an axiom of membership in the proposed sense. Alleged axioms of membership of this kind would never be conclusively recognizable as such, and proofs based on them would thus be impossible. What is wanted rather is a characterization of axioms of membership according to which those axioms will always be recognizable. The condition imposed upon ϕ and α, in order that $\ulcorner \hat{\alpha}\phi \epsilon \mathrm{V} \urcorner$ be an axiom of membership, must be a condition which we can check.

So the best we can do is seek some testable condition on ϕ and α which will serve to include among the axioms of membership as many harmless statements $\ulcorner \hat{\alpha}\phi \epsilon \mathrm{V} \urcorner$ as is conveniently possible, while excluding any actually refutable ones. Looking to (1) for a suggestion of the forms to be excluded, we are led to suspect that the essence of the bad cases of $\ulcorner\hat{\alpha}\phi \in \mathrm{V}\urcorner$ is the presence in ϕ of the cyclic type of clause

$$\ulcorner\alpha \in \gamma_1 \,.\, \gamma_1 \in \gamma_2 \,.\, \ldots \,.\, \gamma_n \in \alpha\urcorner$$

from α around to α. We are tempted to take as axioms of membership all statements $\ulcorner\hat{\alpha}\phi \in \mathrm{V}\urcorner$ such that ϕ is free from that cyclic feature. Actually, however, this restriction is too mild; bad cases turn up which are free from the cyclic feature. Consider, e.g., the abstract:

(2) $$\hat{x} \sim (y)(x = y \,.\supset.\, x \in y).$$

The quantification therein, viz:

$$(y)(x = y \,.\supset.\, x \in y),$$

says that x is a member of whatever it is identical with; in other words, that x is a member of itself. Thus the class which (2) designates is our familiar non-element $\hat{x}\,(x \bar{\in} x)$; yet (2) does not, like '$\hat{x}(x \bar{\in} x)$', exhibit any cyclic situation. Even when expanded into primitive notation, the formula involved in (2) turns out to contain no atomic formulæ beyond '$(z \in x)$', '$(z \in y)$', and '$(x \in y)$'; and these materials are incapable of being put into any cyclic arrangement.

But there is another characteristic, broader than the cyclic one, which is exhibited by the cyclic sorts of context and by (2) as well; namely, it is impossible to put numerals for the variables in such a way that '$\in$' comes to occur always in the manner '$n \in n + 1$'. This is the characteristic upon which exclusions of statements $\ulcorner\hat{\alpha}\phi \in \mathrm{V}\urcorner$ from the axioms of membership will be based.

Let us speak of a formula as *stratified* if it is possible to put numerals for its variables (the same numeral for all occurrences of the same variable) in such a way that '$\in$' comes to be flanked always by consecutive ascending numerals ('$n \in n + 1$'). To show e.g. that the matrix:

(3) $$(y)(x \in y \,.\supset\, (z)(y \in z \,.\equiv.\, x \in y))$$

is stratified, we simply put suitable numerals for the variables:

(4) $$(1)(0 \in 1 \,.\supset (2)(1 \in 2 \,.\equiv.\, 0 \in 1)).$$

The expression (4) is of course not supposed to *mean* anything; it is not a statement or matrix or name, but only a diagrammatic instrument like the truth table.

The property of stratification can obviously be rephrased in terms of a direct test. Given any formula ϕ, the following procedure will decide whether or not it is stratified. First we put '0' for all occurrences of some one arbitrary variable. Then we put numerals (positive, zero, or negative) for further variables, step by step, according to this rule: if $\ulcorner \epsilon\ \alpha \urcorner$ appears preceded by a numeral, put the next higher numeral for *all* occurrences of α; and if $\ulcorner \beta\ \epsilon \urcorner$ appears followed by a numeral, put the next lower numeral for all occurrences of β. If in the course of this process an occurrence of 'ϵ' comes to be preceded by a numeral and followed by a numeral which is not the next higher, then we stop; ϕ is unstratified. If the process yields no such result, and some variables still remain unsupplanted after the process has been carried as far as possible, then we put '0' for an arbitrary one of these residual variables and continue. Having continued thus until all variables are supplanted, and still not having found ϕ unstratified, we conclude that ϕ is stratified.

The notion of stratification is formulated for application to formulæ in primitive notation; hence we must not undertake to show that a formula ϕ is stratified until we have eliminated all definitional abbreviations which conceal any atomic formulæ $\ulcorner \alpha\ \epsilon\ \beta \urcorner$. The abbreviations introduced by D1–8 need not be disturbed; for the set of atomic formulæ $\ulcorner \alpha\ \epsilon\ \beta \urcorner$ occurring in ϕ is unaffected by those definitions, and it is only on that set of atomic formulæ that the question of stratification of ϕ depends. Thus no definitional expansion was needed in showing stratification of (3). But abbreviations introduced by D9–17 and subsequent definitions would have to be eliminated. E.g., to test the formula in (2) for stratification we must first expand '$x = y$' by D10; the whole becomes:

$$\sim (y)((z)(z \in x \;.\!\equiv\!.\; z \in y) \supset .\; x \in y),$$

which proves unstratified by the test of the preceding paragraph.

Assumption of $\ulcorner \hat{\alpha}\phi\ \epsilon\ \mathrm{V} \urcorner$ for stratified ϕ does not attribute elementhood directly to the classes described in (1) and (2), but we must beware still of indirect consequences. If we were to take as axioms of membership all statements $\ulcorner \hat{\alpha}\phi \in \mathrm{V} \urcorner$ with stratified ϕ, and all closures of matrices $\ulcorner \hat{\alpha}\phi \in \mathrm{V} \urcorner$ with stratified ϕ, then we should encounter trouble. E.g., since '$x \in y$' is stratified, we could assert:

$$(5) \qquad (y) \qquad \hat{x}(x \in y) \in \mathrm{V};$$

i.e., by †189, $(y)(y \in \mathrm{V})$. All classes, including those described in (1) and (2), would thus be admitted as elements after all.

So stratification of ϕ is not to be sufficient for truth of the closure of $\ulcorner \hat{\alpha}\phi \in \mathrm{V} \urcorner$. Where ϕ is stratified, we might try assuming not that the closure of $\ulcorner \hat{\alpha}\phi \in \mathrm{V} \urcorner$ is true but only that $\ulcorner \hat{\alpha}\phi \in \mathrm{V} \urcorner$ is true for all *elements* as values of its free variables; i.e., that

$$(6) \qquad \ulcorner (\beta_1) \ldots (\beta_n)(\beta_1, \beta_2, \ldots, \beta_n \in \mathrm{V} \,.\supset.\, \hat{\alpha}\phi \in \mathrm{V}) \urcorner$$

is true where α, β_1, β_2, . . . , β_n exhaust the free variables of ϕ. Then in particular we would affirm not (5) but:

$$(y) \;\; y \in \mathrm{V} \,.\supset.\, \hat{x}(x \in y) \in \mathrm{V},$$

which is quite harmless (and a mere corollary, indeed, of †189).

In the first edition of this book the axioms of membership, in so far as they concern elementhood, were left as described above; viz., (6) for all stratified ϕ. However, in 1942 Rosser showed, by a deduction too complex for inclusion here, that the resulting system was inconsistent. A recent analysis by Wang suggests that we can put the system to rights if we not only restrict the free variables of $\ulcorner \hat{\alpha}\phi \in \mathrm{V} \urcorner$ to elements as in (6), but also restrict the bound variables of ϕ itself to elements. This can be done by changing each quantification $\ulcorner (\gamma)\chi \urcorner$ within ϕ to $\ulcorner (\gamma)(\gamma \in \mathrm{V} \,.\supset\, \chi) \urcorner$, and each quantification $\ulcorner (\exists\gamma)\chi \urcorner$ to $\ulcorner (\exists\gamma)(\gamma \in \mathrm{V} \,.\, \chi) \urcorner$.

Let us then take as axioms of membership all statements of the form (6), i.e. the closures of all matrices of the form

$$(7) \qquad \ulcorner \beta_1, \beta_2, \ldots, \beta_n \in \mathrm{V} \,.\supset.\, \hat{\alpha}\phi \in \mathrm{V} \urcorner,$$

such that ϕ has no free variables but α, β_1, β_2, . . . , β_n and is formed from a stratified formula by restricting all bound variables to elements. E.g., since the formula:

$$(\exists y)(x \in y \,.\, (z)(y \in z \,.\supset.\, z \in w))$$

is stratified, the following is an axiom of membership:

$$(8) \;\; (w) \;\; w \in \mathrm{V} \,.\supset\, \hat{x}(\exists y)(y \in \mathrm{V} \,.\, x \in y \,.\, (z)(z \in \mathrm{V} \,.\supset:\, y \in z \,.\supset.\, z \in w)) \in \mathrm{V}.$$

Again, since (3) is stratified, this is an axiom of membership:

$$\hat{x}(y)(y \in \mathrm{V} \,.\supset:\, x \in y \,.\supset\, (z)(z \in \mathrm{V} \,.\supset:\, y \in z \,.\equiv.\, x \in y)) \in \mathrm{V}.$$

Here there is nothing to correspond to the antecedent of (7), because there are no free variables; the n of (7) is here 0. In general we are to understand (7) thus as simply losing its antecedent when $n = 0$.

Strictly all formulæ are to be thought of as in primitive notation. But then, since in primitive notation all quantifiers are universal, the restriction of a bound variable to elements consists strictly in changing only $\ulcorner(\gamma)\chi\urcorner$ to $\ulcorner(\gamma)(\gamma \in \mathrm{V} \,.\supset \chi)\urcorner$ and never $\ulcorner(\exists\gamma)\chi\urcorner$ to $\ulcorner(\exists\gamma)(\gamma \in \mathrm{V} \,.\, \chi)\urcorner$. Properly thus the restriction of bound variables turns $\ulcorner(\exists\gamma)\chi\urcorner$, or $\ulcorner\sim(\gamma)\sim\chi\urcorner$, into

(9) $$\ulcorner\sim(\gamma)(\gamma \in \mathrm{V} \,.\supset \sim\chi)\urcorner$$

rather than $\ulcorner(\exists\gamma)(\gamma \in \mathrm{V} \,.\, \chi)\urcorner$. However, (9) in turn becomes $\ulcorner(\exists\gamma)(\gamma \in \mathrm{V} \,.\, \chi)\urcorner$ by *100 and *123 (and D8); so we may in practice conveniently treat (8) and similar cases directly as axioms of membership, the excursion through D8, *100, and *123 being left tacit.

§ 29. *Further Axioms of Membership*

IF x and y are the same thing, then whatever can be said of x is naturally true likewise of y. Closures of conditionals of the sort:

$$x = y \,.\supset:\, \ldots x \ldots \,.\equiv.\, \ldots y \ldots$$

should thus be true, '$\ldots x \ldots$' and '$\ldots y \ldots$' being thought of as formulæ differing only in free occurrences of 'x' and 'y'. The following, accordingly, is wanted as a metatheorem:

(1) *If ϕ' is like ϕ except for containing free occurrences of α' in place of some free occurrences of α, then* $\vdash \ulcorner\alpha = \alpha' \,.\supset.\, \phi \equiv \phi'\urcorner$.

This is the principle of the *substitutivity of identity*, in one of its forms. Broader forms of the principle are likewise wanted, dealing with terms ζ and ζ' in general rather than merely with variables α and α'. Now in order to provide for (1), and its extension to terms ζ and ζ', we must supplement our axioms.

In specifying these new substitutivity axioms it is sufficient to limit our attention to (1); the extension to terms ζ and ζ' can be accomplished afterward (§ 30) without further assumption. Nor do we need to take as axioms all the statements covered by (1); it turns out that we need take only the cases where ϕ is atomic. Also it turns out that the occurrences of α replaced by α' can be limited to a single one; and '$\equiv$' can be weakened to '$\supset$'. Freedom of the occurrences of α and α' in ϕ and ϕ' no longer needs to be stipulated, since an occurrence of a variable in an atomic formula is necessarily free. The axioms are therefore describable simply as the closures of formulæ

(2) $$\ulcorner \alpha = \alpha' \,.\supset .\, \phi \supset \phi' \urcorner$$

such that ϕ is atomic and ϕ' is formed from ϕ by putting α' for an occurrence of α. Thus two of them are:

(3) $$(z)(y)(x) \quad x = y \,.\supset :\, x \in z \,.\supset .\, y \in z,$$
(4) $$(z)(y)(x) \quad x = y \,.\supset :\, z \in x \,.\supset .\, z \in y;$$

and there are further ones corresponding to atomic matrices ϕ of any non-logical forms — forms other than $\ulcorner \beta \in \gamma \urcorner$.

In view of D10, the axiom (4) and all others sharing the same form

(5) $$\ulcorner (\alpha)(\beta)(\gamma)(\alpha = \beta \,.\supset :\, \gamma \in \alpha \,.\supset .\, \gamma \in \beta) \urcorner$$

are actually redundant assumptions; for they are readily seen to be theorems on the basis of *103 and *100. In consequence it can be shown that (3) would suffice as sole substitutivity axiom, if we were disposed to disregard non-logical atomic matrices and provide for only such theorems as are logical formulæ. But the notion of theorem is intended as an approximation to the general notion of logical truth; and it seems desirable, as in Chapter II, to pursue our codification of this notion in such a way that it will apply to statements whose atomic formulæ have any sort of subject matter, logical or otherwise. Accordingly the more generous set of substitutivity axioms is adopted. The redundant cases described in (5) are admitted along with the rest merely because an added clause expressly excluding them would be an idle complication.

The substitutivity axioms are appropriately classifiable as further axioms of membership, supplementary to those of § 28; for the substitutivity of identity appears as a condition on 'ϵ'

when '=' is expanded by D10. We turn next to the consideration of yet a third set of axioms of membership.

When we found that the common-sense idiom 'the class of all entities x such that $\ldots x \ldots$' could not in general be trusted to express a class, we weakened it to read 'the class of all elements x such that $\ldots x \ldots$', symbolically '$\hat{x}(\ldots x \ldots)$'. We now need axioms of membership providing in general for the existence of the class described by this weakened idiom. There is to be a class y whose members are all and only those elements x such that $\ldots x \ldots$; symbolically

$$(6) \qquad (\exists y)(x)(x \in y \,.\!\equiv\!.\, x \in V \,.\, \ldots x \ldots),$$

since '$x \in V$' means that x is an element. All statements of the kind (6), then, and the closures of all matrices of this kind, are to be axioms of membership. More rigorously stated, the axioms of membership are to include the closures of all formulæ

$$(7) \qquad \ulcorner(\exists\beta)(\alpha)(\alpha \in \beta \,.\!\equiv\!.\, \alpha \in V \,.\, \phi)\urcorner$$

such that β is not α nor free in ϕ.

To sum up, then, the axioms of membership comprise the closures of the formulæ described in connection with (6) of § 28 and (2) and (7) of the present section. The notation '$\vdash \phi$' will be used hereafter to mean that the closure of ϕ is a theorem in our newly extended sense: an axiom of quantification *or membership* or a statement derivable from such axioms by one or more steps of *modus ponens*. The totality of theorems ceases to be determined by *100–*104 alone, and comes rather to be determined by *100–*104 plus the following:

***200.** *If ϕ has no free variables beyond $\alpha, \beta_1, \beta_2, \ldots, \beta_n$, and is formed from a stratified formula by restricting all bound variables to elements,*
$\vdash \ulcorner \beta_1, \beta_2, \ldots, \beta_n \in V \,.\!\supset\!.\, \hat{\alpha}\phi \in V\urcorner$ ($\vdash \ulcorner\hat{\alpha}\phi \in V\urcorner$ when $n = 0$).

***201.** *If ϕ is atomic, and ϕ' is formed from ϕ by putting α' for an occurrence of α, then* $\vdash \ulcorner \alpha = \alpha' \,.\!\supset\!.\, \phi \supset \phi'\urcorner$.

***202.** *If β is not α nor free in ϕ,*
$\vdash \ulcorner (\exists\beta)(\alpha)(\alpha \in \beta \,.\!\equiv\!.\, \alpha \in V \,.\, \phi)\urcorner$.

What have hitherto been shown to be theorems continue to be

theorems; but we can now establish further theorems, with help of *200–*202, which were lacking before. Two simple ones are the following, which affirm that V and Λ are elements.

†210. $\mathrm{V} \epsilon \mathrm{V}$

Proof. '$(y)(y \epsilon x .\equiv. y \epsilon x)$' is stratified; hence

*200	$\hat{x}(y)(y \epsilon \mathrm{V} .\supset: y \epsilon x .\equiv. y \epsilon x) \epsilon \mathrm{V}$	(1)
*100, *123 (& D15, 10)	$[1 \equiv]$ 210	

†211. $\Lambda \epsilon \mathrm{V}$ *Proof* similar, with '$\sim(y)$' in place of '(y)'.

With help of †210 and *202 it is now possible to prove a theorem to the effect that, whatever y may be, there is something else; in other words, that there is more than one entity.

†212. $(y) \quad (\exists x)(x \neq y)$

Proof.

*100	$z = \mathrm{V} . z \epsilon \mathrm{V} .\supset \sim (z \epsilon x .\equiv. z \epsilon \mathrm{V} . z \bar{\epsilon} x)$	(1)
*149 (& D12)	$[(z)1 \supset. 210 \supset](\exists z)\sim(z \epsilon x .\equiv. z \epsilon \mathrm{V} . z \bar{\epsilon} x)$	(2)
*130	$\sim (z)(z \epsilon x .\equiv. z \epsilon \mathrm{V} . z \bar{\epsilon} x)[\equiv 2]$	(3)
*121 (& D10)	$x = y .\supset. [3 \equiv]\sim(z)(z \epsilon x .\equiv. z \epsilon \mathrm{V} . z \bar{\epsilon} y)$	(4)
*100	$[4 \supset:]\ (z)(z \epsilon x .\equiv. z \epsilon \mathrm{V} . z \bar{\epsilon} y) \supset. x \neq y$	(5)
*202	$(\exists x)(z)(z \epsilon x .\equiv. z \epsilon \mathrm{V} . z \bar{\epsilon} y)$	(6)
*149	$[(x)5 \supset. 6 \supset]$ 212	

Ways of protecting logic from contradictions of the sort considered at the beginning of § 24 were proposed in 1908 by both Russell and Zermelo. Russell's way, known as the theory of types, is restated in Whitehead and Russell, vol. 1, pp. 37–65. The theory is highly complex and at some points obscure. Some of its complexity was resolved in 1914, when Wiener reduced relations to classes (cf. § 36). The dispensability of the most complex and obscure part of the theory, viz. the part having to do with the so-called orders of propositions and propositional functions, was pointed out by Chwistek in 1921. (See Church's review, also Ramsey, pp. 20–29, and my "Axiom of Reducibility.") Thus simplified, the theory of types is substantially this: each entity is conceived as belonging to one and only one of a hierarchy of so-called types; and any formula which represents membership as holding between entities of other than consecutive ascending types is rejected as meaningless, along with all its contexts. (Cf. my "Theory of Types.") In particular, thus, '$x \epsilon x$' and all its contexts are meaningless; and similarly for '$x \epsilon y . y \epsilon x$', '$x \epsilon y . y \epsilon z . z \epsilon x$', etc. The prefix 'the class of all entities x such that' survives, but all the matrices in connection with which this prefix caused trouble are now banished from the language.

Under Russell's theory there is no one exhaustive class V, but rather an infinite array of analogues each of which exhausts a type; for a class is forbidden to draw its members from more than one type. In similar fashion every other logically definable entity, e.g. Λ, the identity function *I* (§ 42), and indeed each number, loses its unity and gives way to infinitely many replicas. Intuitively all these cleavages and reduplications are of course unwelcome; and technically also, for they call continually for more or less elaborate technical manœuvres by way of restoring severed connections.

Zermelo's scheme, refined in 1930 by Skolem, is in effect as follows. Special axioms are adopted providing for the existence of the null class, the class of all subclasses of a given class, the class of all members of members of a given class, and the class having any one or two given things as sole members. Then a so-called *Aussonderungsaxiom* is added which allows generation of further classes via the idiom 'the class of all entities x such that $x \in y$ and . . . x . . .'. The *Aussonderungsaxiom* is the precursor of *202, with which it is identical except for exhibiting a new variable in place of V. For variants of the Zermelo system see Fraenkel's *Einleitung*, my "Set-Theoretic Foundations," and Ackermann's "Mengentheoretische Begründung."

Zermelo's theory is free from the drawbacks noted in connection with Russell's, but has drawbacks of its own. Whereas under Russell's scheme an abstraction prefix 'the class of all entities x (of given type) such that' can be applied outright to any meaningful formula, under Zermelo's scheme generation of classes tends to be laborious and uncertain. Moreover, there is for Zermelo no class which embraces more than an infinitesimal proportion of the totality of entities; thus there is nothing remotely analogous to an exhaustive class, nor does any class have a complement $\bar{x}$ (cf. § 33).

My "New Foundations" strikes a middle course between Russell and Zermelo. Zermelo's initial axioms of class existence are dropped, and the idiom 'the class of all entities x such that' is construed as effective whenever applied to a stratified formula. There is an echo here of Russell, since a stratified formula is one which could be reconciled with the theory of types by some assignment of types to the variables; but types themselves, and the cleavages and reduplications which they involve, are abandoned. Unstratified formulæ remain meaningful, and can often even be shown by more or less devious reasoning to determine classes — just as, in the system of this book, unstratified formulæ can often be shown to determine elements. In contrast to Zermelo's theory, "New Foundations" countenances an exhaustive class V and a complement $\bar{x}$ of every class x. Instead of repudiating the most inclusive classes, the system countenances a universe which is symmetrical as between small and large — as the existence of complements testifies. Within the lower half of this symmetrical universe, a universe fulfilling Zermelo's theory is contained (provided that two rather appendical axioms of Zermelo, viz. the axiom of infinity and the axiom of choice, be appended also to "New Foundations" or else dropped altogether).

In "New Foundations," unlike Zermelo's system, generation of classes is as convenient as in Russell's theory; for the prefix 'the class of all entities x such that' is now available in connection with formulæ corresponding to all Russell's meaningful formulæ. Nor is the phrase 'all entities x' here subject, as with Russell, to the parenthetical qualification 'of given type'. Moreover, the resulting abstracts may be substituted for variables at will without fear, as in Russell's system, of rendering the context meaningless. "New Foundations" shares with Zermelo's system the virtue of involving no complications in the matter of meaningfulness of formulæ.

The system of theorems determined by *100–*104 and *200–*202 could be converted into the system of "New Foundations" merely by dropping *200, deleting '$\alpha \epsilon \mathrm{V}$.' from *202, and stipulating that the ϕ of *202 be stratified.

There is, however, a technical inconvenience to which "New Foundations" and Zermelo's system are alike subject, and from which Russell's scheme is free; namely, the survival of conditions (e.g. '$x \bar{\epsilon} x$') to which no classes correspond. This divergence from Russell is a matter of liberalization rather than restriction, of course, for it consists in the retention of formulæ which Russell rejected as meaningless; but it nevertheless proves awkward in some cases. One unnatural result is that the principle of mathematical induction (§ 44) cannot be established in its general form unless we adopt special axioms for the purpose; cf. Rosser, "Consistency," "Definition."

The system set forth by von Neumann (1926), and recast by Bernays (1937) in a form more directly comparable with tradition, overcomes this kind of difficulty. Matrices which would otherwise have no classes corresponding to them are now accorded classes of a special sort, incapable of being members. This proscription of membership seems to work as well as the repudiation of the classes themselves so far as the avoidance of contradictions is concerned, and it is a milder and technically much more convenient sort of restriction.

The basic difference between the von Neumann-Bernays system and the system developed in this book resides in the conditions of elementhood; whereas the elements of the von Neumann-Bernays system comprise approximately the classes of Zermelo, the elements of the present system comprise all the classes of "New Foundations." For von Neumann and Bernays, no element embraces as members more than an infinitesimal proportion of the totality of elements; consequently the class V of all elements is not an element, nor is the complement $\bar{x}$ of any element x. For the present system on the other hand V is an element (†210) and so is $\bar{x}$ for each element x (†274); the totality of elements is symmetrical as between small and large.

While the elements of the present system answer to the classes of "New Foundations," it is not clear that von Neumann's elements correspond *exactly* to Zermelo's classes. Thus, whereas the lower half of the symmetrical universe of "New Foundations" contains Zermelo's universe, it is not altogether clear that the lower half of our present symmetrical totality of elements contains all of von Neumann's elements. This relationship could be assured by supplementing *200–*202 with a further axiom of membership:

$$(x) \qquad x \,\epsilon\, \mathrm{V} \,.\!\vee\, (\exists y)(x \subset \mathfrak{r}\, y \,.\, y``x = \mathrm{V})$$

(in the notation of §§ 38, 40), which answers to half of von Neumann's 4.2; but let us beware, for perhaps this axiom is actually refutable on the basis of *100–*104 and *200–*202.

Whereas the present system would appear to be much more liberal than von Neumann's in the matter of elementhood, von Neumann's distinguishes itself on an important point of economy; for his system demands, over and above *100–*104, only a finite number of axioms.

In the present system we regain much of the manipulative convenience which was enjoyed in the days of false security before 1902 and sacrificed in the systems of Russell, Zermelo, and "New Foundations." We can compose formulæ in the simple manner of § 23 without fear of meaninglessness, we can apply abstraction prefixes to any of these at will, and we can substitute the results for variables (*231) without thought of stratification or auxiliary existence theorems. We pay for this in clauses of elementhood $\ulcorner \zeta \in V \urcorner$, but at bargain rates.

The contradictions which instigated this whole series of researches, from Russell and Zermelo onward, were implicit in the inferential methods of uncritical common sense; and the various reformulations of logic which have been proposed for avoiding the contradictions have been correspondingly artificial and foreign to common sense. The least artificial and at the same time the technically most convenient formulation would seem to be that which comes as close as it can to the over-liberal canons of common sense without restoring the contradictions. But the more closely we approach this ideal in point of liberality, the more risk we run of subtly reinstating a contradiction for posterity to discover. If we undertake to prove consistency, moreover, we find ourselves in this curious predicament: the proof would itself have to proceed by logic, and hence would be conclusive only in proportion to our prior confidence in the consistency of the logic used. The most we can hope for, in such a proof, is to show that one theory is consistent if another is; and this is interesting just in case the one theory was more suspect than the other. Intuitive obviousness thus becomes the last arbiter — and a fallible one, in view of the contradiction which Russell was able to draw from common-sense logic. But there are degrees of obviousness; future analysis may enable us to derive logic from a set of principles yet more obvious and natural than those which Russell discredited. Meanwhile we adopt a smooth-running technique which does not appear to be inconsistent. Our risk of inconsistency is, at any rate, no greater here than in "New Foundations"; for Wang has shown that the one system is consistent if the other is.

§ 30. *Substitutivity of Identity*

THE SUBSTITUTIVITY of identity, which has been postulated in a rudimentary form (*201), will now be established in its full generality. Seven metatheorems *221–*227 will be proved which are exactly analogous to the principles *120–*123 and *186–*188 of the substitutivity of the biconditional. As a lemma for *221 it is convenient first to prove the following special case thereof (which is at the same time a case of *223).

***220.** *If ϕ' is formed by putting ζ' for an occurrence of ζ in a quasi-atomic context ϕ, then* $\vdash \ulcorner \zeta = \zeta' \,.\supset.\, \phi \equiv \phi' \urcorner$.

Proof. Case 1: ζ and ζ' are variables.
Case 1a: ϕ is atomic.[1]

*201 (& hp)	$\vdash \ulcorner \zeta = \zeta' \,.\supset.\, \phi \supset \phi' \urcorner$.	(1)
*100, *123 (& D10)	$\vdash \ulcorner \zeta = \zeta' \,.\supset.\, \zeta' = \zeta \urcorner$	
*201 (& hp)	$\supset.\, \phi' \supset \phi \urcorner$.	(2)
*100 (& D5)	$\vdash \ulcorner [1 \,.\, 2 \,.\supset] 220 \urcorner$.	

Case 1b: ϕ is not atomic. Then the expansion of ϕ into primitive notation makes the occurrence of the variable ζ give way to many occurrences, because of the reduplications arising from expanding '$\sim$', '.', and '$\equiv$' into terms of '$\downarrow$'; but these occurrences will have a uniform atomic context ψ, which occurs over and over. Let ψ' be the corresponding part of ϕ'. By Case 1a,

$$\vdash \ulcorner \zeta = \zeta' \,.\supset.\, \psi \equiv \psi' \urcorner. \quad (1)$$

From the fact that ϕ and ϕ' are quasi-atomic contexts of ζ and ζ', it is clear that the specified occurrences of ψ and ψ' in the primitive expansions of ϕ and ϕ' will not fall within any quantifications governed by $\ulcorner(\zeta)\urcorner$ or $\ulcorner(\zeta')\urcorner$; in other words, ζ and ζ' are distinct

[1] From the fact that ζ is a variable we cannot conclude that its quasi-atomic context ϕ is atomic; abstracts may still turn up at other points in ϕ. But ϕ *may* be atomic; the atomic is a special case of the quasi-atomic.

from all the variables $\beta_1, \ldots, \beta_n$ with respect to which the occurrences of ψ and ψ' in question are bound in ϕ and ϕ'. Hence

(1), *117 $\vdash \ulcorner \zeta = \zeta' . \supset (\beta_1) \ldots (\beta_n)(\psi \equiv \psi')\urcorner$
*121 $\supset . \phi \equiv \phi'\urcorner$.

Case 2: ζ and ζ' are abstracts.

Case 2a: ϕ and ϕ' are $\ulcorner \alpha \epsilon \zeta \urcorner$ and $\ulcorner \alpha \epsilon \zeta' \urcorner$. *103 (& D10).

Case 2b: ϕ and ϕ' are of other forms than $\ulcorner \alpha \epsilon \zeta \urcorner$ and $\ulcorner \alpha \epsilon \zeta' \urcorner$. Still, definitional expansion of ϕ and ϕ' reveals that the occurrences of ζ and ζ' in question, in ϕ and ϕ', are free occurrences having the immediate contexts $\ulcorner \alpha \epsilon \zeta \urcorner$ and $\ulcorner \alpha \epsilon \zeta' \urcorner$ where α is foreign to ζ and ζ' (cf. § 26). Thus

*121 $\vdash \ulcorner (\alpha)(\alpha \epsilon \zeta . \equiv . \alpha \epsilon \zeta') \supset . \phi \equiv \phi' \urcorner$, q.e.d. (cf. D10).

Case 3: ζ is a variable and ζ' an abstract.

Case 3a: ϕ and ϕ' are $\ulcorner \alpha \epsilon \zeta \urcorner$ and $\ulcorner \alpha \epsilon \zeta' \urcorner$. *103 (& D10).

Case 3b: ϕ and ϕ' are $\ulcorner \hat{\alpha}\psi \epsilon \zeta \urcorner$ and $\ulcorner \hat{\alpha}\psi \epsilon \zeta' \urcorner$. Let γ be new.

*121 $\vdash \ulcorner (\gamma)(\gamma \epsilon \zeta . \equiv . \gamma \epsilon \zeta') \supset .$
$(\exists \gamma)(\gamma = \hat{\alpha}\psi . \gamma \epsilon \zeta) \equiv (\exists \gamma)(\gamma = \hat{\alpha}\psi . \gamma \epsilon \zeta')\urcorner$,
q.e.d. (cf. D10, 12).

Case 3c: ϕ and ϕ' are $\ulcorner \zeta \epsilon \eta \urcorner$ and $\ulcorner \zeta' \epsilon \eta \urcorner$. Let γ be new.

*134 $\vdash \ulcorner \zeta = \zeta' . \zeta \epsilon \eta . \supset (\exists \gamma)(\gamma = \zeta' . \gamma \epsilon \eta)\urcorner$. (1)
*100 $\vdash \ulcorner [1 .] \zeta = \zeta' . \supset : \zeta \epsilon \eta . \supset \text{R1}\urcorner$. (2)
*220 (Case 1) $\vdash \ulcorner \gamma = \zeta . \supset : \gamma \epsilon \eta . \equiv . \zeta \epsilon \eta \urcorner$. (3)
*100, *163 $\vdash \ulcorner (\exists \gamma)([3 .] \gamma = \zeta . \gamma \epsilon \eta) \supset . \zeta \epsilon \eta \urcorner$. (4)
*121 (& D10) $\vdash \ulcorner \zeta = \zeta' . \supset :. [4 \equiv :] \text{R1} \supset . \zeta \epsilon \eta \urcorner$. (5)
*100 (& D5, 12) $\vdash \ulcorner [2 . 5 . \supset :.] \zeta = \zeta' . \supset : \zeta \epsilon \eta . \equiv . \zeta' \epsilon \eta \urcorner$, q.e.d.

Case 3d: ϕ and ϕ' are extra-logical quasi-atomic contexts, as of Da_1–a_n of § 26. Proof similar.

Case 4: ζ is an abstract and ζ' a variable.

*100, *123 (& D10) $\vdash \ulcorner \zeta = \zeta' . \supset . \zeta' = \zeta \urcorner$
*220 (Case 3) $\supset . \phi' \equiv \phi \urcorner$
*100 $\supset . \phi \equiv \phi' \urcorner$.

***221.** *If ϕ' is like ϕ except for containing ζ' at a place where ϕ contains ζ, and $\alpha_1, \ldots, \alpha_n$ exhaust the variables with respect to which those occurrences of ζ and ζ' are bound in ϕ and ϕ', then*

$$\vdash \ulcorner(\alpha_1) \ldots (\alpha_n)(\zeta = \zeta') \supset . \phi \equiv \phi'\urcorner.$$

Proof. The occurrences of ζ and ζ' in question have as immediate contexts either quasi-atomic[1] contexts ψ and ψ' (*Case 1*) or else quantifiers $\ulcorner(\zeta)\urcorner$ and $\ulcorner(\zeta')\urcorner$ (*Case 2*). In Case 1, ζ and ζ' may be abstracts or variables; in Case 2 they must be variables. Case 2 has little practical significance, but must be dealt with because *221, taken literally, includes it.

Case 1. *114 $\vdash \ulcorner(\alpha_1) \ldots (\alpha_n)(\zeta = \zeta') \supset . \zeta = \zeta'\urcorner$
*220 $\supset . \psi \equiv \psi'\urcorner$. (1)

Let $\ulcorner(\beta_1)\chi_1\urcorner$, $\ulcorner(\beta_2)\chi_2\urcorner, \ldots, \ulcorner(\beta_m)\chi_m\urcorner$ ($m \geqslant 0$) be all the quantifications in ϕ which contain the given occurrence of ζ, and let $\ulcorner(\beta_1)\chi'_1\urcorner, \ldots, \ulcorner(\beta_m)\chi'_m\urcorner$ be the corresponding parts of ϕ'. Then, if β_i is a free variable of ζ (or ζ'), the occurrence of ζ (or ζ') will be bound in ϕ (or ϕ') with respect to β_i; wherefore, by hp 2, β_i will be one of $\alpha_1, \ldots, \alpha_n$. For each i from 1 to m, therefore, β_i is one of $\alpha_1, \ldots, \alpha_n$ or else not free in ζ nor in ζ'. Thus none of $\beta_1, \ldots, \beta_m$ is free in $\ulcorner(\alpha_1) \ldots (\alpha_n)(\zeta = \zeta')\urcorner$. Hence

(1), *117 $\vdash \ulcorner(\alpha_1) \ldots (\alpha_n)(\zeta = \zeta') \supset (\beta_1) \ldots (\beta_m)(\psi \equiv \psi')\urcorner$
*121 $\supset . \phi \equiv \phi'\urcorner$.

Case 2 (see above). If the variables ζ and ζ' are the same then ϕ' is ϕ, whereat $\vdash \ulcorner 221\urcorner$ by *100; if on the other hand they are distinct,

*171 $\vdash \ulcorner[(y)212 \equiv](\zeta')(\exists\zeta)(\zeta \neq \zeta')\urcorner$. (1)
*110 $\vdash \ulcorner[1 \supset](\exists\zeta)(\zeta \neq \zeta')\urcorner$. (2)
*130 $\vdash \ulcorner\sim (\zeta)(\zeta = \zeta')[\equiv 2]\urcorner$. (3)
*114 $\vdash \ulcorner(\alpha_1) \ldots (\alpha_n)(\zeta = \zeta') \supset . \zeta = \zeta'\urcorner$. (4)

The occurrence of ζ in ϕ which we are considering is bound to the occurrence of $\ulcorner(\zeta)\urcorner$ which contains it (cf. § 14) — an occurrence of $\ulcorner(\zeta)\urcorner$ which is in ϕ and not, of course, *in* ζ. Thus the occurrence of ζ is bound in ϕ with respect to ζ, wherefore, by hp 2, ζ is one of $\alpha_1, \ldots, \alpha_n$. Thus

(4), *117 $\vdash \ulcorner(\alpha_1) \ldots (\alpha_n)(\zeta = \zeta') \supset (\zeta)(\zeta = \zeta')\urcorner$
*100 $\supset : [3 \supset .]\phi \equiv \phi'\urcorner$.

[1] Possibly atomic, in case ζ and ζ' are variables. See preceding footnote.

***222.** *If* ϕ' *is like* ϕ *except for containing* ζ' *at some places where* ϕ *contains* ζ, *and* $\alpha_1, \ldots, \alpha_n$ *exhaust the variables with respect to which those occurrences of* ζ *and* ζ' *are bound in* ϕ *and* ϕ', *then*

$$\vdash \ulcorner (\alpha_1) \ldots (\alpha_n)(\zeta = \zeta') \supset . \phi \equiv \phi' \urcorner.$$

Proof from *221 like that of *121 from *120.

***223.** *If* ϕ' *is like* ϕ *except for containing free occurrences of* ζ' *in place of some free occurrences of* ζ, *then* $\vdash \ulcorner \zeta = \zeta' . \supset . \phi \equiv \phi' \urcorner$.

Proof: *222 ($n = 0$).

***224.** *If* $\vdash \ulcorner \zeta = \zeta' \urcorner$ *and* ϕ' *is formed from* ϕ *by putting* ζ' *for some occurrences of* ζ, *then* $\vdash \ulcorner \phi \equiv \phi' \urcorner$.

Proof from *222 like that of *123 from *121.

***225.** *If* η' *is like* η *except for containing* ζ' *at some places where* η *contains* ζ, *and* $\alpha_1, \ldots, \alpha_n$ *exhaust the variables with respect to which those occurrences of* ζ *and* ζ' *are bound in* η *and* η', *then*

$$\vdash \ulcorner (\alpha_1) \ldots (\alpha_n)(\zeta = \zeta') \supset . \eta = \eta' \urcorner.$$

Proof: *222, *117 (& D10).

***226.** *If* η' *is like* η *except for containing free occurrences of* ζ' *in place of some free occurrences of* ζ, *then* $\vdash \ulcorner \zeta = \zeta' . \supset . \eta = \eta' \urcorner$.

Proof: *225 ($n = 0$).

***227.** *If* $\vdash \ulcorner \zeta = \zeta' \urcorner$, *and* η' *is formed from* η *by putting* ζ' *for some occurrences of* ζ, *then* $\vdash \ulcorner \eta = \eta' \urcorner$.

Proof from *225 like that of *123 from *121.

The following variant of *223 is occasionally useful.

***228.** *If* ϕ' *is like* ϕ *except for containing free occurrences of* ζ' *in place of some free occurrences of* ζ, *then*

$$\vdash \ulcorner \zeta = \zeta' . \phi . \equiv . \zeta = \zeta' . \phi' \urcorner.$$

Proof. *100 $\vdash \ulcorner [223 \supset]\ 228 \urcorner$.

§ 31. *Substitution for Variables*

SINCE $\hat{x}(\ldots x \ldots)$ is the class of all elements x such that $\ldots x \ldots$, any entity x will belong to the class if and only if it is an element and . . . *x* . . . . That the closures of such biconditionals are theorems is established in the following metatheorem, which will subsequently be generalized (*235).

***230.** $\vdash \ulcorner \alpha \in \hat{\alpha}\phi \,.\!\equiv\!.\, \alpha \in V \,.\, \phi \urcorner$.

Proof. Let β be new.

*171 $\vdash \ulcorner[(x)(y)190 \equiv](\alpha)(\beta)(\alpha \in \beta \,.\!\supset\!.\, \alpha \in V)\urcorner$. (1)

*114 $\vdash \ulcorner[1 \supset :]\alpha \in \beta \,.\!\supset\!.\, \alpha \in V\urcorner$. (2)

*110 $\vdash \ulcorner(\alpha)(\alpha \in \beta \,.\!\supset \phi) \supset : \alpha \in \beta \,.\!\supset \phi\urcorner$

*100 $\supset : [2\,.]\alpha \in \beta \,.\!\supset\!.\, \alpha \in V \,.\, \phi\urcorner$. (3)

*100, *163 $\vdash \ulcorner(\exists\beta)([3\,.]\ \alpha \in \beta \,.\, (\alpha)(\alpha \in \beta \,.\!\supset \phi)) \supset\!.\, \alpha \in V \,.\, \phi\urcorner$. (4)

*100 $\vdash \ulcorner \alpha \in V \,.\, \phi \,.\!\supset \phi\urcorner$. (5)

*135 (& D9) $\vdash \ulcorner \alpha \in \beta \,.\, (\alpha)(\alpha \in \beta \,.\!\supset \phi) \,.\!\supset\!.\, \alpha \in \hat{\alpha}\phi\urcorner$. (6)

*121 $\vdash \ulcorner(\alpha)(\alpha \in \beta \,.\!\equiv\!.\, \alpha \in V \,.\, \phi) \supset :.$

$[6 \equiv :]\ \alpha \in V \,.\, \phi\, [.\ (\alpha)5] \,.\!\supset\!.\, \alpha \in \hat{\alpha}\phi\urcorner$. (7)

(7), *163 $\vdash \ulcorner[202 \supset :]\ \alpha \in V \,.\, \phi \,.\!\supset\!.\, \alpha \in \hat{\alpha}\phi\urcorner$. (8)

*100 (& D9) $\vdash \ulcorner[4 \,.\, 8 \,.\!\supset]\ 230\urcorner$.

Passing mention was made earlier (§ 15) of logical truths of the form $\ulcorner(\alpha)\phi \supset \psi\urcorner$, where ψ is like ϕ except for containing a name ζ in place of the free occurrences of α. The notion of name has since been given an explicit status (§ 27), and now those logical truths are forthcoming as theorems. This is established in the following metatheorem, which is an extension of *103.

***231.** *If ψ is like ϕ except for containing free occurrences of ζ wherever ϕ contains free occurrences of α, then* $\vdash \ulcorner(\alpha)\phi \supset \psi\urcorner$.

Proof. Case 1: ζ is a variable. *103.

Case 2. ζ is an abstract $\ulcorner\hat{\beta}\chi\urcorner$. Let α' be new, and let ϕ' be formed from ϕ by putting α' for the free occurrences of α.

*223 (& hp) $\vdash \ulcorner \alpha' = \hat{\beta}\chi \,.\!\supset\!.\, \phi' \equiv \psi\urcorner$

*122 $\supset : [103 \equiv .](\alpha)\phi \supset \psi\urcorner$. (1)

*202 $\vdash \ulcorner(\exists\alpha')(\beta)(\beta \in \alpha' \,.\!\equiv\!.\, \beta \in V \,.\, \chi)\urcorner$. (2)

*230, *123 $\vdash \ulcorner(\exists\alpha')(\beta)(\beta \in \alpha' \,.\!\equiv\!.\, \beta \in \hat{\beta}\chi)[\equiv 2]\urcorner$. (3)

*171 (& D10) $\vdash \ulcorner[3 \equiv](\exists\alpha')(\alpha' = \hat{\beta}\chi)\urcorner$. (4)

(1), *163 $\vdash \ulcorner[4 \supset]\ 231\urcorner$.

This metatheorem is used constantly in the sequel. It embodies the principle of *application* or *specification* — the principle which leads from a general law, a universal quantification $\ulcorner(\alpha)\phi\urcorner$, to each special case ψ falling under the general law. We get ψ from ϕ by substituting a name or name matrix ζ for the variable of quantification α. It is important to note that the substitution must exhaust the free occurrences of α in ϕ; thus *231 gives us:

$$(x)(x = x) \supset . \Lambda = \Lambda$$

and: (z) $\quad (x)(x = x) \supset . \hat{y}(z \in y) = \hat{y}(z \in y)$

as theorems, but not:

(x) $\quad (x)(x = x) \supset . x = \Lambda.$

The latter is clearly false. In thus requiring exhaustive substitution, *231 resembles *103 and *134 and differs from *120–*124, *186–*188, *220–*228.

The following metatheorem extends *134 as *231 extends *103.

***232.** *If ψ is like ϕ except for containing free occurrences of ζ wherever ϕ contains free occurrences of α, then* $\vdash \ulcorner\psi \supset (\exists\alpha)\phi\urcorner$.

Proof. *231 (& hp) $\quad \vdash \ulcorner[(\alpha)135 \supset] \; 232\urcorner$.

The next extends *231 to the case of many quantifiers.

***233.** *If ψ is like ϕ except for containing free occurrences of ζ_1, $\zeta_2, \ldots, \zeta_n$ in place respectively of all free occurrences of $\alpha_1, \alpha_2, \ldots, \alpha_n$, then*

$$\vdash \ulcorner(\alpha_1)(\alpha_2) \ldots (\alpha_n)\phi \supset \psi\urcorner.$$

Proof. We may assume without loss of generality that $\alpha_1, \alpha_2, \ldots, \alpha_n$ are distinct; for if α_i were α_j then the hp would demand that ζ_i be ζ_j, so that the extra mention of α_j and ζ_j could simply be dropped. Now let ϕ_0 be formed from ϕ by putting *new* distinct variables $\beta_1, \ldots, \beta_n$ for all free occurrences respectively of $\alpha_1, \ldots, \alpha_n$; and, for each i from 1 to n, let ϕ_i be formed from ϕ_{i-1} by putting ζ_i for β_i.

*171	$\vdash \ulcorner(\alpha_1) \ldots (\alpha_n)\phi \supset (\beta_1) \ldots (\beta_n)\phi_0\urcorner$
*231	$\supset (\beta_2) \ldots (\beta_n)\phi_1\urcorner$
*231	$\supset (\beta_3) \ldots (\beta_n)\phi_2\urcorner$
.	.
.	.
.	.
*231	$\supset \phi_n\urcorner$. But ϕ_n is ψ.

The reason for introducing new variables $\beta_1, \ldots, \beta_n$, in the above proof, is as follows. If ϕ_1 were formed rather by putting ζ_1 for α_1 in ϕ, and ϕ_2 by putting ζ_2 for α_2 in ϕ_1, and so on, then the assertion:

$$\vdash \ulcorner(\alpha_1) \ldots (\alpha_n)\phi \supset (\alpha_2) \ldots (\alpha_n)\phi_1\urcorner$$

would not be justified by *231 in case any of $\alpha_2, \ldots, \alpha_n$ happened to be free in ζ_1; and correspondingly for each succeeding step of the proof. The β's obviate this difficulty, foreign as they are to the ζ's.

Note that *233 does not demand a distinct ζ for every distinct α. All of the following, e.g., are theorems according to *233:

$$(z)(w) \quad (x)(y)(x \in z \,.\mathrm{v}.\, z \in y) \supset : w \in z \,.\mathrm{v}.\, z \in \mathrm{V},$$
$$(z)(w) \quad (x)(y)(x \in z \,.\mathrm{v}.\, z \in y) \supset : w \in z \,.\mathrm{v}.\, z \in w,$$
$$(z) \quad (x)(y)(x \in z \,.\mathrm{v}.\, z \in y) \supset : \mathrm{V} \in z \,.\mathrm{v}.\, z \in \mathrm{V},$$
$$(z) \quad (x)(y)(x \in z \,.\mathrm{v}.\, z \in y) \supset : z \in z \,.\mathrm{v}.\, z \in z.$$

Hereafter, when a daggered numeral is used to justify a step of proof, a tacit inference according to *233 will commonly be involved. The following line, e.g., might typically occur in a proof:

†190 $\quad y \in \hat{x}(z \in x) \,.\supset.\, y \in \mathrm{V}.$

Rendered in full, the reasoning here is as follows:

*233 $\quad [(x)(y)190 \supset :]\, y \in \hat{x}(z \in x) \,.\supset.\, y \in \mathrm{V}.$

Similarly the first line in the proof of *234, below, amounts to this:

*233 $\quad \vdash \ulcorner[(x)182 \supset .]\, \zeta = \zeta\urcorner.$

In effect, thus, the matrix following the segregated initial quantifiers of a theorem (the matrix '$x \in y \,.\supset.\, x \in \mathrm{V}$' in the case of †190, or '$x = x$' in the case of †182) is tacitly reconstrued in proofs as containing free occurrences of any desired terms in place of the free variables. Under this procedure the first two lines in the proof of *230 would have given way to the single line:

†190 $\quad \vdash \ulcorner\alpha \in \beta \,.\supset.\, \alpha \in \mathrm{V}\urcorner.$

Note, however, that this innovation relates only to numerals with daggers attached. Divorced of its dagger, the numeral continues to serve as an abbreviation of just the one specific formula which actually appears after the segregated initial quantifiers (if any) in the original statement of the theorem; the undaggered numeral '190', e.g., serves in proofs as an abbreviation of the matrix '$x \in y \mathbin{.\supset.} x \in \mathrm{V}$' and this only.

The following metatheorem depends both upon substitution for variables, as of *231 and *232, and upon the substitutivity of identity as of *223.

***234.** *If α is not free in ζ, and ψ is like ϕ except for containing free occurrences of ζ wherever ϕ contains free occurrences of α, then*

(a) $\vdash \ulcorner \psi \equiv (\alpha)(\alpha = \zeta \mathbin{.\supset} \phi)\urcorner$ *and* (b) $\vdash \ulcorner \psi \equiv (\exists\alpha)(\alpha = \zeta \,.\, \phi)\urcorner$.

Proof.	†182	$\vdash \ulcorner \zeta = \zeta \urcorner$.	(1)
	*231 (& hp)	$\vdash \ulcorner (\alpha)(\alpha = \zeta \mathbin{.\supset} \phi) \supset . [1 \supset]\psi\urcorner$.	(2)
	*232 (& hp)	$\vdash \ulcorner [1\,.]\psi \mathbin{.\supset} (\exists\alpha)(\alpha = \zeta \,.\, \phi)\urcorner$.	(3)
	*223 (& hp2)	$\vdash \ulcorner \alpha = \zeta \mathbin{.\supset.} \phi \equiv \psi \urcorner$.	(4)
	*100, *117 (& hp)	$\vdash \ulcorner \psi \supset (\alpha)([4\,.]\alpha = \zeta \mathbin{.\supset} \phi)\urcorner$.	(5)
	*100, *163 (& hp)	$\vdash \ulcorner (\exists\alpha)([4\,.]\alpha = \zeta \,.\, \phi) \supset \psi \urcorner$.	(6)
	*100	$\vdash \ulcorner [5 \,.\, 2 \mathbin{.\supset}]234\mathrm{a}\urcorner$.	
	*100	$\vdash \ulcorner [3 \,.\, 6 \mathbin{.\supset}]234\mathrm{b}\urcorner$.	

One of the theorems covered by *234b has already been considered separately, in †185.

The stipulation that α not be free in ζ is clearly essential to *234; otherwise we might take ζ as α itself, e.g., thus concluding that

$$\vdash \ulcorner \phi \equiv (\alpha)(\alpha = \alpha \mathbin{.\supset} \phi)\urcorner \quad \text{and} \quad \vdash \ulcorner \phi \equiv (\exists\alpha)(\alpha = \alpha \,.\, \phi)\urcorner$$

and hence that $\vdash \ulcorner \phi \equiv (\alpha)\phi \urcorner$ and $\vdash \ulcorner \phi \equiv (\exists\alpha)\phi\urcorner$ for all choices of ϕ and α.

The principle noted at the beginning of the section, in *230, now achieves its general formulation with help of *231.

***235.** *If ψ is like ϕ except for containing free occurrences of ζ wherever ϕ contains free occurrences of α, then*

$$\vdash \ulcorner \zeta \in \hat{\alpha}\phi \mathbin{.\equiv.} \zeta \in \mathrm{V} \,.\, \psi\urcorner.$$

Proof. *231 (& hp) $\vdash \ulcorner [(\alpha)230 \supset]\, 235\urcorner$.

This principle provides a very useful way of paraphrasing $\ulcorner \zeta \in \hat{\alpha}\phi \urcorner$ so as to eliminate the abstract; a simpler and more convenient way, indeed, than that provided by D9 and D12 themselves. But *235, unlike the definitions, is not always immediately applicable; transformation of $\ulcorner \zeta \in \hat{\alpha}\phi \urcorner$ into an alphabetic variant is sometimes necessary in preparation. E.g., the formula:

(1) $$y \in \hat{x}(\exists y)(y \in x)$$

cannot, as it stands, be paraphrased by *235; for it is impossible to construct a formula ψ which is like '$(\exists y)(y \in x)$' except for containing free occurrences of 'y' wherever '$(\exists y)(y \in x)$' contains free occurrences of 'x'. The alphabetic variant:

$$y \in \hat{x}(\exists z)(z \in x)$$

of (1), on the other hand, can of course be paraphrased by *235; it becomes:

$$y \in \mathrm{V} \,.\, (\exists z)(z \in y).$$

§ 32. *Further Consequences*

THIS SECTION embraces a variety of theorems and metatheorems on membership, identity, and abstraction. The first of these theorems was recognized earlier (§ 23) as the shortest of all logical truths (from the standpoint of primitive notation).

†240. $\sim (x)(x \in x)$ *Proof.* †192 $\Lambda \bar{\in} \Lambda$ (1)
*232 $[1 \supset](\exists x)(x \bar{\in} x)$ (2)
*130 $240\,[\equiv 2]$

†241. $(\exists x)(x \in x)$ *Proof.* *232 $[210 \supset]\ 241$

The next theorem shows that an element x is identical with y if and only if

(1) $$(z)(x \in z \,.\!\supset .\, y \in z).$$

That (1) follows from '$x = y$' is clear, indeed, regardless of whether x is an element (cf. *201); on the other hand the inference of '$x = y$' from (1) does depend on elementhood of x, for if x is not a member of any class then (1) is trivially true independently of y.

†242. $(y)(x)$ $x \in \mathrm{V} \,.\!\supset\,: x = y \,.\!\equiv (z)(x \in z \,.\!\supset .\, y \in z)$

Proof.

*201, *117 $x = y \,.\!\supset (z)(x \in z \,.\!\supset .\, y \in z)$ (1)

*231 $(z)(x \in z \,.\!\supset .\, y \in z) \supset : x \in \hat{w}(x = w) \,.\!\supset .\, y \in \hat{w}(x = w)$

*235, *123 $\supset : x \in \mathrm{V}\,[.182].\supset .\, y \in \mathrm{V} \,.\, x = y$ (2)
*100 $[2\,.]\, x \in \mathrm{V} .\supset : (z)(x \in z .\supset .\, y \in z) \supset .\, y \in \mathrm{V} \,.\, x = y$
*100 $\supset :. [1 \supset :] x = y .\equiv (z)(x \in z .\supset .\, y \in z)$

Note that the third line of the above proof is a compressed rendering of two duplicate steps, as follows:

*235, *123 $\supset : x \in \mathrm{V}\,[.182].\supset .\, y \in \hat{w}(x = w)$
*235, *123 $\supset : x \in \mathrm{V} .\supset .\, y \in \mathrm{V} \,.\, x = y$ (2)

Such telescoping of duplicate steps, in connection with *123 and likewise *224, will be usual hereafter. It occurs again in the second line of the proof of *244, below.

When the '$x \in z .\supset .\, y \in z$' of †242 is strengthened to '$x \in z .\equiv .\, y \in z$', we still have a theorem:

†243. $(y)(x)$ $x \in \mathrm{V} .\supset : x = y .\equiv (z)(x \in z .\equiv .\, y \in z)$

Proof similar, using *223 instead of *201.

The following is the main principle regarding equation of abstracts.

***244.** $\vdash \ulcorner \hat{\alpha}\phi = \hat{\alpha}\psi .\equiv (\alpha)(\alpha \in \mathrm{V} .\supset .\, \phi \equiv \psi)\urcorner$.

Proof. *171 (& D10) $\vdash \ulcorner \hat{\alpha}\phi = \hat{\alpha}\psi .\equiv (\alpha)(\alpha \in \hat{\alpha}\phi .\equiv .\, \alpha \in \hat{\alpha}\psi)\urcorner$
*235, *123 $\equiv (\alpha)(\alpha \in \mathrm{V} \,.\, \phi .\equiv .\, \alpha \in \mathrm{V} \,.\, \psi)\urcorner$
*100, *123 $\equiv (\alpha)(\alpha \in \mathrm{V} .\supset .\, \phi \equiv \psi)\urcorner$.

Various corollaries follow.

***245.** *If* χ *is formed from* ψ *by putting* ϕ *for some free occurrences of* $\ulcorner \alpha \in \hat{\alpha}\phi \urcorner$, *then* $\vdash \ulcorner \hat{\alpha}\psi = \hat{\alpha}\chi \urcorner$.

Proof. *100 $\vdash \ulcorner [230\,.] \alpha \in \mathrm{V} .\supset : \alpha \in \hat{\alpha}\phi .\equiv \phi \urcorner$
*122 (& hp) $\supset .\, \psi \equiv \chi \urcorner$. (1)
*244 $\vdash \ulcorner 245\,[\equiv (\alpha)1] \urcorner$.

***246.** $\vdash \ulcorner \hat{\alpha}\phi = \hat{\alpha}(\alpha \in \mathrm{V} \,.\, \phi)\urcorner$.

Proof. *100 $\vdash \ulcorner \alpha \in \mathrm{V} .\supset : \phi \equiv .\, \alpha \in \mathrm{V} \,.\, \phi \urcorner$. (1)
*244 $\vdash \ulcorner 246\,[\equiv (\alpha)1] \urcorner$.

***247.** $\vdash \ulcorner \hat{\alpha}\phi = \hat{\alpha}(\alpha \in \mathrm{V} .\supset \phi)\urcorner$. *Proof* similar.

***248.** $\vdash \ulcorner \hat{\alpha}\phi = \hat{\alpha}(\alpha \in \mathrm{V} .\equiv \phi)\urcorner$. *Proof* similar.

***249.** $\vdash \ulcorner \hat{\alpha}\phi = V \,.\equiv (\alpha)(\alpha \in V \,.\supset \phi)\urcorner$.

Proof. †189, *224 $\vdash \ulcorner \hat{\alpha}\phi = V \,.\equiv.\, \hat{\alpha}\phi = \hat{\alpha}(\alpha \in V)\urcorner$

*244 $\equiv (\alpha)(\alpha \in V \,.\supset : \phi \equiv .\, \alpha \in V)\urcorner$

*100, *123 $\equiv (\alpha)(\alpha \in V \,.\supset \phi)\urcorner$.

The first line of the above proof involves construing †189 as $\ulcorner V = \hat{\alpha}(\alpha \in V)\urcorner$. This deviation from the original form of †189 turns not only upon use of *233, such as is regularly left tacit in connection with daggered references (cf. § 31), but also upon tacit use of *171 in case the bound variable α happens to be other than 'y'. *233 leads from †189 to '$V = \hat{y}(y \in V)$', and *171 leads thence to $\ulcorner V = \hat{\alpha}(\alpha \in V)\urcorner$. The second line of the following proof is similar. Hereafter, thus, *171 will fare like *233; its use will be left tacit in connection with daggered references.

***250.** $\vdash \ulcorner \hat{\alpha}\phi = \Lambda \,.\equiv (\alpha)(\alpha \in V \,.\supset \sim \phi)\urcorner$.

Proof. †192 $\vdash \ulcorner \alpha \bar{\in} \Lambda\urcorner$. (1)

†189, *224 $\vdash \ulcorner \hat{\alpha}\phi = \Lambda \,.\equiv.\, \hat{\alpha}\phi = \hat{\alpha}(\alpha \in \Lambda)\urcorner$

*244 $\equiv (\alpha)(\alpha \in V \,.\supset : \phi \equiv .\, \alpha \in \Lambda)\urcorner$

*100, *123 $\equiv (\alpha)(\alpha \in V \,.\supset .\sim \phi\,[\equiv 1])\urcorner$.

From *249 and *250 it is clear that the identity $\ulcorner \hat{\alpha}\phi = V\urcorner$ is *not* in general equivalent simply to $\ulcorner(\alpha)\phi\urcorner$, nor $\ulcorner \hat{\alpha}\phi = \Lambda\urcorner$ to $\ulcorner(\alpha)\sim\phi\urcorner$. On the contrary, $\hat{x}(\ldots x \ldots)$ may be V even though '$\ldots x \ldots$' fails for various non-elements x, and $\hat{x}(\ldots x \ldots)$ may be Λ even though '$\ldots x \ldots$' holds for various non-elements x. The '$\supset$' of *193 and *194 cannot in general be strengthened to '$\equiv$'. But '$y = \Lambda$' and '$(x)(x \bar{\in} y)$' do prove equivalent, as seen in †192a.

In the theory of quantification it was found convenient not to require, for significance of $\ulcorner(\alpha)\phi\urcorner$ and $\ulcorner(\exists\alpha)\phi\urcorner$, that α be free in ϕ. In the vacuous case where α is not free in ϕ, such quantifications come out equivalent simply to ϕ. (Cf. pp. 74, 104.) Now it is likewise convenient not to require that α be free in ϕ for significance of $\ulcorner\hat{\alpha}\phi\urcorner$; the definitions of class abstraction already adopted, indeed, impose no such restriction. The significance of abstraction in the vacuous case where α is not free in ϕ is explained in the following two metatheorems, which are corollaries of *249 and *250.

***253.** *If* α *is not free in* ϕ, $\vdash \ulcorner \hat{\alpha}\phi = \mathrm{V} . \equiv \phi \urcorner$.

Proof. *232 $\vdash \ulcorner [210 \supset](\exists\alpha)(\alpha \in \mathrm{V})\urcorner$. (1)
*249 $\vdash \ulcorner \hat{\alpha}\phi = \mathrm{V} . \equiv (\alpha)(\alpha \in \mathrm{V} . \supset \phi)\urcorner$
*161 (& hp) $\equiv . [1 \supset] \phi \urcorner$.

***254.** *If* α *is not free in* ϕ, $\vdash \ulcorner \hat{\alpha}\phi = \Lambda . \equiv \sim \phi \urcorner$.

Proof similar, using *250 instead of *249.

Thus a vacuous abstract $\ulcorner \hat{\alpha}\phi \urcorner$ is capable of designating only V or Λ. It designates V if ϕ is true and Λ if ϕ is false.

We might arbitrarily interpret the two truth values as V and Λ, thus speaking of the truth value of true statements as V and the truth value of falsehoods as Λ. Then the abstraction prefix $\ulcorner \hat{\alpha} \urcorner$, when vacuous, comes to amount to a truth-value operator; $\ulcorner \hat{\alpha}\phi \urcorner$ comes to designate the truth value of ϕ.

From the curious but trivial topic of vacuous abstraction we turn back now, for a short space, to the more serious topic of elementhood.

It has been remarked (§ 28) that *200 provides merely a sufficient condition for elementhood, rather than a necessary one. Other sufficient conditions appear in the following metatheorems.

***256.** $\vdash \ulcorner (\alpha)\phi \supset . \hat{\alpha}\phi \in \mathrm{V}\urcorner$.

Proof. *193 $\vdash \ulcorner (\alpha)\phi \supset . \hat{\alpha}\phi = \mathrm{V}\urcorner$
*223 $\supset : \hat{\alpha}\phi \in \mathrm{V} [. \equiv 210]\urcorner$.

***257.** $\vdash \ulcorner (\alpha) \sim \phi \supset . \hat{\alpha}\phi \in \mathrm{V}\urcorner$.

Proof similar, using *194 and †211.

***258.** *If* α *is not free in* ϕ, $\vdash \ulcorner \hat{\alpha}\phi \in \mathrm{V}\urcorner$.

Proof. *118 (& hp), *123 $\vdash \ulcorner \phi \supset . \hat{\alpha}\phi \in \mathrm{V} [: \equiv 256]\urcorner$. (1)
*118 (& hp), *123 $\vdash \ulcorner \sim \phi \supset . \hat{\alpha}\phi \in \mathrm{V} [: \equiv 257]\urcorner$. (2)
*100 $\vdash \ulcorner [1 . 2 . \supset] 258 \urcorner$.

***259.** $\vdash \ulcorner \hat{\alpha}\phi \in \mathrm{V} . \equiv . \hat{\alpha} \sim \phi \in \mathrm{V}\urcorner$.

Proof. Let β be new. Since $\ulcorner \alpha \bar{\epsilon} \beta \urcorner$ is stratified,
*200 $\vdash \ulcorner \beta \in \mathrm{V} . \supset . \hat{\alpha}(\alpha \bar{\epsilon} \beta) \in \mathrm{V}\urcorner$. (1)
*231 $\vdash \ulcorner [(\beta)1 \supset :] \hat{\alpha}\phi \in \mathrm{V} . \supset . \hat{\alpha}(\alpha \bar{\epsilon} \hat{\alpha}\phi) \in \mathrm{V}\urcorner$
*245, *224 $\supset . \hat{\alpha} \sim \phi \in \mathrm{V}\urcorner$. (2)
Similarly $\vdash \ulcorner \hat{\alpha} \sim \phi \in \mathrm{V} . \supset . \hat{\alpha} \sim \sim \phi \in \mathrm{V}\urcorner$

*100, *123 $\supset . \hat{\alpha}\phi \in V^\urcorner$. (3)
*100 $\vdash^\ulcorner [2 . 3 . \supset]\ 259^\urcorner$.

It has already been noted (§ 24) that the classes

$\hat{x}(x \bar{\epsilon} x)$, $\hat{x}(y)\sim(x \epsilon y . y \epsilon x)$, $\hat{x}(y)(z)\sim(x \epsilon y . y \epsilon z . z \epsilon x)$,

etc. are not elements. By virtue of *259, the same can be proved also of the classes

$\hat{x}(x \epsilon x)$, $\hat{x}(\exists y)(x \epsilon y . y \epsilon x)$, $\hat{x}(\exists y)(\exists z)(x \epsilon y . y \epsilon z . z \epsilon x)$,

etc. All these matters are recorded in the following theorems and metatheorems.

†260. $\hat{x}(x \bar{\epsilon} x) \bar{\epsilon} V$

Proof. *235 $\hat{x}(x \bar{\epsilon} x) \epsilon \hat{x}(x \bar{\epsilon} x) . \equiv . \hat{x}(x \bar{\epsilon} x) \epsilon V . \hat{x}(x \bar{\epsilon} x) \bar{\epsilon} \hat{x}(x \bar{\epsilon} x)$ (1)
*100 $[1 \supset]$ 260

†261. $\hat{x}(x \epsilon x) \bar{\epsilon} V$

Proof. *259, *123 261 [≡ 260]

***262.** *If* $\alpha, \gamma_1, \ldots, \gamma_n$ *are distinct,*
$\vdash^\ulcorner \hat{\alpha}(\gamma_1) \ldots (\gamma_n)\sim(\alpha \epsilon \gamma_1 . \gamma_1 \epsilon \gamma_2 . \ldots . \gamma_n \epsilon \alpha) \bar{\epsilon} V^\urcorner$.

Proof. Let β be new.
*230 $\vdash^\ulcorner \alpha \epsilon \hat{\alpha}(\gamma_1) \ldots (\gamma_n)\sim(\alpha \epsilon \gamma_1 . \ldots . \gamma_n \epsilon \alpha) . \equiv .$
$\alpha \epsilon V . (\gamma_1) \ldots (\gamma_n)\sim(\alpha \epsilon \gamma_1 . \ldots . \gamma_n \epsilon \alpha)^\urcorner$. (1)
*180 $\vdash^\ulcorner (\alpha)(\alpha \epsilon \beta . \equiv R1) \supset . \beta \bar{\epsilon} V^\urcorner$. (2)
*231 $\vdash^\ulcorner [(\beta)2 \supset . (\alpha)1 \supset]\ 262^\urcorner$.

***263.** *If* $\alpha, \gamma_1, \ldots, \gamma_n$ *are distinct,*
$\vdash^\ulcorner \hat{\alpha}(\exists\gamma_1) \ldots (\exists\gamma_n)(\alpha \epsilon \gamma_1 . \gamma_1 \epsilon \gamma_2 . \ldots . \gamma_n \epsilon \alpha) \bar{\epsilon} V^\urcorner$.

Proof. *133, *123
$\vdash^\ulcorner \hat{\alpha}\sim(\exists\gamma_1) \ldots (\exists\gamma_n)(\alpha \epsilon \gamma_1 . \ldots . \gamma_n \epsilon \alpha) \bar{\epsilon} V [. \equiv 262]^\urcorner$. (1)
*259, *123 $\vdash^\ulcorner 263 [\equiv 1]^\urcorner$.

§ 33. *Logical Product, Sum, Complement*

THE CLASS $\hat{z}(z \epsilon x . z \epsilon y)$, which has as members the common members of x and y, is called the *logical product* of x and y and designated by the abbreviated symbolism '$x \cap y$'.

D18. $\ulcorner(\zeta \cap \eta)\urcorner$ *for* $\ulcorner\hat{\alpha}(\alpha \in \zeta \,.\, \alpha \in \eta)\urcorner$.

The choice of α here is subject to the usual understanding: any variable foreign to ζ and η.

The connective '$\cap$' is of a new kind. Whereas '$\downarrow$', '.', 'v', '$\supset$', and '$\equiv$' join formulæ to make formulæ, and '$\in$' and '$=$' join terms to make formulæ, '$\cap$' joins terms to make terms. Like all binary connectives, '$\cap$' is to be construed as carrying with it a pair of parentheses; and D18 has been fashioned accordingly. In practice, however, the parentheses will be dropped when there is no danger of confusion.

The class $\hat{z}(z \in x \,.\mathrm{v}.\, z \in y)$, which has as members all the members of x together with all the members of y, is called the *logical sum* of x and y. The class $\hat{z}(z \bar{\in} x)$, which has as members all elements except the members of x, is called the *complement* or *negate* of x. Symbolically the sum and complement are rendered '$x \cup y$' and '$\bar{x}$'.

D19. $\ulcorner(\zeta \cup \eta)\urcorner$ *for* $\ulcorner\hat{\alpha}(\alpha \in \zeta \,.\mathrm{v}.\, \alpha \in \eta)\urcorner$,

D20. $\ulcorner\bar{\zeta}\urcorner$ *for* $\ulcorner\hat{\alpha}(\alpha \bar{\in} \zeta)\urcorner$.

Thus, where x is the class of Americans and y the class of politicians, $x \cap y$ is the class of American politicians; $x \cup y$ is the class which comprises all Americans (politicians and otherwise) together with all foreign politicians; and $\bar{x}$ is the class of un-American elements.

†270. $(z)(y)(x)$ $\quad z \in x \cap y \,.\!\equiv\!.\, z \in x \,.\, z \in y$

Proof. *230 (& D18) $\quad z \in x \cap y \,.\!\equiv\!.\, z \in \mathrm{V} \,.\, z \in x \,.\, z \in y$

†191, *123 $\quad \equiv.\, z \in x \,.\, z \in y$

†271. $(z)(y)(x)$ $\quad z \in x \cup y \,.\!\equiv:\, z \in x \,.\mathrm{v}.\, z \in y$

Proof. *230 (& D19) $\quad z \in x \cup y \;\equiv:\, z \in \mathrm{V} :\, z \in x \,.\mathrm{v}.\, z \in y$

*100 $\quad \equiv:\, z \in \mathrm{V} \,.\, z \in x \,.\mathrm{v}.\, z \in \mathrm{V} \,.\, z \in y$

†191, *123 $\quad \equiv:\, z \in x \,.\mathrm{v}.\, z \in y$

The analogue of †270 and †271 for negation, viz.:

$$(z)(x) \qquad z \in \bar{x} \,.\!\equiv\!.\, z \bar{\in} x$$

is not forthcoming; on the contrary, z is a member neither of x nor of $\bar{x}$ (nor of anything else) in case z is a non-element. It is to

'$z \in \mathrm{V} \,.\, z \bar{\in} x$', and not just '$z \bar{\in} x$', that '$z \in \bar{x}$' is equivalent (cf. *230, D20).

In the proof of †270 the conjunction '$z \in \mathrm{V} \,.\, z \in x \,.\, z \in y$' is construed first as '$z \in \mathrm{V} : z \in x \,.\, z \in y$' and then as '$z \in \mathrm{V} \,.\, z \in x : z \in y$'. A tacit transformation according to *100 is thus involved. It is natural to leave such reassociative transformations tacit in the case of conjunction, thus treating $\ulcorner \phi \,.\, \psi \,.\, \chi \urcorner$ in effect as $\ulcorner \phi \,.\, \psi : \chi \urcorner$ or $\ulcorner \phi : \psi \,.\, \chi \urcorner$ at convenience; and similarly for alternation.

Products, sums, and complements of elements prove to be elements:

†272. $(y)(x)$ $\quad x, y \in \mathrm{V} \,.\!\supset\!.\, x \cap y \in \mathrm{V}$

Proof: *200 (& D18), since '$z \in x \,.\, z \in y$' is stratified (witness '$0 \in 1 \,.\, 0 \in 1$').

†273. $(y)(x)$ $\quad x, y \in \mathrm{V} \,.\!\supset\!.\, x \cup y \in \mathrm{V}$ *Proof* similar.

†274. (x) $\quad x \in \mathrm{V} \,.\!\equiv\!.\, \bar{x} \in \mathrm{V}$

Proof. †189, *224 $\quad x \in \mathrm{V} \,.\!\equiv\!.\, \hat{y}(y \in x) \in \mathrm{V}$

*259 (& D20) $\quad \equiv .\, \bar{x} \in \mathrm{V}$

Corresponding to all tautologous biconditionals involving '$\sim$', '.', and 'v', it is easy with help of *245 to derive theorems of identity involving '$^{-}$', '$\cap$', and '$\cup$'. The first seven of the following theorems, e.g., correspond to the tautologous forms (2)–(4), (7), (8), (10), and (11) of § 11.

†275. (x) $\quad x = \bar{\bar{x}}$

†276. (x) $\quad x = x \cap x$

†277. (x) $\quad x = x \cup x$

†278. $(y)(x)$ $\quad x \cap y = y \cap x$

†279. $(y)(x)$ $\quad x \cup y = y \cup x$

†280. $(y)(x)$ $\quad \overline{x \cap y} = \bar{x} \cup \bar{y}$

†281. $(y)(x)$ $\quad \overline{x \cup y} = \bar{x} \cap \bar{y}$

†282. $(y)(x)$ $\quad \overline{\bar{x} \cap y} = x \cup \bar{y}$

†283. $(y)(x)$ $\quad \overline{\bar{x} \cup y} = x \cap \bar{y}$

†284. $(y)(x)$ $\quad x \cap y = (x \cup \bar{y}) \cap y$

†285. $(y)(x)$ $\quad x \cup y = (x \cap \bar{y}) \cup y$

†286. $(z)(y)(x)$ $\quad (x \cap y) \cap z = x \cap (y \cap z)$

†287. $(z)(y)(x) \quad (x \cup y) \cup z = x \cup (y \cup z)$

†288. $(z)(y)(x) \quad x \cap (y \cup z) = (x \cap y) \cup (x \cap z)$

†289. $(z)(y)(x) \quad x \cup (y \cap z) = (x \cup y) \cap (x \cup z)$

The proofs of such identities follow a fixed pattern. It will suffice to exhibit two samples.

Proof of †275:	†189	$x = \hat{y}(y \in x)$
	*100, *188	$= \hat{y}\sim\sim(y \in x)$
	*245 (& D20)	$= \bar{\bar{x}}$
Proof of †283:	*245 (& D19, 20)	$\overline{\bar{x} \cup y} = \hat{z}\sim(z \in \bar{x} .\vee. z \in y)$
	*245 (& D20)	$= \hat{z}\sim(z \bar{\in} x .\vee. z \in y)$
	*100, *188	$= \hat{z}(z \in x \,.\, z \bar{\in} y)$
	*245 (& D20, 18)	$= x \cap \bar{y}$

The stacking of identity signs, in the above proofs, is subject to conventions exactly parallel to those which govern the stacking of conditional and biconditional signs (§ 17). Just as the inferences involved in the latter two procedures are justified respectively by *112 and *113, so the inference involved in the stacking of identity signs is justified by the principle:

If $\vdash \ulcorner \zeta_1 = \zeta_2 \urcorner, \vdash \ulcorner \zeta_2 = \zeta_3 \urcorner, \ldots, \vdash \ulcorner \zeta_{n-1} = \zeta_n \urcorner$, *then* $\vdash \ulcorner \zeta_1 = \zeta_n \urcorner$.

Proof.	hp 1	$\vdash \ulcorner \zeta_1 = \zeta_2 \urcorner$.	(1)
	hp 2, *224	$\vdash \ulcorner [1 \equiv .]\ \zeta_1 = \zeta_3 \urcorner$.	(2)
	hp 3, *224	$\vdash \ulcorner [2 \equiv .]\ \zeta_1 = \zeta_4 \urcorner$,	and so on.

In metatheorems or theorems cited by starred or daggered numerals in the course of proofs, the distinction between $\ulcorner \phi \,.\, \psi \urcorner$ and $\ulcorner \psi \,.\, \phi \urcorner$ will hereafter be disregarded; and likewise the distinctions between $\ulcorner \phi \vee \psi \urcorner$ and $\ulcorner \psi \vee \phi \urcorner$, between $\ulcorner \phi \equiv \psi \urcorner$ and $\ulcorner \psi \equiv \phi \urcorner$, between $\ulcorner \zeta = \eta \urcorner$ and $\ulcorner \eta = \zeta \urcorner$, between $\ulcorner \zeta \cap \eta \urcorner$ and $\ulcorner \eta \cap \zeta \urcorner$, and between $\ulcorner \zeta \cup \eta \urcorner$ and $\ulcorner \eta \cup \zeta \urcorner$. This procedure involves tacit transformations according to *100 (and *123) in the case of '.', 'v', and '≡'; according to †183 (and *123) in the case of '='; and according to †278 or †279 (and *224) in the case of '∩' or '∪'. Instances of this practice appear in the last lines of the above proofs of †275 and †283, where the $\ulcorner \hat{\alpha}\psi = \hat{\alpha}\chi \urcorner$ of *245 is treated as $\ulcorner \hat{\alpha}\chi = \hat{\alpha}\psi \urcorner$. This innovation is to apply only to citations by starred or daggered numeral; divorced of its star or dagger and used as an abbreviation, the numeral continues to stand for just the original expression without modification.

†275 is the law of *double negation;* †280–†283 embody the class version of *De Morgan's law;* and †276–†279 and †286–†289, by a natural extension of the terminology applied earlier (§ 11) to statement composition, are said to affirm the idempotence, commutativity, associativity, and mutual distributivity of logical multiplication and addition.[1]

The next two theorems, which have the form of biconditionals rather than identities, are analogues of the tautologous forms (17) and (18) of § 11; analogues in a different way, however, from the way in which †275–†289 are analogues of (2)–(4), (7), etc.

†290. $(y)(x)$ $\quad x = y \,.{\equiv}.\, \bar{x} = \bar{y}$

Proof.

†191, *123 (& D10) $\quad x = y \,.{\equiv}\, (z)(z \in \mathrm{V} \,.\, z \in x \,.{\equiv}.\, z \in \mathrm{V} \,.\, z \in y)$

*100, *123 $\quad \equiv (z)(z \in \mathrm{V} \,.{\supset}{:}\, z \mathrel{\bar{\in}} x \,.{\equiv}.\, z \mathrel{\bar{\in}} y)$

*244 (& D20) $\quad \equiv .\, \bar{x} = \bar{y}$

†291. $(y)(x)$ $\quad x = \bar{y} \,.{\equiv}.\, \bar{x} = y$

Proof. †290 $\quad x = \bar{y} \,.{\equiv}.\, \bar{x} = \bar{\bar{y}}$

†275, *224 $\quad \equiv .\, \bar{x} = y$

Unlike the identities, however, not all the statements which are analogous to tautologous forms in this broader fashion are true. The analogue:

$$(y)(x) \qquad x \neq y \,.{\equiv}.\, x = \bar{y}$$

of the tautologous form (13), e.g., is false. Its truth would require, for every choice of y, that the only entity other than y be $\bar{y}$; whereas we can easily prove e.g. that $\hat{z}(\Lambda \in z)$ is neither Λ nor $\bar{\Lambda}$.

The next eight theorems bring logical addition, multiplication, and negation into relation with V and Λ.

†292. $\bar{\Lambda} = \mathrm{V}$

Proof. *193 (& D20) $\quad [(x)192 \supset]\ 292$

[1] Historically it is rather the application to statement composition that constitutes an extension. The terminology first developed in application to term connectives as in the present section.

†293. $\overline{V} = \Lambda$

Proof. †291 [292 ≡] 293

†294. (x) $x \frown \bar{x} = \Lambda$

Proof. *230 (& D20) $y \epsilon \bar{x} .\equiv. y \epsilon V . y \bar{\epsilon} x$ (1)
*100 $[1 \supset] \sim (y \epsilon x . y \epsilon \bar{x})$ (2)
*194 (& D18) $[(y)2 \supset]$ 294

†295. (x) $x \smile \bar{x} = V$

Proof. †282 $x \smile \bar{x} = \overline{x \frown \bar{x}}$
†294, *227 $= \bar{\Lambda}$
†292 $= V$

†296. (x) $x \frown V = x$

Proof. *246 (& D18) $x \frown V = \hat{y}(y \epsilon x)$
†189 $= x$

†297. (x) $x \smile \Lambda = x$

Proof. †192 $y \bar{\epsilon} \Lambda$ (1)
*100, *123 $[189 \equiv.] x = \hat{y}(y \epsilon x .\text{v}. y \epsilon \Lambda [.1])$
(q.e.d.; cf. D19)

†298. (x) $x \frown \Lambda = \Lambda$

Proof. †192 $y \bar{\epsilon} \Lambda$ (1)
*100 $[1 \supset] \sim (y \epsilon x . y \epsilon \Lambda)$ (2)
*194 (& D18) $[(y)2 \supset]$ 298

†299. (x) $x \smile V = V$

Proof. *100 $y \epsilon V .\supset: y \epsilon x .\text{v}. y \epsilon V$ (1)
*249 (& D19) 299 $[\equiv (y)1]$

The notions of logical sum, product, and complement and the theorems †275–†299 belong to the old "algebra of logic" mentioned in § 1. The above method of deriving this class algebra from the theory of statement composition is essentially Peano's. When inclusion and abstraction are taken as primitive (cf. § 23), on the other hand, the theory of statement composition becomes simply a case of the algebra of classes; cf. my *System*, Ch. XII. For postulational analyses of this algebra as a segregated system see particularly Huntington ("Sets," "New Sets") and Stone.

§ 34. *Inclusion*

IF ALL members of x are members of y, then x is said to be *included* in y; symbolically, $x \subset y$. The definition is thus:

D21. $\ulcorner(\zeta \subset \eta)\urcorner$ *for* $\ulcorner(\alpha)(\alpha \in \zeta \,.\supset .\, \alpha \in \eta)\urcorner$,

where α is foreign to ζ and η. The sign '$\subset$' answers to the word 'are', in such contexts as 'Cats are animals' (i.e., 'The class of cats is included in the class of animals').

Inclusion must be distinguished carefully from membership. Where $x \subset y$ it may or may not be the case that $x \in y$, and where $x \in y$ it may or may not be the case that $x \subset y$. The class of cats, e.g., is included in the class of animals but is not a member of it; i.e., each cat is an animal but the class of cats is not. On the other hand the class of Mormons is a member of the class of Christian sects but is not included in it; it is true that the class of Mormons is a sect but false that each Mormon is a sect. Cases do arise, however, where $x \subset y$ and at the same time $x \in y$; e.g., any element x is both included in V (cf. †333) and a member of V.

Like identity, inclusion is *reflexive* and *transitive*.

†310. (x) $x \subset x$ *Proof:* *100 (& D21).

†311. $(z)(y)(x)$ $x \subset y \,.\, y \subset z \,.\supset .\, x \subset z$

*Proof.**100 $w \in x \,.\supset .\, w \in y : w \in y \,.\supset .\, w \in z :\supset: w \in x \,.\supset .\, w \in z$ (1)
*101 (& D21) $[(w)1 \supset :](w)(w \in x \,.\supset .\, w \in y : w \in y \,.\supset .\, w \in z) \supset .\, x \subset z$ (2)
*140, *123 (& D21) $[2 \equiv]$ 311

But inclusion is not, like identity, commutative (symmetrical). Far from being equivalent, '$x \subset y$' and '$y \subset x$' are incompatible except in the reflexive case where y is x. This appears in the following theorem, according to which mutual inclusion is identity.

†312. $(y)(x)$ $x = y \,.\equiv .\, x \subset y \,.\, y \subset x$ *Proof:* *140 (& D10, 5, 21)

The following metatheorems relate inclusion to quantification and abstraction:

***313.** *If* α *is not free in* ζ, $\vdash \ulcorner(\alpha)(\alpha \in \zeta \,.\supset \phi) \equiv .\, \zeta \subset \hat{\alpha}\phi\urcorner$.

Proof. †190 $\vdash \ulcorner \alpha \epsilon \zeta . \supset . \alpha \epsilon V \urcorner$. (1)
*100, *123 $\vdash \ulcorner (\alpha)(\alpha \epsilon \zeta . \supset \phi [:1]) \equiv (\alpha)(\alpha \epsilon \zeta . \supset . \alpha \epsilon V . \phi)\urcorner$
*230, *123 $\equiv (\alpha)(\alpha \epsilon \zeta . \supset . \alpha \epsilon \hat{\alpha}\phi)\urcorner$
*171 (& hp & D21) $\equiv . \zeta \subset \hat{\alpha}\phi \urcorner$.

***314.** $\vdash \ulcorner (\alpha)(\phi \supset \psi) \supset . \hat{\alpha}\phi \subset \hat{\alpha}\psi \urcorner$.
Proof. *100 $\vdash \ulcorner \phi \supset \psi . \supset : \alpha \epsilon V . \phi . \supset \psi \urcorner$
*230, *123 $\supset : \alpha \epsilon \hat{\alpha}\phi . \supset \psi \urcorner$. (1)
*101 $\vdash \ulcorner [(\alpha)1 \supset .](\alpha)(\phi \supset \psi) \supset (\alpha)(\alpha \epsilon \hat{\alpha}\phi . \supset \psi)\urcorner$
*313 $\supset . \hat{\alpha}\phi \subset \hat{\alpha}\psi \urcorner$.

***315.** *If* α *is not free in* ζ, $\vdash \ulcorner (\alpha)(\phi \supset . \alpha \epsilon \zeta) \supset . \hat{\alpha}\phi \subset \zeta \urcorner$.
Proof. *314 $\vdash \ulcorner (\alpha)(\phi \supset . \alpha \epsilon \zeta) \supset . \hat{\alpha}\phi \subset \hat{\alpha}(\alpha \epsilon \zeta)\urcorner$
†189, *224 (& hp)[1] $\supset . \hat{\alpha}\phi \subset \zeta \urcorner$.

The conditionals described in *314 do not in general hold as biconditionals; $\ulcorner (\alpha)(\phi \supset \psi)\urcorner$ is stronger than $\ulcorner \hat{\alpha}\phi \subset \hat{\alpha}\psi \urcorner$. Truth of $\ulcorner (\alpha)(\phi \supset \psi)\urcorner$ requires that all *entities* fulfilling the condition ϕ fulfill ψ, whereas truth of $\ulcorner \hat{\alpha}\phi \subset \hat{\alpha}\psi \urcorner$ requires merely that all *elements* fulfilling ϕ fulfill ψ. Similar reasoning accounts for the conditional character of *315. Note on the other hand that *313 carries the biconditional.

The next two theorems, analogues of the tautologous forms (5) and (6) (§ 11), show that logical products are included in their factors and logical sums include their summands.

†316. $(y)(x)$ $x \cap y \subset x$
Proof. *100 $z \epsilon x . z \epsilon y . \supset . z \epsilon x$ (1)
*315 (& D18) $[(z)1 \supset]316$

†317. $(y)(x)$ $x \subset x \cup y$
Proof. *100 $z \epsilon x . \supset : z \epsilon x . \vee . z \epsilon y$ (1)
*313 (& D19) $[(z)1 \equiv]317$

The next three theorems, embodying the class version of the law of *transposition*, are analogues of the tautologous forms (14)–(16).

†318. $(y)(x)$ $x \subset \bar{y} . \equiv . y \subset \bar{x}$
Proof. *313 (& D20) $x \subset \bar{y} . \equiv (z)(z \epsilon x . \supset . z \bar{\epsilon} y)$

[1] The hypothesis is needed along with †189 to justify $\ulcorner \zeta = \hat{\alpha}(\alpha \epsilon \zeta)\urcorner$.

*100, *123 $\equiv (z)(z \in y \,.\supset .\, z \bar{\in} x)$
*313 (& D20) $\equiv .\, y \subset \bar{x}$

†319. $(y)(x)$ $x \subset y \,.\equiv .\, \bar{y} \subset \bar{x}$

Proof. †275, *224 $x \subset y \,.\equiv .\, x \subset \bar{\bar{y}}$
†318 $\equiv .\, \bar{y} \subset \bar{x}$

†320. $(y)(x)$ $\bar{x} \subset y \,.\equiv .\, \bar{y} \subset x$ *Proof* similar.

The '$x \subset \bar{y}$' and '$\bar{x} \subset y$' of †318 and †320 amount respectively to saying that the classes x and y are mutually exclusive and that they jointly exhaust all elements. This is apparent from the following two theorems.

†321. $(y)(x)$ $x \subset \bar{y} \,.\equiv .\, x \cap y = \Lambda$

Proof. *313 (& D20) $x \subset \bar{y} \,.\equiv (z)(z \in x \,.\supset .\, z \bar{\in} y)$
*100, *123 $\equiv (z){\sim}(z \in x \,.\, z \in y)$
†270, *123 $\equiv (z)(z \bar{\in} x \cap y)$
†192a $\equiv .\, x \cap y = \Lambda$

†322. $(y)(x)$ $\bar{x} \subset y \,.\equiv .\, x \cup y = V$

Proof. *230, *123 (& D20, 21) $\bar{x} \subset y \,.\equiv (z)(z \in V \,.\, z \bar{\in} x \,.\supset .\, z \in y)$
*100, *123 $\equiv (z)(z \in V \,.\supset : z \in x \,.\vee .\, z \in y)$
*249 (& D19) $\equiv .\, x \cup y = V$

The next two theorems, analogues of the tautologous forms (21) and (22), show ways of paraphrasing an inclusion as an identity.

†323. $(y)(x)$ $x \subset y \,.\equiv .\, x = x \cap y$

Proof. *100, *123 (& D21) $x \subset y \equiv (z)(z \in x \,.\equiv .\, z \in x \,.\, z \in y)$
†270, *123 (& D10) $\equiv .\, x = x \cap y$

†324. $(y)(x)$ $x \subset y \,.\equiv .\, y = x \cup y$ *Proof* similar.

The following variants are occasionally of use.

†325. $(y)(x)$ $x \subset y \,.\equiv .\, x = (x \cup \bar{y}) \cap y$

Proof. †284, *224 [323 ≡] 325

†326. $(y)(x)$ $x \subset y \,.\equiv .\, y = (\bar{x} \cap y) \cup x$

Proof. †285, *224 [324 ≡]326

†327. $(y)(x)$ $x \subset \bar{y} \,.\equiv .\, y = (y \cup x) \cap \bar{x}$

Proof. †318 $x \subset \bar{y} \,.\equiv .\, y \subset \bar{x}$

†325 $\equiv . y = (y \cup \bar{\bar{x}}) \cap \bar{x}$
†275, *224 $\equiv . y = (y \cup x) \cap \bar{x}$

†328. $(y)(x)$ $x \subset \bar{y} . \equiv . x = (y \cup x) \cap \bar{y}$

Proof similar, minus the first step.

The next four are analogues of the tautologous forms (28), (29), (42), and (47).

†329. $(z)(y)(x)$ $x \subset y . \supset . x \cap z \subset y \cap z$

Proof. *100 $w \in x . \supset . w \in y : \supset : w \in x . w \in z . \supset . w \in y . w \in z$
†270, *123 $\supset : w \in x \cap z . \supset . w \in y \cap z$ (1)
*101 (& D21) $[(w)1 \supset]329$

†330. $(z)(y)(x)$ $x \subset y . \supset . x \cup z \subset y \cup z$ *Proof* similar.

†331. $(z)(y)(x)$ $x \subset y \cap z . \equiv . x \subset y . x \subset z$

Proof. †270, *123 (& D21)
$x \subset y \cap z \equiv (w)(w \in x . \supset . w \in y . w \in z)$
*100, *123 $\equiv (w)(w \in x . \supset . w \in y : w \in x . \supset . w \in z)$
*140 (& D21) $\equiv . x \subset y . x \subset z$

†332. $(z)(y)(x)$ $x \cup y \subset z . \equiv . x \subset z . y \subset z$ *Proof* similar.

Note, on the other hand, that the analogues:

$(z)(y)(x)$ $x \subset y \cup z . \equiv : x \subset y . \vee . x \subset z,$
$(z)(y)(x)$ $x \cap y \subset z . \equiv : x \subset z . \vee . y \subset z$

of (43) and (46) are false.

V and Λ have the respective peculiarities of including and being included in every class.

†333. (x) $x \subset \mathrm{V}$ *Proof:* †190 (& D21).

†334. (x) $\Lambda \subset x$

Proof. †323 334 [≡ 298]

Thence we have two corollaries.

†335. (x) $\mathrm{V} \subset x . \equiv . x = \mathrm{V}$

Proof. †312 $[333 .] \mathrm{V} \subset x . \equiv . x = \mathrm{V}$

†336. (x) $x \subset \Lambda . \equiv . x = \Lambda$

Proof. †312 $x \subset \Lambda [. 334] . \equiv . x = \Lambda$

Although many of the classes included in a class x may be members of x, the following theorem shows that they cannot all be members.

†337. (x) $(\exists y)(y \subset x \,.\, y \mathbin{\bar{\epsilon}} x)$

Proof. *100	$z \epsilon x \,.\, z \mathbin{\bar{\epsilon}} z \,.\!\supset\!.\, z \epsilon x$	(1)
*315	$[(z)1 \supset.]\ \hat{z}(z \epsilon x \,.\, z \mathbin{\bar{\epsilon}} z) \subset x$	(2)
*235	$L2 \epsilon L2 \,.\!\equiv\!.\, L2 \epsilon V \,.\, L2 \epsilon x \,.\, L2 \mathbin{\bar{\epsilon}} L2$	
†191, *123	$\equiv.\, L2 \epsilon x \,.\, L2 \mathbin{\bar{\epsilon}} L2$	(3)
*100	$[3 \supset.]\ L2 \mathbin{\bar{\epsilon}} x$	(4)
*232	$[2 \,.\, 4 \,.\!\supset]\ 337$	

Λ, for example, is included in Λ but is not a member of Λ; again any non-element is included in V but is not a member of V.

The sign '$\supset$' was used for inclusion by Gergonne as far back as 1816; Russell has reversed it to distinguish it from the conditional sign. Most of the theorems of the present section were known to Peirce and Schröder.

The distinction between inclusion and membership became fully clear only with the advent of quantification theory; it was clear to Frege in 1879, to Peano in 1889, to Peirce in 1885 (3.396). But in a tentative form the distinction existed in traditional logic as a distinction between "distributive" and "collective predication," drawn to resolve the fallacies of composition and division (e.g. Peter is an Apostle, the Apostles are twelve, therefore Peter is twelve).

§ 35. *Unit Classes*

THE CLASS whose sole member is x is called the *unit class* of x, and designated by the abbreviated symbolism 'ιx'.[1]

D22. $\ulcorner \iota\zeta \urcorner$ *for* $\ulcorner \hat{\alpha}(\alpha = \zeta) \urcorner$.

The first theorem shows that x is a member of y if and only if ιx and y are not mutually exclusive. Two corollaries follow it.

†340.[2] $(y)(x)$ $x \epsilon y \,.\!\equiv\!.\, \iota x \cap y \neq \Lambda$

Proof. *234b $x \epsilon y \,.\!\equiv (\exists z)(z = x \,.\, z \epsilon y)$

[1] The notation is Peano's; 'ι' is for '$ἴσος$'. The distinction between ιx and x seems first to have been drawn explicitly by Frege in 1893 and Peano in 1894. See end of § 25.

[2] Note that $\iota x \cap y$ is the product of ιx and y, whereas $\iota(x \cap y)$ is the unit class

†191, *123 $\equiv (\exists z)(z = x \,.\, z \in V \,.\, z \in y)$
*230, *123 (& D22) $\equiv (\exists z)(z \in \iota x \,.\, z \in y)$
†270, *123 $\equiv (\exists z)(z \in \iota x \cap y)$
†192b $\equiv.\ \iota x \cap y \neq \Lambda$

†341. $(y)(x)$ $x \bar{\epsilon} y \,.\!\equiv.\ \iota x \subset \bar{y}$

Proof. *100 [340 $\supset$:] $x \bar{\epsilon} y \,.\!\equiv.\ \iota x \cap y = \Lambda$
†321 $\equiv.\ \iota x \subset \bar{y}$

†342. $(y)(x)$ $x \in y \,.\!\equiv.\ x \in V \,.\, \iota x \subset y$

Proof. *230 (& D20) $x \in \bar{\bar{y}} \,.\!\equiv.\ x \in V \,.\, x \bar{\epsilon} \bar{y}$
†341, *123 $\equiv.\ x \in V \,.\, \iota x \subset \bar{\bar{y}}$ (1)
†275, *224 342 [$\equiv$ 1]

If x is not an element, and hence a member of no class, there is of course no class whose sole member is x. In this trivial case then, the ιx defined in D22 must turn out to be something else; and it turns out in fact to be Λ.

†343. (x) $x \bar{\epsilon} V \,.\!\equiv.\ \iota x = \Lambda$

Proof. †341 $x \bar{\epsilon} V \,.\!\equiv.\ \iota x \subset \bar{V}$
†293, *224 $\equiv.\ \iota x \subset \Lambda$
†336 $\equiv.\ \iota x = \Lambda$

If on the other hand x is capable of membership at all, ιx has x as member.

†344. (x) $x \in V \,.\!\equiv.\ x \in \iota x$

Proof. *235 (& D22) $x \in V$ [. 182] $.\!\equiv.\ x \in \iota x$

Membership in ιx involves not only identity with x, but elementhood of x.

†345. $(y)(x)$ $y \in \iota x \,.\!\equiv.\ x \in V \,.\, x = y$

Proof. *230 (& D22) $y \in \iota x \,.\!\equiv.\ y \in V \,.\, y = x$
*228 $\equiv.\ x \in V \,.\, x = y$

The composite connective '$\epsilon\,\iota$' thus expresses element-identity; it is stronger than '$=$'. Its commutativity is seen as follows.

---

of $x \cap y$. Just as '$\sim$' is understood as governing the shortest statement following it (cf. §§ 2, 7), so 'ι' is to be understood always as governing the shortest term following it; and similarly for any other singulary operator.

†346. $(y)(x)$ $\quad x \epsilon \iota y \,.\!\equiv\!.\, y \epsilon \iota x$

Proof. *230 (& D22) $\quad x \epsilon \iota y \,.\!\equiv\!.\, x \epsilon \mathrm{V} \,.\, x = y$

†345 $\quad \equiv\!.\, y \epsilon \iota x$

The next theorem shows that ιx is the class whose sole member is x if and only if x is an element.

†347. (x) $\quad (y)(y \epsilon \iota x \,.\!\equiv\!.\, x = y) \equiv .\, x \epsilon \mathrm{V}$

Proof. *100, *123 $\quad \mathrm{L}347 \equiv (y)([345 \equiv\!:]\, x = y \,.\!\supset\!.\, x \epsilon \mathrm{V})$

*234a $\quad \equiv\!.\, x \epsilon \mathrm{V}$

The following theorem, an extension of the foregoing one, shows that $\iota x \cup \iota y$ is the class whose sole members are x and y if and only if x and y are elements.

†348. $(y)(x)$ $\quad (z)(z \epsilon \iota x \cup \iota y \,.\!\equiv\!:\, z = x \,.\!\vee\!.\, z = y) \equiv\!.\, x, y \epsilon \mathrm{V}$

Proof.[1]

†271, *123 $\quad \mathrm{L}348 \equiv (z)(z \epsilon \iota x \,.\!\vee\!.\, z \epsilon \iota y \,:\!\equiv\!:\, z = x \,.\!\vee\!.\, z = y)$

*235, *123 (& D22) $\quad \equiv (z)(z \epsilon \mathrm{V} \,.\, z = x \,.\!\vee\!.\, z \epsilon \mathrm{V} \,.\, z = y \,:\!\equiv\!:\, z = x \,.\!\vee\!.\, z = y)$

*100, *123 $\quad \equiv (z)(z \epsilon \mathrm{V} \,:\, z = x \,.\!\vee\!.\, z = y \,:\!\equiv\!:\, z = x \,.\!\vee\!.\, z = y)$

*100, *123 $\quad \equiv (z)(z = x \,.\!\vee\!.\, z = y \,:\!\supset\!.\, z \epsilon \mathrm{V})$

*100, *123 $\quad \equiv (z)(z = x \,.\!\supset\!.\, z \epsilon \mathrm{V} \,:\, z = y \,.\!\supset\!.\, z \epsilon \mathrm{V})$

*140 $\quad \equiv\!.\, (z)(z = x \,.\!\supset\!.\, z \epsilon \mathrm{V}) \,.\, (z)(z = y \,.\!\supset\!.\, z \epsilon \mathrm{V})$

*234a, *123 $\quad \equiv\!.\, x, y \epsilon \mathrm{V}$

The next two theorems show that the words 'y is the class whose sole member is x' can be put into symbols in three ways:

$$(z)(z \epsilon y \,.\!\equiv\!.\, z = x), \qquad y = \iota x \,.\, x \epsilon \mathrm{V}, \qquad y \subset \iota x \,.\, x \epsilon y.$$

†349. $(y)(x)$ $\quad (z)(z \epsilon y \,.\!\equiv\!.\, z = x) \equiv\!.\, y = \iota x \,.\, x \epsilon \mathrm{V}$

Proof. †189 $\quad y = \hat{z}(z \epsilon y)$ (1)

*121 (& D22) $\quad \mathrm{L}349 \supset\!:[1 \equiv\!.]\, y = \iota x$ (2)

*100 $\quad [2 \supset\!:]\,\mathrm{L}349 \equiv\!.\, y = \iota x \,.\, \mathrm{L}349$

*228 $\quad \equiv\!.\, y = \iota x \,.\, (z)(z \epsilon \iota x \,.\!\equiv\!.\, z = x)$

†347, *123 $\quad \equiv\!.\, y = \iota x \,.\, x \epsilon \mathrm{V}.$

†350. $(y)(x)$ $\quad (z)(z \epsilon y \,.\!\equiv\!.\, z = x) \equiv\!.\, y \subset \iota x \,.\, x \epsilon y$

[1] This proof, neater than my own, is due to N. D. Gautam of Herbert College in India.

Proof. †312, *123 $[349 \equiv :]$ L350 $\equiv .\ y \subset \iota x \,.\, \iota x \subset y \,.\, x \in V$
†342, *123 $\equiv .\ y \subset \iota x \,.\, x \in y$

A peculiarity of ιx is that it is included in any class which it overlaps at all.

†351. $(y)(x)$ $\iota x \subset y \,.\mathsf{v}.\, \iota x \cap y = \Lambda$

Proof. *100 $[342\,.]\ x \in y \,.\supset.\, \iota x \subset y$ (1)
*100 $[1\,.\,340\,.\supset]$ 351

In other words, ιx is included in each class or its complement.

†352. $(y)(x)$ $\iota x \subset y \,.\mathsf{v}.\, \iota x \subset \bar{y}$

Proof. †321, *123 352 $[\equiv 351]$

Inclusion of ιx is analyzable as follows.

†353. $(y)(x)$ $\iota x \subset y \,.\equiv:\, x \in V \,.\supset.\, x \in y$

Proof. †345, *123 (& D21) $\iota x \subset y \,.\equiv\, (z)(x \in V \,.\, x = z \,.\supset.\, z \in y)$
*100, *123 $\equiv (z)(x = z \,.\supset:\, x \in V \,.\supset.\, z \in y)$
*234a $\equiv:\, x \in V \,.\supset.\, x \in y$

Two consequences follow.

†354. $(z)(y)(x)(w)$ $x \in V \,.\, \iota x \cup y = \iota z \cup \iota w \,.\supset:\, x = z \,.\mathsf{v}.\, x = w$

Proof. †317 $\iota x \subset \iota x \cup y$ (1)
*223 $\iota x \cup y = \iota z \cup \iota w \,.\supset:\, [1 \equiv .]\ \iota x \subset \iota z \cup \iota w$
†353 $\supset:\, x \in V \,.\supset.\, x \in \iota z \cup \iota w$ (2)
*100 $[2\,.]$ L354 $.\supset.\, x \in \iota z \cup \iota w$
†271 $\supset:\, x \in \iota z \,.\mathsf{v}.\, x \in \iota w$
*230, *123 (& D22) $\supset:\, x \in V \,.\, x = z \,.\mathsf{v}.\, x \in V \,.\, x = w$
*100 $\supset:\, x = z \,.\mathsf{v}.\, x = w$

†355. $(z)(y)(x)$ $x \in V \,.\, \iota x \cup y = \iota z \,.\supset.\, x = z$

Proof. †277, *224 L355 $\supset.\, x \in V \,.\, \iota x \cup y = \iota z \cup \iota z$
†354 $\supset:\, x = z \,.\mathsf{v}.\, x = z$
*100 $\supset.\, x = z$

The next theorem shows that ιx has one member at most.

†356. $(z)(y)(x)$ $y, z \in \iota x \,.\supset.\, y = z$

Proof. †345,*123 $y, z \in \iota x \,.\supset.\, x \in V \,.\, x = y \,.\, x \in V \,.\, x = z$
*100 $\supset.\, x = y \,.\, x = z$
*100 $\supset:\, [184\supset .]\ y = z$

The next shows that ιx includes two classes at most.

†357. $(y)(x)$ $y \subset \iota x .\equiv: y = \iota x .\vee. y = \Lambda$

Proof. *223

$\iota x \cap y = y .\supset:. \iota x \cap y = \iota x .\vee. \iota x \cap y = \Lambda :\equiv \text{R357}$ (1)

†323, *123 $y \subset \iota x .\supset. [351 \equiv] \text{R357} [:\equiv 1]$ (2)

†312 $y = \iota x .\supset. y \subset \iota x . \iota x \subset y$

*100 $\supset. y \subset \iota x$ (3)

†334 $\Lambda \subset \iota x$ (4)

*223 $y = \Lambda .\supset: y \subset \iota x [.\equiv 4]$ (5)

*100 $[3 . 5 .] \text{R357} .\supset. y \subset \iota x$ (6)

*100 $[2 . 6 .\supset] 357$

The next shows that distinct elements have distinct unit classes.

†358. $(y)(x)$ $x \in V .\supset: \iota x = \iota y .\equiv. x = y$

Proof. *226 $x = y .\supset. \iota x = \iota y$ (1)

*223 $\iota x = \iota y .\supset:. [344 \equiv:] x \in V .\equiv. x \in \iota y$

*230 (& D22), *123 $\supset: x \in V .\equiv. x \in V . x = y$ (2)

*100 $x \in V .\supset:. [2 \supset:] \iota x = \iota y .\supset. x = y$

*100 $\supset:. [1 \supset:] \iota x = \iota y .\equiv. x = y$

Regardless of whether x is an element, ιx proves to be an element.

†359. (x) $\iota x \in V$

Proof. Since '$(z)(z \in y .\equiv. z \in x)$' is stratified,

*200 $x \in V .\supset. \hat{y}(z)(z \in V .\supset: z \in y .\equiv. z \in x) \in V$

*100, *123 $\supset. \hat{y}(z)(z \in V . z \in y .\equiv. z \in V . z \in x) \in V$

†191, *123 (& D10, 22) $\supset. \iota x \in V$ (1)

*223 $\iota x = \Lambda .\supset: \iota x \in V [.\equiv 211]$ (2)

*100 $[1 . 2 . 343 .\supset] 359$

†360. $(y)(x)$ $\iota x \cup \iota y \in V$

Proof. †359 $\iota y \in V$ (1)

†273 $[359 . 1 .\supset] 360$

†361. $(y)(x)$ $\iota x \cap y \in V$

Proof. †323 $\iota x \subset y .\supset. \iota x = \iota x \cap y$

*223 $\supset: [359 \equiv.] \iota x \cap y \in V$ (1)

*223 $\iota x \cap y = \Lambda .\supset: \iota x \cap y \in V [.\equiv 211]$ (2)

*100 $[1 . 2 . 351 .\supset] 361$

In '$y \cap \overline{\iota x}$' we have a notation for the class of all members of y except x.

†362. $(z)(y)(x) \qquad z \in y \cap \overline{\iota x} .\equiv. z \in y . z \neq x$

Proof. *245 (& D20, 22) $\overline{\iota x} = \hat{z}(z \neq x)$ (1)

*223 $[1 \supset :]\ z \in y \cap \overline{\iota x} .\equiv. z \in y \cap \hat{z}(z \neq x)$

†270 $\equiv. z \in y . z \in \hat{z}(z \neq x)$

*230, *123 $\equiv. z \in y . z \in V . z \neq x$

†191, *123 $\equiv. z \in y . z \neq x$

That class turns out to be an element if and only if y itself is.

†363. $(y)(x) \qquad y \in V .\equiv. y \cap \overline{\iota x} \in V$

Proof. †274 $[359 \equiv .]\ \overline{\iota x} \in V$ (1)

†272 $y \in V [.1] .\supset. y \cap \overline{\iota x} \in V$ (2)

†273 $y \cap \overline{\iota x} \in V [.359] .\supset. (y \cap \overline{\iota x}) \cup \iota x \in V$ (3)

†326 $\iota x \subset y .\supset. y = (y \cap \overline{\iota x}) \cup \iota x$

*223 $\supset :. y \cap \overline{\iota x} \in V .\supset. y \in V [: \equiv 3]$

*100 $\supset . [2 \supset]\ 363$ (4)

†318 $\iota x \subset \bar{y} .\supset. y \subset \overline{\iota x}$

†323 $\supset . y = y \cap \overline{\iota x}$

*223 $\supset 363$ (5)

*100 $[4 . 5 . 352 .\supset]\ 363$

From †363 it is apparent that every non-element differs from every element in point of infinitely many members; for if we could reach z from an element w by n steps of deleting or inserting single members, i.e. by n transformations of the type "y to $y \cap \overline{\iota x}$" or "$y \cap \overline{\iota x}$ to y", then we could establish elementhood of z by n applications of †363. In particular it follows that every non-element differs from both Λ and V in point of infinitely many members; hence every finite class is an element, and every class which is so large as to exclude only a finite number of elements is an element.

CHAPTER FIVE

RELATIONS

§ 36. *Pairs and Relations*

IT IS natural to regard any condition on one variable as determining a class: the class of all entities x satisfying the given condition. Correspondingly it is natural to regard any condition on two variables as determining a *relation:* the relation of any entity x to any entity y such that x and y satisfy the given condition. We have seen (§ 24), however, that this point of view is untenable in the case of classes; and the same is readily seen in the case of relations. For, suppose z is the relation of x to y such that x and y satisfy the condition 'x does not bear the relation x to y'. I.e.,

(1) $(y)(x)$ x bears z to y .≡ ∼(x bears x to y).

Thence we have the self-contradictory consequence:

(2) (y) z bears z to y .≡ ∼(z bears z to y).[1]

In the case of classes the analogous difficulty was obviated by barring certain entities, the so-called non-elements, from membership; in other words, by countenancing as classes only classes of so-called elements. A similar expedient suffices here: countenancing as relations only relations of elements. Instead of asking in general that there be such a thing as the relation of any *entity* x to any *entity* y such that . . . x . . . y . . . , then, we ask only that there be such a thing as the relation of any *element* x to any *element* y such that . . . x . . . y . . .. The difficulty encountered in (2) now disappears; for, where z is the relation of any element x to any element y such that ∼(x bears x to y), we have:

(3) $(y)(x)$ x bears z to y .≡. $x, y \in$ V . ∼(x bears x to y)

rather than (1). From (3) we obtain no contradiction, but only the harmless consequence:

(y) z bears z to y .≡. $z, y \in$ V . ∼(z bears z to y),

and the corollary '$z \bar{\in}$ V'.

The new abstraction prefix 'the relation of any element x to any element y such that' will subsequently be defined, and the appro-

[1] Pointed out by Russell, *Principles*, Appendix A.

priate laws will be established (§ 37); but first we must look to the general notion of *relation*. It is characteristic of a relation to pair elements off with elements according to one or another plan. The numerical relation *less than*, e.g., pairs 0 off with 1, also it pairs 0 off with 2, also it pairs 1 off with 2, and in general it pairs x off with y wherever $x < y$. The relation *father of*, again, pairs Abraham off with Isaac, and in general it pairs x off with y wherever x is father of y. It is thus convenient to think of each relation as a class of pairs of elements. To say that an element x bears a relation z to an element y is to say, then, that x paired with y forms a member of z.

This notion of *pair* now needs explicit formulation. It will be rendered symbolically in the fashion '$x;y$', read 'x paired with y'; and we have now to frame a satisfactory definition of this notation. Since relations are to be classes of pairs of elements, these pairs must themselves be elements; otherwise we could not have classes of them. Here, then, is one demand which our definition must meet. Further, pairs must be conceived in a non-commutative fashion, as *ordered* pairs; they must be distinguished not only when they differ as to their constituent elements, but when they are formed from the same elements in reverse order. We must distinguish e.g. between Abraham paired with Isaac and Isaac paired with Abraham, for the former pair is wanted as a member of *father of* whereas the latter pair is not. But any notion of pair which meets these demands will serve our purpose. Now it will be shown that the demands are met by the following definition.

D23. $\ulcorner(\zeta;\eta)\urcorner$ *for* $\ulcorner(\iota\iota\zeta \cup \iota(\iota\zeta \cup \iota\eta))\urcorner$.

According to this definition, $x;y$ is the class whose sole members are ιx and $\iota x \cup \iota y$.

†410. $(z)(y)(x)$ $z \in x;y \,.\!\equiv:\, z = \iota x \,.\vee.\, z = \iota x \cup \iota y$

Proof. †348 (& D23) $(z)410\,[\equiv.\, 359 \,.\, 360]$

These members ιx and $\iota x \cup \iota y$ of $x;y$ are respectively the class whose sole member is x and the class whose sole members are x and y, in case x and y are elements (cf. †347, †348.)

The first of our demands on $x;y$, elementhood, turns out to be met independently of whether x and y are elements.

†411. $(y)(x)$ $x;y \in V$ *Proof:* †360 (& D23)

The other demand was that pairs of elements $x;y$ and $z;w$ always be distinct except where x and y are the same *respectively* as z and w. †417, which shows that this demand is met, will be reached after five preliminary theorems.

†412. $(z)(y)(x)(w)$ $x;y = z;w .\supset: \iota x = \iota z .\vee. \iota x = \iota z \cup \iota w$

Proof. †182 $\iota x = \iota x$ (1)

*100 $[1\supset:]\ \iota x = \iota x .\vee. \iota x = \iota x \cup \iota y$ (2)

*223 $x;y = z;w .\supset: \iota x \in x;y .\equiv. \iota x \in z;w$

†410, *123 $\supset. [2 \equiv]$ R412

†413. $(z)(y)(x)(w)$ $x;y = z;w .\supset: \iota x \cup \iota y = \iota z .\vee. \iota x \cup \iota y = \iota z \cup \iota w$

Proof similar.

†414. $(z)(y)(x)(w)$ $x \in V . x;y = z;w .\supset. x = z$

Proof. †358 $x \in V .\supset: \iota z = \iota x .\equiv. x = z$ (1)

†412 $x;y = z;w .\supset: \iota z = \iota x .\vee. \iota z = \iota x \cup \iota y$ (2)

†355 $x \in V . \iota z = \iota x \cup \iota y .\supset. x = z$ (3)

*100 $[1 . 2 . 3 .\supset]$ 414

Hitherto the use of *100 was restricted to cases where the complexity of the relevant tautologous form did not exceed a certain arbitrary limit (cf. § 17). Hereafter, for brevity, this restriction will be waived; it is already waived, indeed, in the last step of the above proof. The reader has observed the use of *100 through so many proofs that he will now usually be able, even in these more complex cases, to recognize tautology on reasonably brief examination; and when there is doubt the method of tables is always available.

†415. $(z)(y)(x)(w)$ $y \in V . x;y = z;w .\supset: y = z .\vee. y = w$

Proof. †355 $y \in V . \iota x \cup \iota y = \iota z .\supset. y = z$ (1)

†354 $y \in V . \iota x \cup \iota y = \iota z \cup \iota w .\supset: y = z .\vee. y = w$ (2)

*100 $[413 . 1 . 2 .\supset]$415

†416. $(z)(y)(x)(w)$ $x, y, w \in V . x;y = z;w .\supset. y = w$

Proof. †415 $w \in V . x;y = z;w .\supset: w = x .\vee. y = w$ (1)

*223 $w = x .\supset:.$ L414 $\supset. w = z\ [:\equiv 414]$ (2)

*223 $w = z .\supset::$ L415 $\supset: y = w .\vee. y = w\ [.:\equiv 415]$ (3)

*100 $[1 . 2 . 3 .\supset]$416

†417. $(z)(y)(x)(w)$ $x, y, w \epsilon \mathrm{V} .\supset : x;y = z;w .\equiv. x = z . y = w$

Proof. *226 $y = w .\supset. x;y = x;w$ (1)

*223 $x = z .\supset :. [1 \equiv :]\, y = w .\supset. x;y = z;w$ (2)

*100 [414 . 416 . 2 .⊃] 417

Two minor variants of †417 follow as corollaries.

†418. $(z)(y)(x)(w)$ $x,y \epsilon \mathrm{V} .\supset : w \epsilon \mathrm{V} . x;y = z;w .\equiv. x = z . y = w$

Proof. *223 $y = w .\supset : y \epsilon \mathrm{V} .\equiv. w \epsilon \mathrm{V}$ (1)

*100 [417 . 1 .⊃]418

†419. $(z)(y)(x)(w)$ $x,y \epsilon \mathrm{V} .\supset : z,w \epsilon \mathrm{V} . x;y = z;w .\equiv. x = z . y = w$

Proof. *223 $x = z .\supset : x \epsilon \mathrm{V} .\equiv. z \epsilon \mathrm{V}$ (1)

*100 [418 . 1 .⊃]419

Where x is a relation, the statement that y bears x to z may conveniently be rendered in symbols as '$x(y,z)$'. The way to define this notation is clear from foregoing considerations: it is to mean that y and z are elements whose pair $y;z$ is a member of x.

D24. $\ulcorner(\zeta(\eta, \theta))\urcorner$ *for* $\ulcorner(\eta, \theta \epsilon \mathrm{V} . \eta;\theta \epsilon \zeta)\urcorner$.

(The outer parentheses will be retained only after a circumflex.)

Thus '$x(y,z)$' is short for '$y, z \epsilon \mathrm{V} . y;z \epsilon x$'. It may at first appear that the clause '$y, z \epsilon \mathrm{V}$' to the effect that y and z are elements is superfluous here, since '$y;z \epsilon x$' already implies elementhood of $y;z$. But actually the clause is not superfluous, for $y;z$ is an element even when y and z are not (cf. †411). Where y is not an element, we readily see from †343 and †297 that $y;z$ reduces to $\iota\Lambda \cup \iota\iota z$; yet from '$\iota\Lambda \cup \iota\iota z \epsilon x$' we should not want to infer that every non-element y bears x to z, even though it is indeed trivially the case that $y;z \epsilon x$. Again, where z is not an element, $y;z$ reduces to $\iota\iota y$; yet from '$\iota\iota y \epsilon x$' we should not want to infer that y bears x to every non-element z. Again, where neither y nor z is an element, $y;z$ reduces to $\iota\Lambda$; yet from '$\iota\Lambda \epsilon x$' we should not want to infer that every non-element y bears x to every non-element z. Trivial results are thus excluded by incorporating '$y, z \epsilon \mathrm{V}$' into the definition of '$x(y,z)$'.

†420. $(y)(x)$ $\mathrm{V}(x,y) \equiv. x, y \epsilon \mathrm{V}$

Proof. *100 (& D24) $\mathrm{V}(x,y) \equiv. x, y \epsilon \mathrm{V}$ [. 411]

†421. $(z)(y)(x)(w) \quad x \in z \,.\, y \in w \,.\supset V(x,y)$

Proof. †190 $\quad x \in z \,.\supset.\, x \in V$ (1)

†190 $\quad y \in w \,.\supset.\, y \in V$ (2)

*100 $\quad [1 \,.\, 2 \,.\, 420 \,.\supset]\ 421$

†422. $(z)(y)(x)(w) \quad x \subset y \,.\supset.\, x(z,w) \supset y(z,w)$

Proof. *231 (& D21) $\quad x \subset y \,.\supset:\, z;w \in x \,.\supset.\, z;w \in y$

*100 (& D24) $\quad \supset.\, x(z,w) \supset y(z,w)$

†423. $(z)(y)(x)(w) \quad (x \cap y)(z,w) \equiv.\, x(z,w) \,.\, y(z,w)$

Proof. †270, *123 (& D24)

$(x \cap y)(z,w) \equiv.\, z,\, w \in V \,.\, z;w \in x \,.\, z;w \in y$

*100 (& D24) $\quad \equiv.\, x(z,w) \,.\, y(z,w)$

†424. $(z)(y)(x)(w) \quad (x \cup y)(z,w) \equiv.\, x(z,w) \vee y(z,w)$

Proof similar.

Relations in the sense here considered are known, more particularly, as *dyadic* relations; they relate elements in pairs. The relation of giving (y gives z to w) or betweenness (y is between z and w), on the other hand, is triadic; and the relation of paying (x pays y to z for w) is tetradic. But the theory of dyadic relations provides a convenient basis for the treatment also of such polyadic cases. A triadic relation among elements y, z, and w might be conceived as a dyadic relation borne by y to $z;w$. Actually there is some advantage in conceiving it rather as a dyadic relation borne by ιy to $z;w$, for this proves to simplify the conditions of elementhood for triadic relations. The analogue of D24 for triadic relations would therefore assume this form:

$$\ulcorner(\zeta(\eta,\, \theta,\, \theta'))\urcorner \quad \textit{for} \quad \ulcorner(\eta,\, \theta,\, \theta' \in V \,.\, \zeta(\iota\eta,\, \theta;\theta'))\urcorner.$$

Tetradic relations could be handled on the basis of triadic ones in similar fashion:

$$\ulcorner(\zeta(\eta,\, \eta',\, \theta,\, \theta'))\urcorner \quad \textit{for} \quad \ulcorner(\eta,\, \eta',\, \theta,\, \theta' \in V \,.\, \zeta(\iota\eta,\, \iota\eta',\, \theta;\theta'))\urcorner.$$

Similarly for pentadic relations, hexadic ones, and so on. However, we shall have no occasion to proceed beyond the dyadic case.

The notion of a relation as a class of ordered pairs goes back to Peirce (3.220, 329), and the notation '$x;y$' to Frege and Peano. But Wiener (1914) was the first to show that the ordered pair could be defined within the theory of classes. He took $x;y$ as $\iota(\iota\iota x \cup \iota\Lambda) \cup \iota\iota y$, though if it had not been for Russell's theory of types he would obviously have adopted the simpler version $\iota(\iota x \cup \iota\Lambda) \cup \iota y$. The alternative version adopted in D23 is due to Kuratowski. In the present system simplicity would have been gained in some directions, and lost in others, by construing $x;y$ in still a third way: as $\hat{z}(\exists w)(w \in x \,.\, z = \iota w \,.\!\vee\!.\, w \in y \,.\, z = \overline{\iota w})$. This version has the advantage of rendering the antecedent 'x, y, $w \in$ V' superfluous in †417.

§ 37. *Abstraction of Relations*

JUST AS the prefix '$\hat{x}$' of class abstraction corresponds to the words 'the class of all elements x such that', so the prefix '$\hat{x}\hat{y}$' of *relational abstraction* is to correspond to the words 'the relation of any element x to any element y such that'.[1] The relation *factor of*, e.g., can be expressed in terms of relational abstraction as

$$\hat{x}\hat{y}(\exists z)(z \in \text{integer} \,.\, x \times z = y).$$

The form of notation '$\hat{x}\hat{y}(\ldots x \ldots y \ldots)$' is readily defined if we reflect that this relation is to comprise as members just those pairs $x;y$ of elements such that $\ldots x \ldots y \ldots$. Anything z is to belong to $\hat{x}\hat{y}(\ldots x \ldots y \ldots)$ just in case

$$(\exists x)(\exists y)(x, y \in \text{V} \,.\, z = x;y \,.\, \ldots x \ldots y \ldots).$$

The definition adopted is thus the following.

D25. $\ulcorner\hat{\alpha}\hat{\beta}\phi\urcorner$ *for* $\ulcorner\hat{\gamma}(\exists\alpha)(\exists\beta)(\alpha, \beta \in \text{V} \,.\, \gamma = \alpha;\beta \,.\, \phi)\urcorner$.

Accordingly we have the following metatheorem.

***430.** *If α and β are free in neither ζ nor η,*
$\vdash \ulcorner\hat{\alpha}\hat{\beta}\phi\,(\zeta,\eta) \equiv . \,\zeta, \eta \in \text{V} \,.\, (\exists\alpha)(\exists\beta)(\alpha = \zeta \,.\, \beta = \eta \,.\, \phi)\urcorner$.

Proof. †411 $\vdash \ulcorner\zeta;\eta \in \text{V}\urcorner$. (1)

*235 (& D25 & hp)

$\vdash \ulcorner\zeta;\eta \in \hat{\alpha}\hat{\beta}\phi \,.\!\equiv\!.\, [1\,.]\ (\exists\alpha)(\exists\beta)(\alpha, \beta \in \text{V} \,.\, \zeta;\eta = \alpha;\beta \,.\, \phi)\urcorner$. (2)

†419, *117 (& hp)

$\vdash \ulcorner\zeta, \eta \in \text{V} \,.\!\supset\! (\alpha)(\beta)(\alpha, \beta \in \text{V} \,.\, \zeta;\eta = \alpha;\beta \,.\!\equiv\!.\, \alpha = \zeta \,.\, \beta = \eta)\urcorner$

[1] This notation was adapted from Frege by Russell. My reference to elementhood, in interpreting the notation, is of course a departure.

$$*121 \qquad \supset :[2 \equiv .]\ \mathrm{L}2 \equiv (\exists\alpha)(\exists\beta)(\alpha = \zeta \,.\, \beta = \eta \,.\, \phi)\urcorner. \quad (3)$$

$$*100\ (\&\ \mathrm{D}24) \qquad \vdash \ulcorner [3 \supset]\ 430\urcorner.$$

In practice the notation $\ulcorner\hat{\alpha}\hat{\beta}\phi\urcorner$ will appear only where α and β are distinct variables. No such explicit restriction needs be imposed, however, because the desired metatheorems governing $\ulcorner\hat{\alpha}\hat{\beta}\phi\urcorner$ turn out not to be violated by the case $\ulcorner\hat{\alpha}\hat{\alpha}\phi\urcorner$. We find that $\hat{x}\hat{x}(\ldots x \ldots)$ is the relation which each member of $\hat{x}(\ldots x \ldots)$ bears to itself.

***431.** $\vdash \ulcorner \hat{\alpha}\hat{\alpha}\phi\,(\zeta,\eta) \equiv .\ \zeta = \eta \,.\, \zeta \in \hat{\alpha}\phi\urcorner.$

Proof. Let α' and β be new and distinct, and let ϕ' be formed from ϕ by putting α' for all free occurrences of α.

$$\begin{array}{lll}
*430 & \vdash \ulcorner \hat{\alpha}\hat{\alpha}\phi\,(\alpha',\beta) & \equiv .\ \alpha', \beta \in \mathrm{V} \,.\, (\exists\alpha)(\exists\alpha)(\alpha = \alpha' \,.\, \alpha = \beta \,.\, \phi)\urcorner \\
*137,\ *123 & & \equiv .\ \alpha', \beta \in \mathrm{V}.\ (\exists\alpha)(\alpha = \alpha' \,.\, \alpha = \beta \,.\, \phi)\urcorner \\
*234\mathrm{b},\ *123 & & \equiv .\ \alpha', \beta \in \mathrm{V} \,.\, \alpha' = \beta \,.\, \phi'\urcorner \\
*228,\ *123 & & \equiv .\ \alpha', \alpha' \in \mathrm{V} \,.\, \alpha' = \beta \,.\, \phi'\urcorner \\
*100 & & \equiv .\ \alpha' = \beta \,.\, \alpha' \in \mathrm{V} \,.\, \phi'\urcorner \\
*235,\ *123 & & \equiv .\ \alpha' = \beta \,.\, \alpha' \in \hat{\alpha}\phi\urcorner. \qquad (1) \\
*233 & \vdash \ulcorner [(\alpha')(\beta)1 \supset] 431\urcorner. &
\end{array}$$

Thence we have the following corollary.

***432.** $\vdash \ulcorner \hat{\alpha}\hat{\alpha}\phi\,(\zeta,\zeta) \equiv .\ \zeta \in \hat{\alpha}\phi\urcorner.$

Proof.

$$\begin{array}{ll}
\dagger 182 & \vdash \ulcorner \zeta = \zeta\urcorner. \qquad (1) \\
*431 & \vdash \ulcorner \hat{\alpha}\hat{\alpha}\phi\,(\zeta,\zeta) \equiv .\ [1\,.]\ \zeta \in \hat{\alpha}\phi\urcorner.
\end{array}$$

Since $\hat{x}\hat{y}(\ldots x \ldots y \ldots)$ is the relation of any element x to any element y such that $\ldots x \ldots y \ldots$, any entity x will bear the relation to any entity y if and only if x and y are elements and $\ldots x \ldots y \ldots$. That the closures of such biconditionals are theorems is established in the following metatheorem. It is the analogue, for relations, of *230.

***433.** $\vdash \ulcorner \hat{\alpha}\hat{\beta}\phi\,(\alpha,\beta) \equiv .\ \alpha, \beta \in \mathrm{V} \,.\, \phi\urcorner.$

Proof. *Case 1:* α is β.

$$\begin{array}{lll}
*432 & \vdash \ulcorner \hat{\alpha}\hat{\alpha}\phi\,(\alpha,\alpha) & \equiv .\ \alpha \in \hat{\alpha}\phi\urcorner \\
*230 & & \equiv .\ \alpha \in \mathrm{V} \,.\, \phi\urcorner \\
*100 & & \equiv .\ \alpha, \alpha \in \mathrm{V} \,.\, \phi\urcorner,\ \text{q.e.d.}
\end{array}$$

Case 2: α is not β. Let α' and β' be new and distinct, let ϕ' be formed from ϕ by putting α' for all free occurrences of α, and let ϕ'' be formed from ϕ' by putting β' for all free occurrences of β.

*430 $\vdash \ulcorner \hat{\alpha}\hat{\beta}\phi\ (\alpha',\beta') \equiv . \alpha', \beta' \epsilon \mathrm{V} . (\exists\alpha)(\exists\beta)(\alpha = \alpha' . \beta = \beta' . \phi)\urcorner$
*158, *123 $\equiv . \alpha', \beta' \epsilon \mathrm{V} . (\exists\alpha)(\alpha = \alpha' . (\exists\beta)(\beta = \beta' . \phi))\urcorner$
*234b, *123 $\equiv . \alpha', \beta' \epsilon \mathrm{V} . (\exists\beta)(\beta = \beta' . \phi')\urcorner$
*234b, *123 $\equiv . \alpha', \beta' \epsilon \mathrm{V} . \phi''\urcorner$. (1)
*233 $\vdash \ulcorner[(\alpha')(\beta')1 \supset]\ 433\urcorner$.

The principle embodied in *433 has the following as its more general formulation. This is the analogue of *235.

***434.** *If ψ is like ϕ except for containing free occurrences of ζ and η in place respectively of all free occurrences of α and β, then*

$$\vdash \ulcorner \hat{\alpha}\hat{\beta}\phi\ (\zeta,\eta) \equiv . \zeta, \eta \epsilon \mathrm{V} . \psi\urcorner.$$

Proof. *233 (& hp) $\vdash \ulcorner[(\alpha)(\beta)433 \supset]434\urcorner$.

The following is the analogue, for relations, of *314.

***435.** $\vdash \ulcorner(\alpha)(\beta)(\phi \supset \psi) \supset . \hat{\alpha}\hat{\beta}\phi \subset \hat{\alpha}\hat{\beta}\psi\urcorner$.

Proof. Let γ be new.

*100 $\vdash \ulcorner \phi \supset \psi . \supset : \alpha, \beta \epsilon \mathrm{V} . \gamma = \alpha;\beta . \phi . \supset . \alpha, \beta \epsilon \mathrm{V} . \gamma = \alpha;\beta . \psi\urcorner$. (1)
*101 $\vdash \ulcorner[(\beta)1 \supset .]\ (\beta)(\phi \supset \psi) \supset (\beta)\mathrm{R}1\urcorner$
*149 $\supset . (\exists\beta)\mathrm{LR}1 \supset (\exists\beta)\mathrm{RR}1\urcorner$. (2)
Similarly, from (2),
$\vdash \ulcorner(\alpha)(\beta)(\phi \supset \psi) \supset . (\exists\alpha)(\exists\beta)\mathrm{LR}1 \supset (\exists\alpha)(\exists\beta)\mathrm{RR}1\urcorner$. (3)
(3), *117 $\vdash \ulcorner(\alpha)(\beta)(\phi \supset \psi) \supset (\gamma)((\exists\alpha)(\exists\beta)\mathrm{LR}1 \supset (\exists\alpha)(\exists\beta)\mathrm{RR}1)\urcorner$
*314 (& D25) $\supset . \hat{\alpha}\hat{\beta}\phi \subset \hat{\alpha}\hat{\beta}\psi\urcorner$.

The iterated use of 'L' and 'R', as above, is self-explanatory; 'LR1', e.g., stands as an abbreviation for the matter:

$$\alpha, \beta \epsilon \mathrm{V} . \gamma = \alpha;\beta . \phi$$

which stands to the left of the main connective in the expression:

$$\alpha, \beta \epsilon \mathrm{V} . \gamma = \alpha;\beta . \phi . \supset . \alpha, \beta \epsilon \mathrm{V} . \gamma = \alpha;\beta . \psi$$

for which 'R1' stands. (Cf. § 24.)

The analogue of †189, viz. '$(x)(x = \hat{y}\hat{z}(x(y,z)))$', is readily seen to be false; for the relation $\hat{y}\hat{z}(x(y,z))$ will have only pairs of elements as members, whereas x itself may have some non-pairs as members besides. The relation $\hat{y}\hat{z}(x(y,z))$ is rather the *relational part* of x

— the class of all those members of x which are pairs of elements. It is the largest relation included in x, and it coincides with x only in case x is itself a relation. It will be referred to briefly as $\dot{x}$. When we are concerned with a complex expression in place of 'x' the superior dot may conveniently be placed at the beginning; e.g., '$\dot{}(x \cap y)$'. The following definition is accordingly adopted.

D26. $\ulcorner\dot{\zeta}\urcorner$ *or* $\ulcorner\dot{}\zeta\urcorner$ *for* $\ulcorner\hat{\alpha}\hat{\beta}(\zeta(\alpha,\beta))\urcorner$.

The following theorem and metatheorem reveal positions from which the dot is suppressible.

†436. $(z)(y)(x) \quad \dot{x}(y,z) \equiv x(y,z)$

Proof. *433 (& D26) $\quad \dot{x}(y,z) \equiv .\ y, z \in \mathrm{V} \,.\, x(y,z)$

*100 (& D24) $\quad \equiv x(y,z)$

***437.** $\vdash \ulcorner\dot{}\hat{\alpha}\hat{\beta}\phi = \hat{\alpha}\hat{\beta}\phi\urcorner$.

Proof. Case 1: α is not β.

*433, *188 $\quad \vdash \ulcorner\hat{\alpha}\hat{\beta}$ L433 $= \hat{\alpha}\hat{\beta}$ R433$\urcorner$

*100, *188 (& D25) $\quad = \hat{\alpha}\hat{\beta}\phi\urcorner$. (1)

*171 (& D26) $\quad \vdash \ulcorner[1 \equiv]437\urcorner$.

Case 2: α is β. Let γ and δ be new and distinct.

*137, *188 (& D25) $\quad \vdash \ulcorner\hat{\alpha}\hat{\alpha}\phi = \hat{\delta}(\exists\alpha)(\alpha, \alpha \in \mathrm{V} \,.\, \delta = \alpha;\alpha \,.\, \phi)\urcorner$

*234b, *188 (& D25) $\quad = \hat{\alpha}\hat{\gamma}(\phi \,.\, \alpha = \gamma)\urcorner$ (1)

*437 (Case 1) $\quad = \dot{}\hat{\alpha}\hat{\gamma}(\phi \,.\, \alpha = \gamma)\urcorner$

(1), *227 $\quad = \dot{}\hat{\alpha}\hat{\alpha}\phi\urcorner$.

Use of †436 and *437 will be tacit; dots will simply be dropped, without comment, from the positions $\ulcorner\dot{\zeta}(\eta,\theta)\urcorner$ and $\ulcorner\dot{}\hat{\alpha}\hat{\beta}\phi\urcorner$. (Cf. proofs of †446, †447.) This practice is still followed when the relevant context $\ulcorner\dot{\zeta}(\eta,\theta)\urcorner$ or $\ulcorner\dot{}\hat{\alpha}\hat{\beta}\phi\urcorner$ is concealed under definitional abbreviations. E.g., the second dot in '$\ddot{x}$' would be dropped by tacit use of *437 on the ground that it has the position '$\dot{}\hat{y}\hat{z}\, x(y,z)$' when '$\dot{x}$' is expanded by D26. (Cf. proof of †450; also § 38.)

The relational part $\dot{\mathrm{V}}$ of V is the universal relation — the relation which every element bears to every element. It is the class of all pairs of elements.

†438. $\dot{\mathrm{V}} = \hat{x}(\exists y)(\exists z)(y, z \in \mathrm{V} \,.\, x = y;z)$

Proof. †420, *188 (& D26) $\dot{\mathrm{V}} = \hat{y}\hat{z}(y, z \in \mathrm{V})$
*100, *188 (& D25) $= \mathrm{R}438$

The following two theorems provide alternative renderings of $\dot{x}$.

†439. (x) $\dot{x} = \hat{y}\hat{z}(y;z \in x)$ *Proof:* *100, *188 (& D24–26).

†440. (x) $\dot{x} = x \cap \dot{\mathrm{V}}$

Proof. *228, *123 (& D25)

$[439 \equiv .]\dot{x} = \hat{w}(\exists y)(\exists z)(y, z \in \mathrm{V} \,.\, w = y;z \,.\, w \in x)$

*158, *188 $= \hat{w}(\exists y)(w \in x \,.\, (\exists z)(y, z \in \mathrm{V} \,.\, w = y;z))$

*158, *188 $= \hat{w}(w \in x \,.\, (\exists y)(\exists z)(y, z \in \mathrm{V} \,.\, w = y;z))$

*245 $= \hat{w}(w \in x \,.\, w \in \hat{w}(\exists y)(\exists z)(y, z \in \mathrm{V} \,.\, w = y;z))$

†438, *227 (& D18) $= x \cap \dot{\mathrm{V}}$

The relational part $\dot{\Lambda}$ of Λ is simply Λ. The null class and the null relation are the same.

†441. $\dot{\Lambda} = \Lambda$

Proof. †440 $\dot{\Lambda} = \Lambda \cap \dot{\mathrm{V}}$
†298 $= \Lambda$

To say that x is a relation is to say that all members of x are pairs of elements. Thus a convenient symbolic rendering of 'x is a relation' is '$x \subset \dot{\mathrm{V}}$'. Another is '$x = \dot{x}$'. The following theorem establishes the equivalence of the two.

†442. (x) $x \subset \dot{\mathrm{V}} \,.\!\equiv.\, x = \dot{x}$

Proof. †323 $x \subset \dot{\mathrm{V}} \,.\!\equiv.\, x = x \cap \dot{\mathrm{V}}$
†440, *224 $\equiv.\, x = \dot{x}$

Since a relation is merely a class of pairs, the notion of class inclusion applies to relations in obvious fashion. The relation of father, e.g., is included in that of parent. The following four theorems deal with inclusion of a relation $\dot{x}$.

†443. (x) $\dot{x} \subset x$

Proof. †316 $x \cap \dot{\mathrm{V}} \subset x$ (1)
†440, *224 $443 [\equiv 1]$

†444. (x) $\dot{x} \subset \dot{\mathrm{V}}$ *Proof* similar.

†445. $(y)(x)$ $\dot{x} \subset y \,.\!\equiv.\, \dot{x} \subset \dot{y}$

Proof. †331 $\dot{x} \subset y$ [. 444] .≡. $\dot{x} \subset y \cap \dot{V}$
†440, *224 ≡. $\dot{x} \subset \dot{y}$

†446. $(y)(x)$ $\dot{x} \subset y$.≡ $(z)(w)(x(z, w) \supset y(z, w))$

Proof. †422, *117 $\dot{x} \subset y$.⊃ R446 (1)
*435 (& D26) R446 ⊃. $\dot{x} \subset \dot{y}$ (2)
*100 [1 . 2 . 445 .⊃] 446

In the first line of the above proof, note the tacit transition from '$\dot{x}(z,w)$' to '$x(z,w)$'. This illustrates the convention which was adopted in connection with †436. The first line of the next proof is similar.

†447. $(y)(x)$ $\dot{x} = \dot{y}$.≡ $(z)(w)(x(z, w) \equiv y(z, w))$

Proof. *223, *117 $\dot{x} = \dot{y}$.⊃ R447 (1)
*186 (& D26) R447 ⊃. $\dot{x} = \dot{y}$ (2)
*100 [1 . 2 .⊃] 447

Since each relation x is identical with $\dot{x}$ (cf. †442), theorems containing dotted variables may be read simply as theorems about all relations. The dot on a variable serves, in effect, merely to confine a law to relations. †444, e.g., may be read as saying that every relation is included in $\dot{V}$; and †447 may be read as saying that relations are identical if and only if they hold between just the same things.

The notions of logical sum and product, like that of class inclusion, apply to relations in obvious fashion. The sum of the relations *father of* and *mother of*, e.g., is the relation *parent of*; and the product of the relations *fond of* and *parent of* is the relation *fond parent of*. The first and third of the following theorems show that the sum of the relational parts of two classes is the relational part of the sum and that the product of the relational parts is the relational part of the product.

†448. $(y)(x)$ $\dot{x} \cup \dot{y} = {}^{\cdot}(x \cup y)$

Proof. †288 $(\dot{V} \cap x) \cup (\dot{V} \cap y) = \dot{V} \cap (x \cup y)$ (1)
†440, *224 448 [≡ 1]

†449. $(y)(x)$ $\dot{x} \cap y = {}^{\cdot}(x \cap y)$

Proof similar, using †286.

†450. $(y)(x)$ $\qquad \dot{x} \cap \dot{y} = {}^{\cdot}(x \cap y)$

Proof. †449 $\qquad \dot{x} \cap \dot{y} = {}^{\cdot}(\dot{x} \cap y)$

†449, *227 $\qquad = {}^{\cdot}(x \cap y)$

In this last step, note the tacit transition from '${}^{\cdot\cdot}(x \cap y)$' to '${}^{\cdot}(x \cap y)$'. This illustrates the convention explained earlier.

The last two of the above theorems show that the dot may be applied to a product or to either or both factors indifferently. But the analogue of †449 for the logical sum does not hold.

The definition D25 of relational abstraction might equivalently have been formulated in the fashion:

D25′. $\ulcorner \hat{\alpha}\hat{\beta}\phi \urcorner$ *for* $\ulcorner \hat{\gamma}(\exists\alpha)(\exists\beta)(\iota\gamma\ (\alpha,\beta) \,.\, \phi)\urcorner$,

in view of the theorem:

$(z)(y)(x) \qquad \iota z\ (x,y) \equiv .\ x, y \in V \,.\, z = x;y.$

Proof. *235, *123 (& D22, 24)

$\iota z\ (x,y) \equiv .\ x, y \in V\ [.\ 411]\ .\ x;y = z$

The variant D25′ recommends itself if one is seeking a model for definitions of abstraction of polyadic relations. The following definitions would serve for the polyadic cases (see end of § 36).

$\ulcorner \hat{\alpha}\hat{\beta}\hat{\gamma}\phi \urcorner$ *for* $\ulcorner \hat{\delta}(\exists\alpha)(\exists\beta)(\exists\gamma)(\iota\delta\ (\alpha, \beta, \gamma) \,.\, \phi)\urcorner$,

$\ulcorner \hat{\alpha}\hat{\beta}\hat{\gamma}\hat{\delta}\phi \urcorner$ *for* $\ulcorner \hat{\alpha}'(\exists\alpha)(\exists\beta)(\exists\gamma)(\exists\delta)(\iota\alpha'(\alpha, \beta, \gamma, \delta) \,.\, \phi)\urcorner$,

etc.

§ 38. *Converse, Image, Relative Product*

THE CONVERSE $\breve{x}$ of a relation x is the relation which y bears to z just in case $x(z,y)$. Where x is the relation *parent of*, $\breve{x}$ is *child of;* where x is *greater than*, $\breve{x}$ is *less than;* where x is *above*, $\breve{x}$ is *below*. The pairs belonging to $\breve{x}$ are those of x with their order reversed.

D27. $\ulcorner \breve{\zeta} \urcorner$ *or* $\ulcorner {}^{\smile}\zeta \urcorner$ *for* $\ulcorner \hat{\alpha}\hat{\beta}(\zeta(\beta,\alpha))\urcorner$.

The breve '${}^{\smile}$', like the superior dot of D26, is placed at the beginning in complex cases; e.g., '${}^{\smile}(x \cap y)$'.

We thus have the following theorem.

†460. $(z)(y)(x) \qquad \breve{x}(y,z) \equiv x(z,y)$

Proof. *433 (& D27) $\quad \breve{x}(y,z) \equiv .\ y, z \in \mathrm{V} .\ x(z,y)$
*100 (& D24) $\quad \equiv x(z,y)$

The converse of the converse of anything x is the relational part of x; and thus the converse of the converse of a relation is the relation itself.

†461. $(x) \qquad \breve{\breve{x}} = \dot{x}$ *Proof:* †460, *188 (& D27, 26).

In view of the convention explained in connection with †436 and *437 (§ 37), the dot is regularly dropped from the positions $\ulcorner\dot{\breve{\zeta}}\urcorner$ and $\ulcorner\breve{\dot{\zeta}}\urcorner$ alike; for these contexts are abbreviations respectively of $\ulcorner\hat{\alpha}\hat{\beta}\ \dot{\zeta}(\beta,\alpha)\urcorner$ and $\ulcorner\cdot\ \hat{\alpha}\hat{\beta}\ \zeta(\beta,\alpha)\urcorner$, and thus involve the dot in the conventionally suppressible ways $\ulcorner\dot{\zeta}(\eta,\theta)\urcorner$ and $\ulcorner\cdot\hat{\alpha}\hat{\beta}\phi\urcorner$. Instances of this practice appear in the last lines of the next two proofs.

†462. $(y)(x) \qquad x \subset y .\equiv. \breve{x} \subset \breve{y}$

Proof. *119, *123 $\quad [446 \equiv :]\ \dot{x} \subset y .\equiv (w)(z)(x(z,w) \supset y(z,w))$
†460, *123 $\quad \equiv (w)(z)(\breve{x}(w,z) \supset \breve{y}(w,z))$
†446 $\quad \equiv. \breve{x} \subset \breve{y}$

The above theorem shows that inclusion on the part of relations is equivalent to inclusion on the part of their converses. Three corollaries follow.

†463. $(y)(x) \qquad \dot{x} \subset \breve{y} .\equiv. \breve{x} \subset y$

Proof. †462 $\quad \dot{x} \subset \breve{y} .\equiv. \breve{x} \subset \breve{\breve{y}}$
†461, *224 $\quad \equiv. \breve{x} \subset \dot{y}$
†445 $\quad \equiv. \breve{x} \subset y$

†464. $(y)(x) \qquad x \subset y .\supset. \breve{x} \subset \breve{y}$

Proof. †311 $\quad [443.]\ x \subset y .\supset. \dot{x} \subset y$
†462 $\quad \supset. \breve{x} \subset \breve{y}$

†465. $(y)(x) \qquad x \subset \breve{y} .\supset. \breve{x} \subset y$ *Proof* similar, using †463.

The *image* of a class y by a relation x, or more briefly the *x-image* of y, symbolically $x``y$, is the class of all those elements which bear $x$ to one or more members of $y$. Where $x$ is the parent relation and $y$ is the class of violinists, $x``y$ is the class of parents of violinists. The members of $x``y$ are the first elements of all those element-pairs which belong to x and have as their second elements members of y.

D28. $\ulcorner(\zeta``\eta)\urcorner$ *for* $\ulcorner\hat{\alpha}(\exists\beta)(\zeta(\alpha,\beta) \,.\, \beta \in \eta)\urcorner$.

We thus have the following theorems.

†466. $(z)(y)(x) \qquad z \in x``y \,. \equiv (\exists w)(x(z,w) \,.\, w \in y)$

Proof. *230 (& D28) $\quad z \in x``y \,.\equiv.\, z \in \mathrm{V} \,.\, (\exists w)(x(z,w) \,.\, w \in y)$

*158 $\quad \equiv (\exists w)(z \in \mathrm{V} \,.\, x(z,w) \,.\, w \in y)$

*100 (& D24), *123 $\quad \equiv (\exists w)(x(z,w) \,.\, w \in y)$

†467. $(z)(y)(x)(w) \qquad x(z,w) \,.\, w \in y \,.\supset.\, z \in x``y$

Proof. *135 $\quad x(z,w) \,.\, w \in y \,.\supset (\exists w)(x(z,w) \,.\, w \in y)$

†466 $\quad \supset.\, z \in x``y$

†468. $(z)(y)(x) \qquad x``y \subset z \,.\equiv (w)(w')(x(w,w') \,.\, w' \in y \,.\supset.\, w \in z)$

Proof.

†466, *123 (& D21) $\quad x``y \subset z \,.\equiv (w)((\exists w')(x(w,w') \,.\, w' \in y) \supset.\, w \in z)$

*161, *123 $\quad \equiv$ R468

†469. $(z)(y)(x) \qquad x \subset y \,.\supset.\, x``z \subset y``z$

Proof. †422 $\quad x \subset y \,.\supset.\, x(w,w') \supset y(w,w')$ (1)

*100 $\quad [1\,.]\, x \subset y \,.\, x(w,w') \,.\, w' \in z \,.\supset.\, y(w,w') \,.\, w' \in z$

†467 $\quad \supset.\, w \in y``z$ (2)

*100, *117 $\quad x \subset y \,.\supset (w)(w')([2\,.]\, x(w,w') \,.\, w' \in z \,.\supset.\, w \in y``z)$

†468 $\quad \supset.\, x``z \subset y``z$

†470. $(z)(y)(x) \qquad y \subset z \,.\supset.\, x``y \subset x``z \qquad$ *Proof* similar.

De Morgan remarked that traditional logic was inadequate to proving the statement 'If horses are animals then heads of horses are heads of animals.' This statement is an instance of †470.

The next two theorems are distributive laws. They show e.g. that the parents of violinists comprise the fathers of violinists and the mothers of violinists, and that the wives of classicists comprise the wives of Latinists and the wives of Hellenists.

†471. $(z)(y)(x) \qquad (x \cup y)``z = (x``z) \cup (y``z)$

Proof. †424, *188 (& D28)

$(x \cup y)``z = \hat{w}(\exists w')(x(w,w') \vee y(w,w') \,.\, w' \in z)$

*100, *188 $\quad = \hat{w}(\exists w')(x(w,w') \,.\, w' \in z \,.\vee.\, y(w,w') \,.\, w' \in z)$

*141, *188 $= \hat{w}((\exists w')(x(w,w') \,.\, w' \epsilon z) \vee (\exists w')(y(w,w') \,.\, w' \epsilon z))$

†466, *188 (& D19) $= (x``z) \cup (y``z)$

†472. $(z)(y)(x)$ $x``(y \cup z) = (x``y) \cup (x``z)$ *Proof* similar.

The practice of telescoping duplicate steps, adopted hitherto (§ 32) in connection with *123 and *224, will be usual also in connection with *188, *227, and *245. The last line of the proof of †471, e.g., is a condensation of the following two lines:

†466, *188 $= \hat{w}(w \epsilon x``z \,.\!\vee (\exists w')(y(w,w') \,.\, w' \epsilon z))$
†466, *188 (& D19) $= (x``z) \cup (y``z)$

Distributive laws parallel to †471 and †472 do not hold for the logical product. The statement:

$$(z)(y)(x) \qquad (x \cap y)``z = (x``z) \cap (y``z)$$

is seen to be false by taking x as the relation *older than*, y as *brother of*, and z as the class of sailors; for $(x \cap y)``z$ then becomes the class of elder brothers of sailors, whereas $(x``z) \cap (y``z)$ becomes the class of all persons who are at once older than some sailors and brothers of some sailors. Even the younger brother of a sailor will belong to the latter class, so long as he happens to be older than some other sailor. Again, the statement:

$$(z)(y)(x) \qquad x``(y \cap z) = (x``y) \cap (x``z)$$

is seen to be false by taking x as the relation of benefiting, y as the class of cripples, and z as the class of musicians; for $x``(y \cap z)$ then comprises just the benefactors of crippled musicians, whereas $(x``y) \cap (x``z)$ takes in also any person who happens to have benefited both an unmusical cripple and an uncrippled musician.

The next theorems explain images involving Λ and V.

†473. (x) $\Lambda``x = \Lambda$

Proof. †192 $y;z \,\bar{\epsilon}\, \Lambda$ (1)
*100 (& D24) $[1 \supset] \sim (\Lambda(y,z) \,.\, z \epsilon x)$ (2)
*131 $\sim(\exists z)(\Lambda(y,z) \,.\, z \epsilon x)[\equiv (z)2]$ (3)
*194 (& D28) $[(y)3 \supset]$ 473

†474. (x) $x``\Lambda = \Lambda$ *Proof* similar.

†475. $(x) \qquad \mathrm{V}``x = \hat{y}(x \neq \Lambda)$

Proof. †420, *188 (& D28) $\mathrm{V}``x = \hat{y}(\exists z)(y, z \in \mathrm{V} \,.\, z \in x)$
†191, *188 $= \hat{y}(\exists z)(y \in \mathrm{V} \,.\, z \in x)$
*158, *188 $= \hat{y}(y \in \mathrm{V} \,.\, (\exists z)(z \in x))$
*246 $= \hat{y}(\exists z)(z \in x)$
†192b, *188 $= \hat{y}(x \neq \Lambda)$

†476. $(x) \qquad x``\mathrm{V} = \hat{y}(\exists z)\, x(y,z)$ *Proof:* *100, *188 (& D28, 24).

In view of †475 and *253–*254, $\mathrm{V}``x$ is $\Lambda$ or V according as $x$ is or is not $\Lambda$. On the other hand $x``\mathrm{V}$ is, in view of †476, the class of all elements bearing x to something. This class is called the *domain* of x. Where x is the parent relation, its domain $x``\mathrm{V}$ is the class of all parents.

†477. $(y)(x) \qquad y \in x``\mathrm{V} \,. \equiv (\exists z)\, x(y,z)$

Proof. †466 $y \in x``\mathrm{V} \,. \equiv (\exists z)(x(y,z) \,.\, z \in \mathrm{V})$
*100 (& D24), *123 $\equiv (\exists z)\, x(y,z)$

The images of unit classes play an important rôle. The image by x of ιz is simply the class of all elements bearing x to z.

†478. $(z)(x) \qquad x``\iota z = \hat{y}(x(y,z))$

Proof. †345, *188 (& D28) $x``\iota z = \hat{y}(\exists w)(x(y,w) \,.\, z \in \mathrm{V} \,.\, w = z)$
*234b, *188 $= \hat{y}(x(y,z) \,.\, z \in \mathrm{V})$
*100 (& D24), *188 $= \hat{y}(x(y,z))$

†479. $(z)(y)(x) \qquad y \in x``\iota z \,. \equiv x(y,z)$

Proof. †478, *224 $y \in x``\iota z \,.\equiv.\, y \in \hat{y}(x(y,z))$
*230 $\equiv.\, y \in \mathrm{V} \,.\, x(y,z)$
*100 (& D24) $\equiv x(y,z)$

The image of y by $\breve{x}$, i.e. $\breve{x}``y$, might be called the *converse image* of $y$ by $x$. It is the class of all elements to which members of $y$ bear $x$. Where $x$ is parent and $y$ violinists, $\breve{x}``y$ is the class of all children of violinists.

†480. $(y)(x) \qquad \breve{x}``y = \hat{w}(\exists z)(x(z,w) \,.\, z \in y)$

Proof: †460, *188 (& D28).

†481. $(y)(x)(w) \qquad w \in \breve{x}``y \,. \equiv (\exists z)(x(z,w) \,.\, z \in y)$

Proof. †466 $w \in \breve{x}``y \,. \equiv (\exists z)(\breve{x}(w,z) \,.\, z \in y)$ (1)
†460, *123 [1 ≡] 481

The domain $\breve{x}``V$ of $\breve{x}$ is called the *converse domain* of $x$. It is the class of all elements to which $x$ is borne. Where $x$ is the relation of hate, its converse domain $\breve{x}``V$ is the class of all who are hated.

†482. (x) $\breve{x}``V = \hat{z}(\exists y)x(y,z)$ *Proof* similar, from †476.

†483. $(z)(x)$ $z \in \breve{x}``V .\equiv (\exists y)x(y,z)$ *Proof* similar, from †477.

Similarly $\breve{x}``\iota y$ is the class of all elements to which y bears x.

†484. $(y)(x)$ $\breve{x}``\iota y = \hat{z}(x(y,z))$ *Proof* similar, from †478.

†485. $(z)(y)(x)$ $z \in \breve{x}``\iota y .\equiv x(y,z)$ *Proof* similar, from †479.

The *relative product* of a relation x into a relation x', symbolically $x \mid x'$, is the relation which y bears to w whenever y bears x to something which bears x' to w. Where x is the brother relation and x' the parent relation, $x \mid x'$ is the relation of uncle. Where x is the father relation and x' the mother relation, $x \mid x'$ becomes the relation of maternal grandfather and $x' \mid x$ becomes the relation of paternal grandmother.

D29. $\ulcorner(\zeta \mid \eta)\urcorner$ *for* $\ulcorner\hat{\alpha}\hat{\gamma}(\exists\beta)(\zeta(\alpha, \beta) . \eta(\beta, \gamma))\urcorner$.

We thus have the following theorems.

†486. $(x')(y)(x)(w)$ $(x \mid x')(y,w) \equiv (\exists z)(x(y,z) . x'(z,w))$

Proof from *433 like that of †466 from *230.

†487. $(x')(z)(y)(x)(w)$ $x(y,z) . x'(z,w) .\supset (x \mid x')(y,w)$

Proof from †486 like that of †467 from †466.

†488. $(z)(y)(x)$ $x|y \subset z .\equiv (x')(y')(z')(x(x',y') . y(y',z') .\supset. z(x',z'))$

Proof from †446 and †486 like that of †468 from D21 and †466.

†489. $(z)(y)(x)$ $x \subset y .\supset. x \mid z \subset y \mid z$

Proof from †487 and †488 like that of †469 from †467 and †468.

†490. $(z)(y)(x)$ $y \subset z .\supset. x \mid y \subset x \mid z$ *Proof* similar.

Relative multiplication is not commutative. A maternal grandfather, e.g., is not the same as a paternal grandmother. But relative multiplication is indeed *associative*, as the following theorem shows.

†491. $(z)(y)(x)$ $(x \mid y) \mid z = x \mid (y \mid z)$

Proof. †486, *188 (& D29)

$$(x \mid y) \mid z = \hat{x}'\hat{w}'(\exists z')((\exists y')(x(x',y') \,.\, y(y',z')) \,.\, z(z',w'))$$

*158, *188 $= \hat{x}'\hat{w}'(\exists z')(\exists y')(x(x',y') \,.\, y(y',z') \,.\, z(z',w'))$

*138, *188 $= \hat{x}'\hat{w}'(\exists y')(\exists z')(x(x',y') \,.\, y(y',z') \,.\, z(z',w'))$

*158, *188 $= \hat{x}'\hat{w}'(\exists y')(x(x',y') \,.\, (\exists z')(y(y',z') \,.\, z(z',w')))$

†486, *188 (& D29) $= x \mid (y \mid z)$

A null factor nullifies the product.

†492. (x) $\Lambda \mid x = \Lambda$ *Proof* similar to that of †473.

†493. (x) $x \mid \Lambda = \Lambda$ *Proof* similar.

The converse of the relative product is the relative product of the converses in opposite order.

†494. $(y)(x)$ $\breve{\ }(x \mid y) = \breve{y} \mid \breve{x}$

Proof. †486, *188 (& D27) $\breve{\ }(x \mid y) = \hat{w}\hat{z}(\exists x')(y(x', w) \,.\, x(z, x'))$

†460, *188 (& D29) $= \breve{y} \mid \breve{x}$

E.g., since the converse of *teacher of* is *pupil of* and the converse of *child of* is *parent of*, the above theorem tells us that the converse of the relative product *teacher of child of* is the relative product *parent of pupil of*.

The concluding theorem exhibits an important connection between the image and relative product. It says e.g. that the uncles of violinists are the brothers of the parents of violinists.

†495. $(z)(y)(x)$ $(x \mid y)``z = x``(y``z)$

Proof like that of †491, using D28 instead of D29 and †466 instead of the last use of †486.

The notions of converse and relative product, along with the identity function *I* (cf. § 42), are familiar to students of group theory. In this connection they were clearly formulated and investigated by Cayley (1854). But in group theory the notions of converse and relative product occur only in application to a special sort of relations, viz. functions (in the sense of § 40) whose converses are functions. Converses and relative products of relations generally are dealt with by DeMorgan (1860), and so also is the notion of image. But DeMorgan does not distinguish explicitly between the relative product and the image; the selfsame notation designates $x \mid y$ or $x``y$ according as y is or is not a relation. The same practice recurs in Peirce; but in Frege (1879) the distinction becomes explicit. The present notation is Russell's.

§ 39. *The Ancestral*

A CLASS y is said to be *closed* with respect to a relation x if whatever bears x to a member of y belongs in turn to y; i.e., if $x``y \subset y$. The class of humans, e.g., is closed with respect to the nephew relation; for all nephews of humans are human. The class of males is likewise closed with respect to the nephew relation. The class of even numbers, again, is closed with respect to the relation *square of;* for the squares of even numbers are even. On the other hand the class of males is not closed with respect to the parent relation; for not all parents of males are male. Again, the class of even numbers is not closed with respect to the relation *half of;* for the halves of even numbers are not always even.

An element may belong to an infinite variety of classes each of which is closed with respect to a given relation x. Napoleon, e.g., belongs to an infinite variety of classes each of which is closed with respect to the parent relation. One of these is the class of all organisms, since parents of organisms are organisms. Even broader classes of the kind can be formed, by supplementing the class of organisms with any parentless sort of objects — say desks; for no such trivial supplementation will alter the fact that all parents of members are members. Moreover, there are classes narrower than that of organisms which still contain Napoleon and are closed with respect to the parent relation. One such is the class of all organisms born before 1800; Napoleon belongs to this, and so do the parents of all the members. Still narrower classes of the kind can be got by casting out all childless persons, or all persons of Chinese descent; for no parents of the residual members will thereby be extruded, nor presumably will Napoleon himself. Certain members, however, will be common to this whole infinite variety of classes. One common member, of course, is Napoleon. Two more are the parents of Napoleon; clearly they must belong to any class which contains Napoleon and is closed with respect to parent. Four others are the parents of Napoleon's parents; and so on. Napoleon and all his ancestors are thus distinguished from the rest of the world by this formal characteristic: they belong to *every* class which contains Napoleon and is closed with respect to the parent relation. If we so construe 'ancestor' as to reckon a person among his own ancestors, then 'z is an ancestor of Napoleon' can be explained as meaning that z belongs to every class which contains Napoleon and is closed with respect to parent; symbolically,

$$(y)(\text{parent}``y \subset y \,.\, \text{Napoleon} \in y \,.\!\supset\!.\, z \in y).$$

The relation of ancestor can be defined, in terms of that of parent, as

$$\hat{z}\hat{w}(y)(\text{parent}``y \subset y \,.\, w \in y \,.\!\supset\!.\, z \in y).$$

More generally, given any relation x instead of parent, the relation

(1) $$\hat{z}\hat{w}(y)(x``y \subset y \,.\, w \in y \,.\!\supset\!.\, z \in y)$$

will be called the *ancestral* of x; symbolically, $*x$. Where x is the parent relation, $*x$ is the genealogical relation of ancestor; where x is the offspring relation, $*x$ is the relation of descendant. In general, $*x$ is the relation of z to w such that z is w or else bears x to w or else bears x to something which bears x to w or else etc. The rigorous formulation of $*x$, viz. (1), is important in that it constitutes a logical analysis of the essential notion involved in the idiom 'etc.'.

We thus adopt the following definition.

D30. $\ulcorner *\zeta \urcorner$ *for* $\ulcorner \hat{\alpha}\hat{\beta}(\gamma)(\zeta``\gamma \subset \gamma \,.\, \beta \in \gamma \,.\!\supset\!.\, \alpha \in \gamma)\urcorner$.

The following theorem is forthcoming.

†510. $(z)(y)(x)(w)$ $\quad x``y \subset y \,.\, *x(z,w) \,.\, w \in y \,.\!\supset\!.\, z \in y$

Proof.

*110 $\quad (y)(x``y \subset y \,.\, w \in y \,.\!\supset\!.\, z \in y) \supset\!:\, x``y \subset y \,.\, w \in y \,.\!\supset\!.\, z \in y$ (1)

*433 (& D30) $\quad *x(z,w) \equiv .\, z, w \in V \,.\, L1$ (2)

*100 $\quad [1 \,.\, 2 \,.\!\supset] \; 510$

Every element bears the ancestral of every relation to itself.

†511. $(y)(x)$ $\quad *x(y,y) \equiv .\, y \in V$

Proof. *100 $\quad x``z \subset z \,.\, y \in z \,.\!\supset\!.\, y \in z$ (1)

*434 (& D30) $*x(y,y) \equiv . \; y, y \in V$ [. (z)1]
*100 $\equiv . \; y \in V$

Two corollaries follow.

†512.[1] $(y)(x)$ $\quad \dot{y} \subset *x \mid y$

Proof. *100 (& D24) $y(z,w) \supset . \; z \in V \, . \; y(z,w)$
†511, *123 $\supset . \; *x(z,z) \, . \; y(z,w)$
†487 $\supset (*x \mid y)(z,w)$ (1)
†446 512 [≡ (z)(w)1]

†513. $(y)(x)$ $\quad \dot{y} \subset y \mid *x$ *Proof* similar.

Every relation is included in its ancestral.

†514. (x) $\quad \dot{x} \subset *x$

Proof. *110 (& D21) $x``y \subset y \, . \supset : z \in x``y \, . \supset . \; z \in y$ (1)
*100, *117 $x(z,w) \supset (y)$([1 . 467 .] $x``y \subset y \, . \; w \in y \, . \supset . \; z \in y)$ (2)
*435 (& D26, 30) [(z)(w)2 ⊃] 514

A class is closed with respect to x if and only if it is closed with respect to $*x$.

†515. $(y)(x)$ $\quad x``y \subset y \, . \equiv . \; *x``y \subset y$

Proof.
*100, *117 $x``y \subset y \, . \supset (z)(w)$([510 .] $*x(z,w) \, . \; w \in y \, . \supset . \; z \in y)$
†468 $\supset . \; *x``y \subset y$ (1)
†469 [514 ⊃.] $x``y \subset *x``y$ (2)
†311 [2 .] $*x``y \subset y \, . \supset . \; x``y \subset y$ (3)
*100 [1 . 3 .⊃] 515

The ancestral is its own ancestral, and its own relative self-product.

†516. (x) $\quad *x = **x$ *Proof:* †515, *188 (& D30).

†517. (x) $\quad *x = *x \mid *x$

Proof. †510 $x``y \subset y \, . \; *x(z',z) \, . \; z \in y \, . \supset . \; z' \in y$ (1)
*100, *117
$*x(z',z) \, . \; *x(z,w) \, . \supset (y)$([1 . 510 .] $x``y \subset y \, . \; w \in y \, . \supset . \; z' \in y)$ (2)
*161 [(z)2 ≡.] $(\exists z)$L2 ⊃ R2 (3)

[1] Note that the star in '$*x \mid y$' applies to 'x', not to '$x \mid y$.' See second footnote of § 35.

*435 (& D29, 30)	$[(z')(w)\ 3\ \supset .]\ *x \mid *x \subset *x$	(4)
†512	$*x \subset *x \mid *x$	(5)
†312	$517\ [\equiv .\ 5\ .\ 4]$	

The next two theorems show e.g. that the parents of one's ancestors and the ancestors of one's parents figure among one's ancestors.

†518. (x) $x \mid *x \subset *x$

Proof. †489 $[514 \supset .]\ x \mid *x \subset *x \mid *x$ (1)

†517, *224 $518\ [\equiv 1]$

†519. (x) $*x \mid x \subset *x$ *Proof* similar, using †490.

The class $*x``y$ comprises all the objects which bear $*x$ to members of $y$. Thus, where $x$ is the parent relation and $y$ the class of lawyers, $*x``y$ is the class of all ancestors of lawyers. The following theorem says that $*x``y$ is closed with respect to x.

†520. $(y)(x)$ $x``(*x``y) \subset *x``y$

Proof. †469 $[518 \supset .]\ (x \mid *x)``y \subset *x``y$ (1)

†495, *224 $[1 \equiv]\ 520$

The following sort of inference is characteristic of the ancestral. From the premisses:

(z)(w)(parent(z,w) . w was born before 1800
.⊃. z was born before 1800)

and:

ancestor (Fiorecchio, Napoleon)

and:

Napoleon was born before 1800,

e.g., we infer that Fiorecchio was born before 1800. In general, from premisses of the forms:

$(z)(w)(x(z,w)\ \ldots .\ w\ \ldots .\supset .\ \ldots .\ z\ \ldots),$
$*x\ (A,B),$
$\ldots B \ldots$

we infer that $\ldots A \ldots .$ The metatheorem governing such inferences is as follows.

***521.** *If α and α' are distinct and not free in ζ, and α' is not free in*

ϕ, and ϕ', ψ, and ψ' are like ϕ except for containing free occurrences respectively of α', η, and η' wherever ϕ contains free occurrences of α, then

$$\vdash \ulcorner(\alpha')(\alpha)(\zeta(\alpha', \alpha) \,.\, \phi \,.\supset \phi') \,.\, *\zeta(\eta', \eta) \,.\, \psi \,.\supset \psi'\urcorner.$$

Proof. †510 $\vdash\ulcorner\zeta``\hat{\alpha}\phi \subset \hat{\alpha}\phi \,.\, *\zeta(\eta', \eta) \,.\, \eta \in \hat{\alpha}\phi \,.\supset.\, \eta' \in \hat{\alpha}\phi\urcorner$. (1)

*100 (& D24)

$\vdash\ulcorner\zeta(\alpha', \alpha) \,.\, \phi \,.\supset \phi' :\equiv: \zeta(\alpha', \alpha) \,.\, \alpha \in \mathrm{V} \,.\, \phi \,.\supset.\, \alpha' \in \mathrm{V} .\, \phi'\urcorner$

*235 (& hp3), *123 $\equiv: \zeta(\alpha', \alpha) \,.\, \alpha \in \hat{\alpha}\phi \,.\supset.\, \alpha' \in \hat{\alpha}\phi\urcorner$. (2)

Similarly $\vdash\ulcorner\zeta``\hat{\alpha}\phi \subset \hat{\alpha}\phi \,.\, *\zeta(\eta', \eta) \,.\, \psi \,.\supset \psi' [:\equiv 1]\urcorner$. (3)

†468 (& hp1–2) $\vdash\ulcorner\zeta``\hat{\alpha}\phi \subset \hat{\alpha}\phi \,.\equiv (\alpha')(\alpha)\mathrm{R}2\urcorner$

(2), *123 $\equiv (\alpha')(\alpha)\mathrm{L}2\urcorner$. (4)

(4), *123 $\vdash\ulcorner[3 \equiv]\ 521\urcorner$.

Inferences of the above kind remain valid when the first of the three premisses, viz.:

$$(z)(w)(x(z,w) \ldots .\, w \ldots .\supset. \ldots z \ldots),$$

is weakened to read:

$$(z)(w)(x(z,w) \,.\, *x(w,B) \ldots .\, w \ldots .\supset. \ldots z \ldots).$$

The metatheorem to this effect is as follows.

***522.** *If α and α' are distinct and free in neither ζ nor η, and* [etc. as in *521], *then*

$$\vdash\ulcorner(\alpha')(\alpha)(\zeta(\alpha', \alpha) \,.\, *\zeta(\alpha, \eta) \,.\, \phi \,.\supset \phi') \,.\, *\zeta(\eta', \eta) \,.\, \psi \,.\supset \psi'\urcorner.$$

Proof. †518 $\vdash\ulcorner\zeta \mid *\zeta \subset *\zeta\urcorner$. (1)

†511 $\vdash\ulcorner*\zeta(\eta, \eta) \equiv.\, \eta \in \mathrm{V}\urcorner$. (2)

†422 $\vdash\ulcorner[1 \supset.]\ (\zeta \mid *\zeta)(\alpha', \eta) \supset *\zeta(\alpha', \eta)\urcorner$. (3)

†487 $\vdash\ulcorner\zeta(\alpha', \alpha) \,.\, *\zeta(\alpha, \eta) \,.\supset (\zeta \mid *\zeta)(\alpha', \eta)\urcorner$. (4)

*100 $\vdash\ulcorner[4 \,.\, 3 \,.\supset :.]\ \mathrm{L}4 \,.\, \phi \,.\supset \phi' :\equiv: \mathrm{L}4 \,.\, \phi \,.\supset.\, *\zeta(\alpha', \eta) \,.\, \phi'\urcorner$. (5)

(5), *123 $\vdash\ulcorner\mathrm{L}522 \supset.\, (\alpha')(\alpha)\mathrm{R}5 \,.\, *\zeta(\eta', \eta) \,.\, \psi\urcorner$

*100 (& D24) $\supset: [2 \supset.]\ (\alpha')(\alpha)\mathrm{R}5 \,.\, *\zeta(\eta', \eta) \,.\, *\zeta(\eta, \eta) \,.\, \psi\urcorner$

*521 (& hp)[1] $\supset.\, *\zeta(\eta', \eta) \,.\, \psi'\urcorner$

*100 $\supset \psi'\urcorner$.

Typical use of *521 occurs in the proofs of the next two theorems.

†523. (x) $\quad \breve{*x} = *\breve{x}$

Proof. †487 $*x(w,z) \,.\, x(z,y) \,.\supset (*x \mid x)(w,y)$ (1)

†422 $[519 \supset.]\ (*x \mid x)(w,y) \supset *x(w,y)$ (2)

[1] In this step $\ulcorner*\zeta(\alpha, \eta) \,.\, \phi\urcorner$ plays the rôle of the ϕ of *521.

*100	$[460\,.\,1\quad 2\,.]\breve{x}(y,z) \,.\, {*x}(w,z) \,.\!\supset {*x}(w,y)$	(3)
*521	$[(y)(z)3\,.]\, {*\breve{x}}(y,w) \,.\, {*x}(w,w) \,.\!\supset {*x}(w,y)$	(4)
†511	${*x}(w,w) \equiv .\; w \in \mathrm{V}$	(5)
*100 (& D24)	$[4\,.\,5\,.]\, {*\breve{x}}(y,w) \,.\!\supset {*x}(w,y)$	
†460	$\supset \breve{}{*x}(y,w)$	(6)
†446	${*\breve{x}} \subset \breve{}{*x}\;[.\!\equiv (y)(w)6]$	(7)
†463	$[7 \equiv .]\, \breve{}{*\breve{x}} \subset {*x}$	(8)
*231	$[(x)8 \supset .]\, \breve{}{*\breve{\breve{x}}} \subset {*\breve{x}}$	(9)
†461, *224	$[9 \equiv .]\, \breve{}{*x} \subset {*\breve{x}}$	(10)
†312	$523\,[\equiv .\, 10\,.\,7]$	

†524. (x)	${*x} \mid x = x \mid {*x}$	
Proof. †513	$\dot{x} \subset x \mid {*x}$	(1)
†422	$[1 \supset .]\, x(z', w) \supset (x \mid {*x})(z', w)$	(2)
†422	$[518 \supset .]\, (x \mid {*x})(z,w) \supset {*x}(z,w)$	(3)
†487	$x(y,z) \,.\, {*x}(z,w) \,.\!\supset (x \mid {*x})(y,w)$	(4)
*100	$[3\,.\,4\,.]\, x(y,z) \,.\, x(z',w) \supset \mathrm{L}3 \,.\!\supset .\, x(z',w) \supset \mathrm{R}4$	(5)
*521	$[(y)(z)\,5\,.]\, {*x}(y, z')[.\,2] \,.\!\supset .\, x(z', w) \supset \mathrm{R}4$	(6)
*100	$[6\,.]\, {*x}(y, z') \,.\, x(z', w) \,.\!\supset \mathrm{R}4$	(7)
†488	${*x} \mid x \subset x \mid {*x}\;[.\!\equiv (y)(z')(w)7]$	(8)
†462	$[8 \equiv .]\, \breve{}({*x} \mid x) \subset \breve{}(x \mid {*x})$	(9)
†494, *224	$[9 \equiv .]\, \breve{x} \mid \breve{}{*x} \subset \breve{}{*x} \mid \breve{x}$	(10)
†523, *224	$[10 \equiv .]\, \breve{x} \mid {*\breve{x}} \subset {*\breve{x}} \mid \breve{x}$	(11)
*231	$[(x)11 \supset .]\, \breve{\breve{x}} \mid {*\breve{\breve{x}}} \subset {*\breve{\breve{x}}} \mid \breve{\breve{x}}$	(12)
†461, *224	$[12 \equiv .]\, x \mid {*x} \subset {*x} \mid x$	(13)
†312	$524\,[\equiv .\, 8\,.\,13]$	

Where x is the parent relation, we have seen that the ancestral ${*x}$ is the ancestor relation — provided that we so construe 'ancestor' as to count every person and indeed every element trivially as an ancestor of itself. The relations $x \mid {*x}$ and ${*x} \mid x$, then, are respectively the relations "parent of ancestor of" (in the above sense of 'ancestor') and "ancestor of parent of". †518 and †519 showed that these relations are included in the ancestor relation; and the above theorem †524 shows further that they are the same. This relation ${*x} \mid x$ or $x \mid {*x}$ is in fact simply the ancestor relation in the *narrower*, more usual sense; it is the ancestor relation when we construe the latter in such a way as not to reckon one among his own ancestors. It is the relation which y bears to z when y is a parent of z or a parent of a parent of z or . . ..

For any x, in general, the relation $*x \mid x$ (or $x \mid *x$) will be called the *proper ancestral* of x. The ancestral $*x$ of x is the relation of an element y to an element z such that y *is* z *or* bears x to z or bears x to something which bears x to z or . . .; on the other hand the proper ancestral $*x \mid x$ is rather the relation which y bears to z just in case y bears x to z or bears x to something which bears x to z or . . ..

The line of reasoning used in D30 was first set forth by Frege in 1879 (*Begriffschrift*, p. 60), in defining what I have called the proper ancestral. In 1884 he introduced the ordinary ancestral, defining it in precisely the manner of D30 (cf. *Grundlagen*, p. 94; also *Grundgesetze*, p. 60).

Whitehead and Russell's notation for the ancestral is $\ulcorner \zeta_* \urcorner$; but I have modified it to $\ulcorner *\zeta \urcorner$ because of a disinclination to accord parentheses to singulary operators. Whitehead and Russell's notation would have to be construed as involving implicit parentheses in the fashion $\ulcorner (\zeta_*) \urcorner$ in order to deal with ambiguous combinations such as 'ιx_*'.

Whitehead and Russell's version of the ancestral diverges from the present one (Frege's) in this minor respect: y does not bear their x_* to itself unless y bears x to something or something bears x to y (cf. *Principia*, *90.01). The practical advantage of this complication is not apparent. In any case the two versions are interdefinable in obvious fashion:

$$x_* = \hat{y}\hat{z}(*x(y, z) \;.\; (\exists w)(x(y, w) \vee x(w, y))),$$
$$*x = \hat{y}\hat{z}(x_*(y, z) \;\vee .\; y = z).$$

The proper ancestral is accorded a separate definition by Whitehead and Russell, under the notation $\ulcorner \zeta_{\text{po}} \urcorner$. (Cf. *91·01·52).

§ 40. *Functions*

INSTANCES of the notion of *function* are expressed by such locutions as 'quadruple of', 'double of', 'square of'; and the *application* of a function is expressed by completing such locutions, in the fashion 'double of x', 'square of x'. The notation of functional application is $\ulcorner \zeta`\eta \urcorner$, or more strictly $\ulcorner (\zeta`\eta) \urcorner$. Thus $z`3$ is 12 where $z$ is *quadruple of;* $z`3$ is 9 where z is *square of;* and, to take a non-numerical example, $z`V$ is ιV where z is *unit class of*.

A suitable way of formally defining the notation $\ulcorner \zeta`\eta \urcorner$ suggests itself when we reflect that a function is simply a relation; *quadruple of*, e.g., is the relation of the quadruple to the quarter. The result $z`x$ of applying a function z to an element x is the element which bears the relation z to x, symbolically $(\iota y)\, z(y,x)$. Thus

D31.[1] $\ulcorner(\zeta`\eta)\urcorner$ *for* $\ulcorner(\iota\alpha)\, \zeta(\alpha,\eta)\urcorner$.

In the usual terminology, $z`x$ is the *value* of the function $z$ for the *argument* $x$; 12, e.g., is the value of quadruple for the argument 3. More briefly, $z`x$ is spoken of as *the z of x*.

A function is, as observed, a relation. But it is a relation of a special sort, having the peculiarity that no two elements bear it to the same element.[2] No two numbers, e.g., are quadruples of the same number. Common mathematical usage allows the term 'function' a broader scope, indeed, and countenances so-called many-valued functions as well as single-valued ones; but we may conveniently waive this usage, for the inclusive notion of function which it involves is already at hand in the general notion of relation. It is only where a unique element bears x to y, symbolically

(1) $$(\exists w)(z)(z = w \,.\!\equiv x(z,y)),$$

that we can speak non-trivially of *the* x of y; where (1) fails, $x`y$ reduces trivially to Λ (cf. *197, D31).

Functionality of x, then, is conveniently explained as single-valuedness of x; i.e., as truth of (1) for all elements y to which the relation x is borne at all. But there are further relations x which, though not functions in this sense, do fulfill (1) for certain choices of y. Such relations may, by a natural extension, be spoken of as functions in a relative sense; viz., as functions with respect to this or that appropriate argument y. Functionality of x with respect to y receives its symbolic expression in (1).

The class of those elements y with respect to which x is a function, in this sense, will be called the *range of functionality* of x, symbolically $\mathfrak{r}x$. Thus $\mathfrak{r}x$ is the class of all elements y for which (1) holds.

D32. $\ulcorner\mathfrak{r}\zeta\urcorner$ *for* $\ulcorner\hat{\beta}(\exists\gamma)(\alpha)(\alpha = \gamma \,.\!\equiv \zeta(\alpha,\beta))\urcorner$.

[1] The notation and the definition are both Russell's ("Mathematical Logic," p. 253).

[2] Here I follow Peano, "Sulla definizione di funzione."

The range of functionality of the relation *author of*, e.g., is the class of uncollaborated writings; the range of functionality of the relation *member of* is the class of all unit classes of elements; and the range of functionality of the relation *included in* has Λ as sole member.

Ways of paraphrasing '$y \in \mathfrak{r}x$' are shown in the following two theorems.

†530. $(y)(x)$ $\quad y \in \mathfrak{r}x \,.\!\equiv (\exists w)(z)(z = w \,.\!\equiv x(z,y))$

Proof. †182	$w = w$	(1)
*103	$(z)(z = w \,.\!\equiv x(z,y)) \supset . [1 \equiv]\, x(w,y)$	(2)
*100 (& D24), *163	$(\exists w)([2 .]\ \mathrm{L}2) \supset . y \in \mathrm{V}$	(3)
*230 (& D32)	$y \in \mathfrak{r}x \,. \equiv . y \in \mathrm{V} \,.\ \mathrm{R}530$	
*100	$\equiv . [3 .]\ \mathrm{R}530$	

†531. $(y)(x)$ $\quad y \in \mathfrak{r}x \,.\!\equiv (z)(z = x\text{‘}y \,.\!\equiv x(z,y))$

Proof. *196 (& D31)	$(z)(z = w \,.\!\equiv x(z,y)) \supset . w = x\text{‘}y$	(1)
*223	$w = x\text{‘}y \,. \supset .\ \mathrm{L}1 \equiv \mathrm{R}531$	(2)
*100, *163	$(\exists w)(\mathrm{L}1\ [.\ 1\ .\ 2]) \supset \mathrm{R}531$	(3)
*232	$\mathrm{R}531 \supset (\exists w)\mathrm{L}1$	(4)
*100	$[3\ .\ 4\ .\ 530\ . \supset]\ 531$	

If y is in the range of x, then $x\text{‘}y$ bears x to y; and otherwise $x\text{‘}y$ is Λ.

†532. $(y)(x)$ $\quad y \in \mathfrak{r}x \,. \supset x(x\text{‘}y,\ y)$

Proof. †182	$x\text{‘}y = x\text{‘}y$	(1)
†531	$y \in \mathfrak{r}x \,. \supset \mathrm{R}531$	
*231	$\supset . [1 \equiv]\ x(x\text{‘}y,\ y)$	

†533. $(y)(x)$ $\quad y \,\tilde{\in}\, \mathfrak{r}x \,. \supset . x\text{‘}y = \Lambda$

Proof. *100	$[530 .]\ y \,\tilde{\in}\, \mathfrak{r}x \,. \supset \sim \mathrm{R}530$	
*197 (& D31)	$\supset . x\text{‘}y = \Lambda$	

From †533 it follows that $x\text{‘}y$ is Λ unless y is an element.

†534. $(y)(x)$ $\quad y \,\tilde{\in}\, \mathrm{V} \,. \supset . x\text{‘}y = \Lambda$

Proof. †190	$y \in \mathfrak{r}x \,. \supset . y \in \mathrm{V}$	(1)
*100	$[1\ .\ 533\ . \supset]\ 534$	

Thus y is an element if $x\text{‘}y$ has a member.

†535. $(z)(y)(x)$ $\quad z \in x`y \,.\supset.\, y \in V$

Proof. †192 $\quad z \,\bar{\epsilon}\, \Lambda$ (1)

*223 $\quad x`y = \Lambda \,.\supset:\, z \,\bar{\epsilon}\, x`y \,[.\equiv 1]$ (2)

*100 $\quad [2 \,.\, 534 \,.\supset]\, 535$

But $x`y$ is an element regardless of x and y.

†536. $(y)(x)$ $\quad x`y \in V$

Proof. *223 $\quad x`y = \Lambda \,.\supset.\, 536 \,[\equiv 211]$ (1)

*100 (& D24) $\quad [532 \,.\, 533 \,.\, 1 \,.\supset]\, 536$

The paraphrases of '$y \in \mathfrak{r}x$' provided by †530 and †531 can now be supplemented with a third.

†537. $(y)(x)$ $\quad y \in \mathfrak{r}x \,.\equiv.\, \iota(x`y) = x``\iota y$

Proof.

†345, *123 (& D10) $\quad R537 \equiv (z)([536\,.]z = x`y \,.\equiv.\, z \in x``\iota y)$

†479, *123 $\quad \equiv R531$ (1)

*100 $\quad [1 \,.\, 531 \,.\supset]\, 537$

The following theorem exhibits a certain analogy to †495.

†538. $(z)(y)(x)$ $\quad z \in \mathfrak{r}y \,.\supset.\, (x \mid y)`z = x`(y`z)$

Proof. †537 $\quad z \in \mathfrak{r}y \,.\supset.\, y``\iota z = \iota(y`z)$

*223 $\quad \supset:\, w \in x``(y``\iota z) \,.\equiv.\, w \in x``\iota(y`z)$

†495, *224 $\quad \supset:\, w \in (x \mid y)``\iota z \,.\equiv.\, w \in x``\iota(y`z)$

†479, *123 $\quad \supset.\, (x \mid y)(w,z) \equiv x(w, y`z)$ (1)

*159 (& D4) $\quad [(w)1 \equiv:]\, z \in \mathfrak{r}y \,.\supset\, (w)R1$

*186 (& D31) $\quad \supset R538$

If x is the father relation, y is the son relation, and z is Eve, then $(x \mid y)`z$ is Adam whereas $x`(y`z) = x`\Lambda = \Lambda$. Hence the need of the antecedent '$z \in \mathfrak{r}y$' in the above theorem.

The next concerns the ranges of functionality of relative products.

†539. $(y)(x)$ $\quad \mathfrak{r}x = V \,.\supset.\, \mathfrak{r}y \subset \mathfrak{r}(x \mid y)$

Proof. †536 $\quad y`z \in V$ (1)

*223 $\quad \mathfrak{r}x = V \,.\supset:\, y`z \in \mathfrak{r}x \,[.\equiv 1]$

†537 $\quad \supset.\, \iota(x`(y`z)) = x``\iota(y`z)$ (2)

†537 $\quad z \in \mathfrak{r}y \,.\supset.\, \iota(y`z) = y``\iota z$

*223 $\supset :.\, [2 \equiv :]\, \mathfrak{r}x = \mathrm{V} \,.\!\supset.\, \iota(x\text{‘}(y\text{‘}z)) = x\text{“}(y\text{“}\iota z)$

†495, *224 $\supset :\, \mathfrak{r}x = \mathrm{V} \,.\!\supset.\, \iota(x\text{‘}(y\text{‘}z)) = (x \mid y)\text{“}\iota z$ (3)

*223 R538 $\supset ::\, z \epsilon \mathfrak{r}y \,.\!\supset :\, \mathfrak{r}x = \mathrm{V} \,.\!\supset.\, \iota((x \mid y)\text{‘}z) = (x \mid y)\text{“}\iota z[.\!:\equiv 3]$

†537, *123 $\supset :.\, z \epsilon \mathfrak{r}y \,.\!\supset :\, \mathfrak{r}x = \mathrm{V} \,.\!\supset.\, z \epsilon \mathfrak{r}(x \mid y)$ (4)

*100, *117 $\mathfrak{r}x = \mathrm{V} \,.\!\supset (z)(z \epsilon \mathfrak{r}y\, [.\, 4\, .\, 538] \,.\!\supset.\, z \epsilon \mathfrak{r}(x \mid y))$

(q.e.d.; cf. D21)

Besides functions of the sort here considered, which apply to arguments one at a time, there are those which apply to arguments two or more at a time. An example which applies to arguments two at a time is the power function, whose value for arguments x and y is x^y. A non-numerical example is logical sum, whose value for arguments x and y is $x \cup y$. Just as a function of one argument is a dyadic relation, so a function of two arguments is a triadic relation; the power function, e.g., is the triadic relation which holds among any numbers z, x, and y (in that order) when $z = x^y$. A function of three arguments, similarly, is a tetradic relation; and so on. In terms of the notions introduced at the end of § 36, the definition D31 of application is readily extended to functions of two or more arguments.

$$\ulcorner(\zeta\text{‘}(\eta, \theta))\urcorner \quad \textit{for} \quad \ulcorner(\imath\alpha)\, \zeta(\alpha, \eta, \theta)\urcorner,$$
$$\ulcorner(\zeta\text{‘}(\eta, \theta, \theta'))\urcorner \quad \textit{for} \quad \ulcorner(\imath\alpha)\, \zeta(\alpha, \eta, \theta, \theta')\urcorner,$$

etc. The notion of range admits of extension in similar fashion, in terms of the notions introduced at the end of § 37. A range of n-argument functionality, in general, is itself an n-adic relation.

$$\ulcorner\mathfrak{r}_2\, \zeta\urcorner \quad \textit{for} \quad \ulcorner\hat{\beta}\hat{\gamma}(\exists\delta)(\alpha)(\alpha = \delta \,.\!\equiv \zeta(\alpha, \beta, \gamma))\urcorner,$$
$$\ulcorner\mathfrak{r}_3\, \zeta\urcorner \quad \textit{for} \quad \ulcorner\hat{\beta}\hat{\gamma}\hat{\delta}(\exists\alpha')(\alpha)(\alpha = \alpha' \,.\!\equiv \zeta(\alpha, \beta, \gamma, \delta))\urcorner,$$

etc.

§ 41. *Abstraction of Functions*

THE FUNCTION *quadruple of* may be thought of as abstracted from the name matrix '$4 \times x$' in somewhat the same way in which the class $\hat{x}(4 < x)$ is abstracted from the statement matrix '$4 < x$'. The class $\hat{x}(4 < x)$ has, we recall, this characteristic feature: an element x is a member of the class if and only if $4 < x$. The parallel feature of the function *quadruple of* is this: application of the function to an element x yields the element $4 \times x$.

The notation of functional abstraction is $\ulcorner\lambda_\alpha \zeta\urcorner$. Thus *quadruple of* is $\lambda_x(4 \times x)$; *double of* is $\lambda_x(2 \times x)$; *square of* is $\lambda_x(x^2)$; *unit class of* is $\lambda_x \iota x$. The appropriate formal definition of $\ulcorner\lambda_\alpha \zeta\urcorner$ suggests itself when we reflect, e.g., that *quadruple of* is simply the relation of quadruple to quarter, and *square of* is the relation of square to root; i.e.,

$$\lambda_x(4 \times x) = \hat{y}\hat{x}(y = 4 \times x),$$
$$\lambda_x(x^2) = \hat{y}\hat{x}(y = x^2).$$

Thus

D33. $\ulcorner\lambda_\alpha \zeta\urcorner$ *for* $\ulcorner\hat{\beta}\hat{\alpha}(\beta = \zeta)\urcorner$.

In view of the general convention governing the choice of odd variables in definitions (cf. § 24), the β of D33 must be distinct from α and the variables of ζ; on the other hand α may, and in non-trivial cases will, occur in ζ.

Mathematicians tend in practice to confuse a name matrix ζ with the name $\ulcorner\lambda_\alpha \zeta\urcorner$ of the corresponding function. Such expressions as '$4 + x$', 'x^2', '$x^2 + 2x - 1$', etc., though commonly treated as if they designated functions, actually designate nothing; they are not names at all, because of the free variable (cf. § 27). When their free variables are supplanted by numerals, moreover, these expressions come to designate not functions but numbers. The functions which '$4 \times x$', 'x^2', and '$x^2 + 2x - 1$' suggest are properly designated rather by the functional abstracts '$\lambda_x(4 \times x)$', '$\lambda_x(x^2)$, and '$\lambda_x(x^2 + 2x - 1)$'; these expressions are straightforward names, devoid of free variables.

Note further that the distinction in meaning between ζ and $\ulcorner\lambda_\alpha \zeta\urcorner$ survives even in the case where ζ itself *is* a name, and thus devoid of free occurrences of α. It is in this case that the so-called "constant functions" appear; $\lambda_x 5$, e.g., is the function whose value is uniformly 5 for all arguments. The names '5' and '$\lambda_x 5$' are altogether different in meaning; whereas 5 is a number, the "constant function" $\lambda_x 5$ is rather the relation which 5 bears to all elements (cf. *540).

Abstraction, defined in D33 for functions of one argument, is extensible to the case of many arguments in obvious fashion:

$$\ulcorner\lambda_{\alpha\beta}\,\zeta\urcorner \quad \textit{for} \quad \ulcorner\hat{\gamma}\hat{\alpha}\hat{\beta}(\gamma = \zeta)\urcorner,$$
$$\ulcorner\lambda_{\alpha\beta\gamma}\,\zeta\urcorner \quad \textit{for} \quad \ulcorner\hat{\delta}\hat{\alpha}\hat{\beta}\hat{\gamma}(\delta = \zeta)\urcorner,$$

etc. (cf. end of § 40). $\lambda_{xy}(x^y)$, e.g., is the power function; and $\lambda_{xy}(x^y)\,{}^{\backprime}\,(5,3)$ is 125. In what follows, however, attention will be limited to the one-argument case.

The following metatheorem deals with the functional abstract $\ulcorner\lambda_\alpha\,\zeta\urcorner$ in the context $\ulcorner\lambda_\alpha\zeta\,(\eta,\alpha)\urcorner$ of relational predication (D24). Note that the occurrences of α in $\ulcorner\lambda_\alpha\,\zeta\urcorner$ are bound, whereas the occurrence of α which terminates $\ulcorner\lambda_\alpha\zeta\,(\eta,\alpha)\urcorner$ is free.

***540.** $\vdash \ulcorner\lambda_\alpha\zeta\,(\eta,\alpha) \equiv.\ \alpha, \zeta \in \mathrm{V}\,.\,\eta = \zeta\urcorner.$

Proof. *434 (& D33) $\vdash \ulcorner\lambda_\alpha\zeta\,(\eta,\alpha) \equiv.\ \alpha, \eta \in \mathrm{V}\,.\,\eta = \zeta\urcorner$
*228 $\equiv.\ \alpha, \zeta \in \mathrm{V}\,.\,\eta = \zeta\urcorner.$

As explained, $\lambda_x(4 \times x)$ is supposed to be the function whose value, for any element x as argument, is $4 \times x$. In general, $\lambda_x(\ldots x \ldots)$ is supposed to be the function whose value, for any element x as argument, is $\ldots x \ldots$ (granted elementhood in turn of $\ldots x \ldots$). This is provided by the following metatheorem.

***541.**[1] $\vdash \ulcorner\alpha, \zeta \in \mathrm{V}\,.\!\supset\!.\ \lambda_\alpha\zeta\,{}^{\backprime}\alpha = \zeta\urcorner.$

Proof. Let β be new.
*100, *117 $\vdash \ulcorner\alpha, \zeta \in \mathrm{V}\,.\!\supset (\beta)(\beta = \zeta\,.\!\equiv\!.\ \alpha, \zeta \in \mathrm{V}\,.\,\beta = \zeta)\urcorner$
*540, *123 $\supset (\beta)(\beta = \zeta\,.\!\equiv \lambda_\alpha\zeta\,(\beta,\alpha))\urcorner$
*196 (& D31) $\supset.\ \lambda_\alpha\zeta\,{}^{\backprime}\alpha = \zeta\urcorner.$

Just as the metatheorems *230 and *433 on the abstraction of classes and relations were generalized in *235 and *434, the metatheorem *541 on functional abstraction is generalized in the following.

[1] So far as parentheses are concerned, the prefix $\ulcorner\lambda\alpha\urcorner$ as a whole is treated as if it were a singulary operator; thus it is that no parentheses are entered in the left side of D33. Accordingly the $\ulcorner\lambda\alpha\zeta{}^{\backprime}\alpha\urcorner$ of *541 is to be read as composed of $\ulcorner\lambda\alpha\zeta\urcorner$ and α, not of $\ulcorner\lambda\alpha\urcorner$ and $\ulcorner\zeta{}^{\backprime}\alpha\urcorner$; cf. second footnote to § 35. The functional abstract of $\ulcorner\zeta{}^{\backprime}\alpha\urcorner$ with respect to α would be rendered rather $\ulcorner\lambda\alpha(\zeta{}^{\backprime}\alpha)\urcorner$. The latter can, as a matter of fact, be shown to obey a law somewhat like *541; namely, if α is not free in ζ then $\vdash \ulcorner\mathfrak{r}\zeta = \mathrm{V}\,.\!\supset\!.\ \lambda\alpha(\zeta{}^{\backprime}\alpha) = \dot{\zeta}\urcorner.$

***542.** *If θ is like ζ except for containing free occurrences of η in place of all free occurrences of α, then*

$$\vdash \ulcorner \eta, \theta \in \mathrm{V} \,.\!\supset .\, \lambda_\alpha \zeta \,`\, \eta = \theta \urcorner.$$

Proof. *231 (& hp) $\vdash \ulcorner [(\alpha)541 \supset]\ 542 \urcorner$.

The next metatheorem explains the notion of range as applied to a functional abstract.

***543.** $\vdash \ulcorner \mathfrak{r}\, \lambda_\alpha \zeta = \hat{\alpha}(\zeta \in \mathrm{V}) \urcorner$.

Proof. Let β be new.

†189	$\vdash \ulcorner \mathfrak{r}\, \lambda_\alpha \zeta = \hat{\alpha}(\alpha \in \mathfrak{r}\, \lambda_\alpha \zeta) \urcorner$
*531, *188	$= \hat{\alpha}(\beta)(\lambda_\alpha \zeta \,`\, \alpha = \beta \,.\!\equiv \lambda_\alpha \zeta\ (\beta, \alpha)) \urcorner$
*540, *188	$= \hat{\alpha}(\beta)(\lambda_\alpha \zeta \,`\, \alpha = \beta \,.\!\equiv.\, \alpha, \zeta \in \mathrm{V} \,.\, \beta = \zeta) \urcorner$
*100, *188	$= \hat{\alpha}(\beta)(\lambda_\alpha \zeta \,`\, \alpha = \beta \,.\!\supset.\, \alpha, \zeta \in \mathrm{V} \,.\, \beta = \zeta :.$ $\beta = \zeta \,.\!\supset\!: \alpha, \zeta \in \mathrm{V} \,.\!\supset.\, \lambda_\alpha \zeta \,`\, \alpha = \beta) \urcorner$
*140, *188	$= \hat{\alpha}((\beta)(\lambda_\alpha \zeta \,`\, \alpha = \beta \,.\!\supset.\, \alpha, \zeta \in \mathrm{V} \,.\, \beta = \zeta) \,.$ $(\beta)(\beta = \zeta \,.\!\supset\!: \alpha, \zeta \in \mathrm{V} \,.\!\supset.\, \lambda_\alpha \zeta \,`\, \alpha = \beta)) \urcorner$
*234a, *188	$= \hat{\alpha}(\alpha, \zeta \in \mathrm{V} \,.\, \lambda_\alpha \zeta \,`\, \alpha = \zeta\ [.\ 541]) \urcorner$
*100, *188	$= \hat{\alpha}(\alpha, \zeta \in \mathrm{V}\ [.\ 541]) \urcorner$
*246	$= \hat{\alpha}(\zeta \in \mathrm{V}) \urcorner$.

The next explains $\ulcorner \lambda_\alpha \zeta \,`\, \eta \urcorner$ in the cases not covered by *542.

***544.** *If θ is like ζ except for containing free occurrences of η in place of all free occurrences of α, then*

$$\vdash \ulcorner \sim (\eta, \theta \in \mathrm{V}) \supset .\, \lambda_\alpha \zeta \,`\, \eta = \Lambda \urcorner.$$

Proof.

*235 (& hp), *123	$\vdash \ulcorner \sim (\eta, \theta \in \mathrm{V}) \supset .\, \eta \mathbin{\bar{\in}} \hat{\alpha}(\zeta \in \mathrm{V}) \urcorner$
*543, *224	$\supset .\, \eta \mathbin{\bar{\in}} \mathfrak{r}\, \lambda_\alpha \zeta \urcorner$
†533	$\supset .\, \lambda_\alpha \zeta \,`\, \eta = \Lambda \urcorner$.

Note that the antecedent in *544 is not $\ulcorner \eta \mathbin{\bar{\in}} \mathrm{V} \,.\, \theta \mathbin{\bar{\in}} \mathrm{V} \urcorner$, but the much weaker condition $\ulcorner \sim (\eta \in \mathrm{V} \,.\, \theta \in \mathrm{V}) \urcorner$, i.e. $\ulcorner \eta \mathbin{\bar{\in}} \mathrm{V} \,.\!\vee\!.\, \theta \mathbin{\bar{\in}} \mathrm{V} \urcorner$. Thus *544 goes farther than †534.

Statements of the form $\ulcorner \lambda_\alpha \zeta\ (\lambda_\alpha \zeta \,`\, \eta, \eta) \urcorner$ are not all true. If $\eta$ designates a non-element, or if the result of putting $\eta$ for $\alpha$ in $\zeta$ designates a non-element, then $\ulcorner \lambda_\alpha \zeta\ (\lambda_\alpha \zeta \,`\, \eta, \eta) \urcorner$ will be false; nor will there be *any* term θ for which $\ulcorner \lambda_\alpha \zeta\ (\theta, \eta) \urcorner$ *is* true. If on the other hand there is such a term θ at all, we can infer $\ulcorner \lambda_\alpha \zeta \,`\, \eta = \theta \urcorner$; this is shown in the following metatheorem.

***545.** $\vdash \ulcorner \lambda_{\alpha}\zeta\ (\theta, \eta) \supset . \lambda_{\alpha}\zeta\ `\eta = \theta \urcorner$.

Proof. Let β be new.

$$
\begin{array}{lll}
{*}223 & \vdash \ulcorner \beta = \zeta . \supset :. \alpha, \zeta \in V . \supset . \lambda_{\alpha}\zeta\ `\alpha = \beta\, [:\equiv 541] \urcorner. & (1) \\
{*}540 & \vdash \ulcorner \lambda_{\alpha}\zeta\ (\beta, \alpha) \supset . \alpha, \zeta \in V . \beta = \zeta \urcorner & \\
{*}100 & \qquad \supset : [1 \supset .]\, \lambda_{\alpha}\zeta\ `\alpha = \beta \urcorner. & (2) \\
{*}233 & \vdash \ulcorner [(\alpha)(\beta) 2 \supset]\, 545 \urcorner. &
\end{array}
$$

The value of the unit-class function $\lambda_x \iota x$ for an argument y proves to be ιy regardless of whether y is an element.

†546. $(y) \qquad \lambda_x \iota x\ `y = \iota y$

$$
\begin{array}{llll}
\textit{Proof.} & \dagger 359 & \iota y \in V & (1) \\
 & {*}542 & y \in V\, [.1] . \supset 546 & (2) \\
 & \dagger 534 & y \bar{\in} V . \supset . \lambda_x \iota x\ `y = \Lambda & (3) \\
 & \dagger 343 & y \bar{\in} V . \supset . \iota y = \Lambda & \\
 & {*}223 & \qquad \supset :. y \bar{\in} V . \supset 546\, [:\equiv 3] & (4) \\
 & {*}100 & [2 . 4 . \supset]\, 546 &
\end{array}
$$

The notion of functional abstraction goes back to Frege (*Grundgesetze*, vol. 1, pp. 14f, 54f). The present notation is Church's. Both Frege and Church construe functional abstraction more broadly, however, to include class abstraction as a special case. Correspondingly relational abstraction becomes, for those authors, a case of the abstraction of functions of many arguments. Special features of the groundwork of Frege's logic and Church's make this assimilation a natural one; and it would be equally natural in the logic based on inclusion and abstraction (cf. § 23).

§ 42. *Identity and Membership as Relations*

THE IDENTITY function I is defined as follows.[1]

D34. 'I' *for* '$\lambda_x x$'.

It is the function whose value, for any element as argument, is that element itself.

†550. $(x) \qquad x \in V . \equiv . I`x = x$

Proof. *541 (& D34) $\qquad x, x \in V . \supset . I`x = x \qquad$ (1)

[1] See end of § 38.

†536	$I\text{‘}x \,\epsilon\, \mathrm{V}$	(2)
*223	$I\text{‘}x = x \,.\!\supset\!: [2 \equiv .]\, x \,\epsilon\, \mathrm{V}$	(3)
*100	$[1 \,.\, 3 \,.\!\supset]550$	

Its range embraces all elements.

†551. $\mathrm{r}I = \mathrm{V}$

Proof. *543 (& D34) $\mathrm{r}I = \hat{x}(x \,\epsilon\, \mathrm{V})$
†189 $= \mathrm{V}$

Like all functions, I is a relation. In view of D34 and D33, indeed, it is the identity relation $\hat{x}\hat{y}(x = y)$. Note, however, that '$I(x,y)$' is not equivalent to '$x = y$'. Since only elements can enter into a relation, '$I(x,y)$' holds only where x and y are the same element; on the other hand '$x = y$' holds wherever x and y are the same entity, whether an element or a non-element.

†552. $(y)(x)$ $I(x,y) \equiv . \, x \,\epsilon\, \mathrm{V} \,.\, x = y$

Proof. *434 (& D34, 33) $I(x,y) \equiv . \, x, y \,\epsilon\, \mathrm{V} \,.\, x = y$
*228 $\equiv . \, x, x \,\epsilon\, \mathrm{V} \,.\, x = y$
*100 $\equiv . \, x \,\epsilon\, \mathrm{V} \,.\, x = y$

†553. $(y)(x)$ $I(x,y) \equiv . \, y \,\epsilon\, \mathrm{V} \,.\, x = y$

Proof. *228, *123 $[552 \equiv]\ 553$

†554. $(y)(x)$ $I(x,y) \equiv . \, x \,\epsilon\, \iota y$

Proof. †345, *123 $554\ [\equiv 553]$

†555. (x) $I(x,x) \equiv . \, x \,\epsilon\, \mathrm{V}$

Proof. †552 $I(x,x) \equiv . \, x \,\epsilon\, \mathrm{V}\, [. \ 182]$

To say that x includes I is to say that every element bears x to itself.

†556. (x) $I \subset x \,.\!\equiv (y)(y \,\epsilon\, \mathrm{V} \,.\!\supset x(y,y))$

Proof. †446 $I \subset x \,.\!\equiv (y)(z)(I(y,z) \supset x(y,z))$
†552, *123 $\equiv (y)(z)(y \,\epsilon\, \mathrm{V} \,.\, y = z \,.\!\supset x(y,z))$
*100, *123 $\equiv (y)(z)(y = z \,.\!\supset\!: y \,\epsilon\, \mathrm{V} \,.\!\supset x(y,z))$
*234a, *123 $\equiv$ R556

I is its own converse.

†557. $I = \breve{I}$

Proof. *437 $I = \dot{I}$
†552, *188 (& D26) $= \hat{y}\hat{z}(y \in V . y = z)$
†553, *188 (& D27) $= \check{I}$

Imaging by I has no effect.

†558. (x) $I``x = x$

Proof. †553, *188 (& D28) $I``x = \hat{y}(\exists z)(y = z . z \in V . z \in x)$
*234b, *188 (& D18) $= V \cap x$
†296 $= x$

Relative multiplication by I leaves a relation unchanged.

†559. (x) $I \mid x = \dot{x}$

Proof. †553, *188 (& D29) $I \mid x = \hat{y}\hat{w}(\exists z)(y = z . z \in V . x(z,w))$
*100 (& D24), *188 $= \hat{y}\hat{w}(\exists z)(y = z . x(z,w))$
*234b, *188 (& D26) $= \dot{x}$

†560. (x) $x \mid I = \dot{x}$ *Proof* similar.

I is included in every ancestral.

†561. (x) $I \subset *x$

Proof. †511 $y \in V . \supset *x(y,y)$ (1)
†556 561 [≡ (y)1]

†562. (x) $I \cup \dot{x} \subset *x$

Proof. †332 562 [≡. 561 . 514]

The next theorem shows that I added to the proper ancestral yields the ancestral.

†563. (x) $I \cup (x \mid *x) = *x$

Proof. †332 L563 $\subset *x$ [.≡. 561 . 518] (1)
†422 [1 ⊃.] L563 $(z,w) \supset *x(z,w)$ (2)
†487 $x(y,z) . *x(z,w) . \supset (x \mid *x)(y,w)$ (3)
*100 [2 . 3 .] $x(y,z)$. L563 $(z,w) . \supset . I(y,w) \vee (x \mid *x)(y,w)$
†424 $\supset$ L563 (y,w) (4)
*521 [(y)(z)4 .] $*x(y,w)$. L563 $(w,w) . \supset$ L563 (y,w) (5)
†424 L563 $(w,w) \equiv . I(w,w) \vee (x \mid *x)(w,w)$
†555, *123 $\equiv : w \in V . \vee (x \mid *x)(w,w)$ (6)
*100 (& D24) [5 . 6 .] $*x(y,w) . \supset$ L563 (y,w) (7)
†446 $*x \subset$ L563 [.≡ (y)(w)7] (8)
†312 563 [≡. 1 . 8]

The terms 'transitive' and 'reflexive', applied hitherto to modes of statement composition (§ 11) and predicates (§ 25), may by extension be applied also to *relations* in obvious fashion; historically this is, indeed, their original application. A relation w is transitive if

$$(x)(y)(z) \qquad w(x, y) \,.\, w(y, z) \,.\supset w(x, z);$$

or, equivalently (by †488), if $w \mid w \subset w$. Thus I is transitive, by †559. A relation w is totally reflexive (see note to § 25) if every element bears w to itself; or, equivalently (by †556), if $I \subset w$. Thus I is totally reflexive. A relation w is reflexive, more generally, if

$$(x)(y) \qquad w(x, y) \vee w(y, x) \,.\supset w(x, x).$$

The term 'commutative' carries over in similar fashion; but the term ordinarily used in connection with relations is rather 'symmetrical.' A relation w is symmetrical if

$$(x)(y) \qquad w(x, y) \equiv w(y, x);$$

or, equivalently, if $w = \breve{w}$. †557 affirms the symmetry of I.

We have attributed transitivity, reflexivity, etc., on the one hand to the connective sign '=', which is a two-place predicate, and on the other hand to the corresponding relation $\hat{x}\hat{y}(x = y)$ or I. By a similar transfer the idempotence, commutativity, etc., which have been attributed to the term connectives '$\cap$' and '$\cup$' (§ 33) might be attributed likewise to the corresponding functions $\lambda_{xy}(x \cap y)$ and $\lambda_{xy}(x \cup y)$. In general, a function w of two arguments is idempotent, commutative, or associative according as

$$w`(x, x) = x,$$
$$w`(x, y) = w`(y, x),$$
or
$$w`(w`(x, y), z) = w`(x, w`(y, z))$$

for all elements x, y, and z; and correspondingly for distributivity.

I is, we have seen, the identity relation $\hat{x}\hat{y}(x = y)$. We turn now to $\mathfrak{E}$, which is the *membership relation* $\hat{x}\hat{y}(x \in y)$.

D35. '$\mathfrak{E}$' *for* '$\hat{x}\hat{y}(x \in y)$'.

'$\mathfrak{E}(x,y)$' diverges from '$x \in y$' in the same way in which '$I(x,y)$' was seen to diverge from '$x = y$'. Since only elements can enter into a relation, '$\mathfrak{E}(x,y)$' holds only where y is an element and $x \in y$.

†564. $(y)(x) \qquad \mathfrak{E}(x,y) \equiv . \, x \in y \,.\, y \in V$

Proof. *433 (& D35) $\quad \mathfrak{E}(x,y) \equiv . \, x \in y \,.\, x, y \in V$

†191, *123 $\quad \equiv . \, x \in y \,.\, y \in V$

But the elementhood clause is suppressible in the reflexive case.

†565. (x) $\mathfrak{E}(x,x) \equiv . \, x \in x$

Proof. †564 $\mathfrak{E}(x,x) \equiv . \, x \in x \, . \, x \in V$

†191 $\equiv . \, x \in x$

Though '$x \in \iota y$' and '$x = y$' are not equivalent (cf. †345), '$\mathfrak{E}(x, \iota y)$' and '$I(x,y)$' are.

†566. $(y)(x)$ $\mathfrak{E}(x, \iota y) \equiv I(x,y)$

Proof. †359 $\iota y \in V$ (1)

†564 $\mathfrak{E}(x, \iota y) \equiv . \, x \in \iota y$ [.1]

†554 $\equiv I(x,y)$

$\mathfrak{E}``x$ is the class of all members of members of x.

†567. (x) $\mathfrak{E}``x = \hat{y}(\exists z)(y \in z \, . \, z \in x)$

Proof. †564, *188 (& D28) $\mathfrak{E}``x = \hat{y}(\exists z)(y \in z \, . \, z \in V \, . \, z \in x)$

†191, *188 $= \text{R567}$

Where x is the class of professional baseball clubs, e.g., $\mathfrak{E}``x$ is the class of professional baseball players. If lines and planes are conceived as classes of points, and $x$ is the class of all lines in a given plane, then $\mathfrak{E}``x$ is the plane itself.

CHAPTER SIX

NUMBER

§43. *Zero, One, Successor*

TO SAY that the Apostles are pious is to attribute a property to each man among the Apostles; and to say that we are unfortunate is to attribute a property to each individual among us. To say that the Apostles are twelve, on the other hand, is to attribute a property rather to the class of Apostles; and to say that we are seven is to attribute a property to us as a class. These properties twelve and seven, symbolically 12 and 7, are properties of classes; or, in keeping with our custom of treating properties as classes (cf. § 22), they may be construed as classes of classes. 7 is the class of all seven-member classes; 12 is the class of all dozens. In order to belong to 7 or 12 or anything else, a class must of course be an element; but this is no restriction, for we know that every class of seven or twelve or a million members is an element (cf. end of § 35).

What has been said of 7 and 12 holds equally of the rest of the so-called *natural numbers*, viz. 0, 1, 2, 3, etc. Thus 3 is the class of all three-member classes; 2 is the class of all two-member classes; 1 is the class of all one-member classes; and 0 is the class of *the* no-member class, i.e.,

D36. '0' *for* '$\iota\Lambda$'.

Given any natural number, say 7, the next (viz. 8) is readily defined; for a class belonging to 8 (an eight-member class) is characterized by the fact that it comes to belong to 7 when a member is dropped. A class x belongs to 8 if and only if the removal of a member y from x leaves a class ($x \cap \overline{\iota y}$; cf. § 35) which belongs to 7.

$$8 = \hat{x}(\exists y)(y \in x \,.\, x \cap \overline{\iota y} \in 7).$$

In general, given any natural number z, the next is definable as

(1) $$\hat{x}(\exists y)(y \in x \,.\, x \cap \overline{\iota y} \in z).$$

In terms of 0, accordingly, which is already at hand (D36), we can define 1 as follows:

D37. '1' *for* '$\hat{x}(\exists y)(y \in x \,.\, x \cap \overline{\iota y} \in 0)$'.

In terms of 1, in turn, we can define 2 in similar fashion:

D38. '2' *for* '$\hat{x}(\exists y)(y \in x \,.\, x \cap \overline{\iota y} \in 1)$'.

Similarly we can define 3, 4, and so on, as far as we like.

Whenever z is an element, (1) designates an element in turn.

†610. (z) $z \in V \,.\!\supset \hat{x}(\exists y)(y \in x \,.\, x \cap \overline{\iota y} \in z) \in V$

Proof. Since '$(\exists y)(y \in x \,.\, (\exists w)(w \in z \,.\, (x')(x' \in w \,.\!\equiv.\, x' \in x \,.\, \sim (y')(y' \in x' \,.\!\equiv.\, y' \in y))))$' is stratified,[1]

*200 $z \in V \,.\!\supset\!.\, \hat{x}(\exists y)(y \in V \,.\, y \in x \,.\, (\exists w)(w \in V \,.\, w \in z \,.\, (x')(x' \in V \,.\!\supset\!:\, x' \in w \,.\!\equiv.\, x' \in x \,.\, \sim(y')(y' \in V \,.\!\supset\!:\, y' \in x' \,.\!\equiv.\, y' \in y))))) \in V$

*100, *123 $\supset\!.\, \hat{x}(\exists y)(y \in V \,.\, y \in x \,.(\exists w)((x')(x' \in V \,.\, x' \in w \,.\!\equiv.\, x' \in V \,.\, x' \in x \,.\, \sim(y')(y' \in V \,.\, y' \in x' \,.\!\equiv.\, y' \in V .\, y' \in y)) \,.\, w \in V \,.\, w \in z)) \in V$

†191, *123 (& D10) $\supset\!.\, \hat{x}(\exists y)(y \in x \,.\, (\exists w)((x')(x' \in w \,.\!\equiv.\, x' \in x \,.\, x' \neq y) \,.\, w \in z)) \in V$

†362, *123 (D10, 12) $\supset\!.\, \hat{x}(\exists y)(y \in x \,.\, x \cap \overline{\iota y} \in z) \in V$

It is now apparent that 0, 1, 2, etc. are elements.

†611. $0 \in V$ *Proof:* †359 (& D36).

†612. $1 \in V$ *Proof.* †610 (& D37) [611 ⊃]612

†613. $2 \in V$ *Proof.* †610 (& D38) [612 ⊃]613

We can extend this series of theorems as far as we like. Each natural number is an element.

The function whose application to any element z yields the class expressed by (1), above, will be called the *successor* function, S.

D39.[2] 'S' *for* '$\lambda_z \hat{x}(\exists y)(y \in x \,.\, x \cap \overline{\iota y} \in z)$'.

[1] Note that the use of numerals in showing stratification does not involve any circular presupposition of the arithmetic which we are beginning to construct; for the numerals so used do not enter into the formulæ themselves, but belong rather to syntax (cf. Ch. VII) along with the words 'assignment', 'matrix', etc. The case is similar with numerical subscripts on Greek letters. Any sort of counters would indeed serve these purposes; the numerical character is not essential.

[2] This formulation of the successor function is due essentially to Frege, who is likewise responsible for the identification of a natural number with the class of all classes having that number of members. See *Grundlagen*, pp. 79f, 89f.

The successor of any element z is the class expressed in (1).

†614. (z) $z \in V .\supset. S`z = \hat{x}(\exists y)(y \in x . x \cap \overline{\iota y} \in z)$

Proof. *100 [610 .] $z \in V .\supset. z \in V$. R610
*541 (& D39) $\supset$ R614

In view then of the elementhood of 0, 1, etc., we have the following theorems.

†615. $1 = S`0$ *Proof.* †614 (& D37) [611 $\supset$] 615

†616. $2 = S`1$ *Proof.* †614 (& D38) [612 $\supset$] 616

We can extend this series of theorems as far as we like.

Though the terminology 'successor' suggests interest in $S`z$ only where z is a natural number, still the range of functionality of S embraces not merely the natural numbers but all elements.

†617. $rS = V$

Proof. *249 $\hat{z}$ R610 $= V$ [.$\equiv$ (z)610] (1)
*543 (& D39), *224 617 [$\equiv$ 1]

x belongs to the successor of an element z just in case x deprived of one of its members belongs to z.

†618. $(z)(x)$ $z \in V .\supset: x \in S`z .\equiv (\exists y)(y \in x . x \cap \overline{\iota y} \in z)$

Proof. †190 $x \cap \overline{\iota y} \in z .\supset. x \cap \overline{\iota y} \in V$
†363 $\supset. x \in V$ (1)
*100, *163 $(\exists y)$([1 .] $y \in x . x \cap \overline{\iota y} \in z) \supset. x \in V$ (2)
*230 $x \in \hat{x}$ L2 .$\equiv$. $x \in V$. L2
*100 $\equiv$. L2 [. 2] (3)
†614 $z \in V .\supset$ R614
*223 $\supset$. R618 [$\equiv$ 3]

Two corollaries follow.

†619. $(z)(x)$ $x \in S`z .\supset (\exists y)(y \in x . x \cap \overline{\iota y} \in z)$.

Proof. †535 $x \in S`z .\supset. z \in V$ (1)
*100 [1 . 618 .$\supset$] 619

†620. $(z)(y)(x)$ $y \in x . x \cap \overline{\iota y} \in z . z \in V .\supset. x \in S`z$

Proof. *135 $y \in x . x \cap \overline{\iota y} \in z .\supset$ RR618 (1)
*100 [1 . 618 .$\supset$] 620

The null class is its own successor.

†621. $\mathrm{S}'\Lambda = \Lambda$

$$
\begin{array}{lll}
\textit{Proof.}\ \dagger 192 & x \cap \overline{\iota y}\ \bar{\epsilon}\ \Lambda & (1) \\
\quad *100 & [1\ \supset]\ \sim (y \,\epsilon\, x \,.\, x \cap \overline{\iota y} \,\epsilon\, \Lambda) & (2) \\
\quad \dagger 614\ (\&\ \mathrm{D}8) & [211\ \supset .]\ \mathrm{S}'\Lambda = \hat{x} \sim (y)2 & (3) \\
\quad *121\ (\&\ \mathrm{D}16) & [(x)((y)2 \equiv 182) \supset .\ 3 \equiv]\ 621\ ^{1} &
\end{array}
$$

0 has one and only one member, viz. Λ.

†622. $\Lambda \,\epsilon\, 0$

Proof. †344 (& D36) $[211 \equiv]\ 622$

†623. $(x) \quad x \,\epsilon\, 0 \,.\!\equiv .\, x = \Lambda$

Proof. †345 (& D36) $x \,\epsilon\, 0 \,.\!\equiv .\, [211\ .]\ x = \Lambda$

0 is not a successor.

†624. $(x) \quad 0 \neq \mathrm{S}'x$

$$
\begin{array}{lll}
\textit{Proof.}\ \dagger 192 & y\ \bar{\epsilon}\ \Lambda & (1) \\
\quad *223 & 0 = \mathrm{S}'x \,.\supset :[622 \equiv .]\ \Lambda \,\epsilon\, \mathrm{S}'x & \\
\quad \dagger 619 & \qquad \supset (\exists y)(y \,\epsilon\, \Lambda \,.\, \Lambda \cap \overline{\iota y} \,\epsilon\, x) & \\
\quad *156\ (\&\ \mathrm{D}8) & \qquad \supset . \sim (y)1 \,.\, (\exists y)(\Lambda \cap \overline{\iota y} \,\epsilon\, x) & (2) \\
\quad *100 & [2 \,.\, (y)1 \,.\supset]\ 624 &
\end{array}
$$

1, the class of all one-member classes, has been expressed in two ways in D37 and †615; and it receives four more expressions in the course of the following theorems.

†625. $1 = \hat{x}(\exists y)(y \,\epsilon\, x \,.\, x \subset \iota y)$

$$
\begin{array}{ll}
\textit{Proof.}\ \dagger 623,\ *188\ (\&\ \mathrm{D}37) & 1 = \hat{x}(\exists y)(y \,\epsilon\, x \,.\, x \cap \overline{\iota y} = \Lambda) \\
\quad \dagger 321,\ *188 & \quad = \hat{x}(\exists y)(y \,\epsilon\, x \,.\, x \subset \overline{\overline{\iota y}}) \\
\quad \dagger 275,\ *227 & \quad = \mathrm{R}625
\end{array}
$$

†626. $1 = \hat{x}(\exists y)(z)(z \,\epsilon\, x \,.\!\equiv .\, z = y)$

Proof. †350, *123 $626\ [\equiv 625]$

†627. $1 = \hat{x}(\exists y)(y \,\epsilon\, \mathrm{V} \,.\, x = \iota y)$

Proof. †349, *123 $[626 \equiv]\ 627$

[1] This is a striking case of compound bracketing. Its step-by-step resolution involves no new idea. First we may bracket out '$(y)2 \equiv$ ', then' $(x)182 \supset .$', then '$3 \equiv$'; or we may follow any of five other orders.

I.e., 1 is the class of unit classes of elements. Also it is the range of functionality of the membership relation.

†628. $1 = \mathrm{r}\,\mathfrak{E}$

Proof. †626 $1 = \hat{x}(\exists y)(z)(z = y \,.\!\equiv\!.\, z \in x)$
*245 $= \hat{x}(\exists y)(z)(z = y \,.\!\equiv\!.\, x \in \hat{x}(z \in x))$
*230, *188 $= \hat{x}(\exists y)(z)(z = y \,.\!\equiv\!.\, z \in x \,.\, x \in \mathrm{V})$
†564, *188 (& D32) $= \mathrm{r}\,\mathfrak{E}$

To have 1 member (or 0, or 2, etc.) is to be a member of 1 (or 0, or 2, etc.). The following theorem says that if x has one member then anything y will belong to x if and only if x is ιy.

†629. $(y)(x)$ $x \in 1 \,.\!\supset\!:\, y \in x \,.\!\equiv\!.\, x = \iota y$

Proof. †358 $z \in \mathrm{V} \,.\!\supset\!:\, y = z \,.\!\equiv\!.\, \iota z = \iota y$ (1)
*100 $[1 \supset:]\, z \in \mathrm{V} \,.\, \iota z = \iota y \,.\!\equiv\!.\, z \in \mathrm{V} \,.\, y = z$
†345 $\equiv\!.\, y \in \iota z$ (2)
*223 $x = \iota z \,.\!\supset\!:.\, z \in \mathrm{V} \,.\, x = \iota y \,.\!\equiv\!.\, y \in x\, [:\equiv 2]$ (3)
*100, *163 $(\exists z)([3.]\, z \in \mathrm{V} \,.\, x = \iota z) \supset \mathrm{R}629$ (4)
†627, *224 $x \in 1 \,.\!\equiv\!.\, x \in \hat{x}\ \mathrm{L}4$
*230 $\equiv\!.\, x \in \mathrm{V} \,.\, \mathrm{L}4$ (5)
*100 $[4 \,.\, 5 \,.\!\supset]\ 629$

§ 44. *Natural Numbers*

THE NATURAL numbers 0, 1, and 2 have been defined (D36–38), and each further natural number is definable by a continuation of the same method. Given any natural number z, the next is

$$\hat{x}(\exists y)(y \in x \,.\, x \cap \overline{\iota y} \in z)$$

or more briefly S‘z (cf. †611–†616, § 43). But the class Nn of all natural numbers has yet to be formally defined as a whole. Roughly, the definition is of course this: Nn comprises 0, 1, 2, etc., i.e., 0, S‘0, S‘(S‘0), etc. The problem is to formalize the ‘etc.’; and the key is provided by the notion of the ancestral.

S‘0 is that which bears the relation S to 0; S‘(S‘0) is that which bears S to that which bears S to 0; and so on. To be a natural number, thus, is to be 0 or to bear S to 0 or to bear S to something which bears S to 0 or etc. In short, to be a natural number is to bear $*$S to 0 (cf. § 39). Thus Nn is definable as the class of all bearers of $*$S to 0; symbolically, $*$S“ι0 (cf. †478).

D40. ‘Nn’ *for* ‘($*$S“ι0)’.

Nn is closed with respect to S.

†630. S“Nn ⊂ Nn *Proof:* †520 (& D40).

I.e., successors of natural numbers are natural numbers.

†631. (x) $x \epsilon$ Nn .⊃. S‘$x \epsilon$ Nn

Proof. †190 $x \epsilon$ Nn .⊃. $x \epsilon$ V
†617, *224 ⊃. $x \epsilon$ ιS
†532 ⊃ S(S‘x, x) (1)
†467 S(S‘x, x) . $x \epsilon$ Nn .⊃. S‘$x \epsilon$ S“Nn (2)
*231 (& D21) [630 ⊃:] S‘$x \epsilon$ S“Nn .⊃. S‘$x \epsilon$ Nn (3)
*100 [1 . 2 . 3 .⊃]631

0, 1, etc. are natural numbers.

†632. 0 ϵ Nn

Proof. †511 $*$S(0,0) [≡ 611] (1)
†479 (& D40) 632 [≡ 1]

†633. 1 ϵ Nn

Proof. †631 [632 ⊃.] S‘0 ϵ Nn (1)
†615, *224 633 [≡ 1]

†634. 2 ϵ Nn *Proof* similar, using †633 and †616.

Any class which is closed with respect to S and contains 0 contains every natural number.

†635. (x) S“x ⊂ x . 0 ϵ x .⊃. Nn ⊂ x

Proof. †510 L635 . $*$S(y,0) .⊃. $y \epsilon x$ (1)
*100, *117 L635 ⊃ (y)([1.] $*$S(y,0) .⊃. $y \epsilon x$)
†479, *123 (& D21, 40) ⊃. Nn ⊂ x

The form of inference which we considered in connection with the ancestral (§ 39) takes on a more familiar aspect when adapted to the theory of natural numbers. From the premisses:

$$(x)(x + x = 2 \times x \,.\supset.\, (\mathrm{S}`x) + (\mathrm{S}`x) = 2 \times (\mathrm{S}`x)),$$
$$0 + 0 = 2 \times 0,$$
$$17 \in \mathrm{Nn},$$

e.g., we infer that $17 + 17 = 2 \times 17$. Or, incorporating the third premiss rather into the conclusion and leaving 17 unspecified, we infer from the first two of the above premisses that

$$(y) \qquad y \in \mathrm{Nn} \,.\supset.\, y + y = 2 \times y.$$

Such inference, called *mathematical induction*, is afforded by the following metatheorem.[1]

***636.** *If ψ, ϕ', and ϕ_0 are like ϕ except for containing free occurrences respectively of ζ, $\ulcorner \mathrm{S}`\alpha \urcorner$, and '0' wherever ϕ contains free occurrences of α, then*

$$\vdash \ulcorner (\alpha)(\phi \supset \phi') \,.\, \phi_0 \,.\, \zeta \in \mathrm{Nn} \,.\supset \psi \urcorner.$$

Proof. Let β be new, and let χ be formed from ϕ by putting β for all free occurrences of α.

*545 (& D39) $\vdash \ulcorner \mathrm{S}(\beta,\alpha) \supset. \beta = \mathrm{S}`\alpha \urcorner.$ (1)

*100 $\vdash \ulcorner [1{:}]\, \beta = \mathrm{S}`\alpha \,.\supset.\, \phi \supset \chi \,{:}\supset{:}\, \mathrm{S}(\beta,\alpha) \,.\, \phi \,.\supset \chi \urcorner.$ (2)

*101 $\vdash \ulcorner [(\beta)2 \supset.](\beta)\mathrm{L}2 \supset (\beta)\mathrm{R}2 \urcorner.$ (3)

*234a, *123 (& hp) $\vdash \ulcorner \phi \supset \phi' \,.\supset (\beta)\mathrm{R}2\, [{:}\equiv 3.] \urcorner.$ (4)

*101 $\vdash \ulcorner [(\alpha)4 \supset.]\, (\alpha)(\phi \supset \phi') \supset (\alpha)(\beta)\mathrm{R}2 \urcorner$
*119 $\supset (\beta)(\alpha)\mathrm{R}2 \urcorner.$ (5)

†479 (& D40) $\vdash \ulcorner \zeta \in \mathrm{Nn} \,.\equiv *\mathrm{S}(\zeta,0) \urcorner.$ (6)

*100 $\vdash \ulcorner [5 \,.\, 6 \,.]\, \mathrm{L}636 \,.\supset.\, (\beta)(\alpha)\mathrm{R}2 \,.\, *\mathrm{S}(\zeta,0) \,.\, \phi_0 \urcorner$
*521 (& hp) $\supset \psi \urcorner.$

Suppose the three premisses in the above example are at hand as theorems, numbered say †811, †812, †813. Then *636 gives:

$$[811 \,.\, 812 \,.\, 813 \,.\supset.]\ 17 + 17 = 2 \times 17$$

and more generally:

$$(y) \qquad [811 \,.\, 812 \,.]\, y \in \mathrm{Nn} \,.\supset.\, y + y = 2 \times y.$$

[1] Mathematical induction was used and explicitly recognized by Pascal in 1654 (vol. 3, p. 298) and Fermat in 1659 (vol. 2, pp. 431ff). But the principle of mathematical induction retained the status of an ultimate arithmetical axiom until 1879, when Frege defined the ancestral and by its means the class of natural numbers.

A stronger form of mathematical induction is provided by the following metatheorem, which is related to *522 as *636 is related to *521.

***637.** *If* [etc. as in *636] *then*

$$\vdash \ulcorner(\alpha)(\alpha \in \mathrm{Nn} \,.\, \phi \,.\!\supset \phi') \,.\, \phi_0 \,.\, \zeta \in \mathrm{Nn} \,.\!\supset \psi\urcorner.$$

Proof similar, from *522.

Illustrations of the use of *637 will appear in the proofs of †638 and †641.

It may happen that a class x deprived of a certain member y belongs to a class z, while x deprived of a different member w does not belong to z. When z is a natural number, however, this will not happen; if x deprived of one member belongs to a natural number z (i.e., has z members) then x deprived of any different member belongs likewise to z. If removal of Peter leaves eleven Apostles, removal of John would have had the same effect. This is shown in the following theorem.

†638. $(z)(y)(x)(w) \qquad z \in \mathrm{Nn} \,.\!\supset : y, w \in x \,.\, x \cap \overline{\iota w} \in z \,.\!\supset .\, x \cap \overline{\iota y} \in z$

Proof. *223 $\quad w = y \,.\!\supset : x \cap \overline{\iota w} \in \mathrm{S}`z \,.\!\equiv .\, x \cap \overline{\iota y} \in \mathrm{S}`z$ (1)

*233

$(w)(x)\mathrm{R638} \supset : y, x' \in x \cap \overline{\iota w} \,.\, (x \cap \overline{\iota w}) \cap \overline{\iota x'} \in z \,.\!\supset .\, (x \cap \overline{\iota w}) \cap \overline{\iota y} \in z$ (2)

*100, *163 $\quad (\exists x')([2.]\ x' \in x \cap \overline{\iota w} \,.\, (x \cap \overline{\iota w}) \cap \overline{\iota x'} \in z)$

$\supset : (w)(x)\mathrm{R638} \,.\, y \in x \cap \overline{\iota w} \,.\!\supset \mathrm{RR2}$

†362, *123 $\quad \supset : (w)(x)\mathrm{R638} \,.\, y \in x \,.\, w \neq y \,.\!\supset \mathrm{RR2}$ (3)

†362, *123

$w \in x \,.\, w \neq y \,.\, \mathrm{RR2} \,.\, z \in \mathrm{V} \,.\!\supset .\, w \in x \cap \overline{\iota y} \,.\, \mathrm{RR2} \,.\, z \in \mathrm{V}$

†286, *224 $\quad \supset .\, w \in x \cap \overline{\iota y} \,.\, x \cap (\overline{\iota w} \cap \overline{\iota y}) \in z \,.\, z \in \mathrm{V}$

†286, *224 $\quad \supset .\, w \in x \cap \overline{\iota y} \,.\, (x \cap \overline{\iota y}) \cap \overline{\iota w} \in z \,.\, z \in \mathrm{V}$

†620 $\quad \supset \mathrm{RR1}$ (4)

†619 $\mathrm{LR1} \supset \mathrm{L3}$

*100 $\quad \supset : [3 \,.\, 4 \,.](w)(x)\mathrm{R638} \,.\, y, w \in x \,.\, w \neq y \,.\, z \in \mathrm{V} \,.\!\supset \mathrm{RR1}$ (5)

†190 $\quad z \in \mathrm{Nn} \,.\!\supset .\, z \in \mathrm{V}$

*100 $\quad \supset : [5 \,.\, 1 \,.]\ (w)(x)\mathrm{R638} \,.\, y, w \in x \,.\, \mathrm{LR1} \,.\!\supset \mathrm{RR1}$ (6)

*100, *117

$z \in \mathrm{Nn} \,.\, (w)(x)\mathrm{R638} \,.\!\supset (w)(x)\ ([6.]\ y, w \in x \,.\, \mathrm{LR1}.\!\supset \mathrm{RR1})$ (7)

†620 $\quad w \in x \,.\, x \cap \overline{\iota w} \in 0\,[.\ 611] \,.\!\supset .\, x \in \mathrm{S}`0$

†615, *224 $\quad \supset .\, x \in 1$ (8)

*100 $[8 . 629 .]\ y, w \in x . x \cap \overline{\iota w} \in 0 .\supset. x = \iota y$

*223 $\supset: [294 \equiv .]\ x \cap \overline{\iota y} = \Lambda$

*223 $\supset: x \cap \overline{\iota y} \in 0\ [. \equiv 622]$ (9)

*637 $[(z)7 . (w)(x)9 .]\ z \in \mathrm{Nn} .\supset (w)(x)\mathrm{R}638$

*114 $\supset \mathrm{R}638$

If z is a natural number and y is any member of x, then x will have S‘z members if and only if x deprived of y has z members. If on the other hand y is a member of $\bar{x}$ (hence any element not belonging to x), then x will have z members if and only if x with y thrown in has S‘z members. Such is the import of the following two theorems.

†639. $(z)(y)(x)$ $z \in \mathrm{Nn} . y \in x .\supset: x \in \mathrm{S}‘z .\equiv. x \cap \overline{\iota y} \in z$

Proof. †190 $z \in \mathrm{Nn} .\supset. z \in \mathrm{V}$ (1)

*100, *163 $(\exists w)([638 .]\ w \in x . x \cap \overline{\iota w} \in z) \supset: \mathrm{L}639 \supset. x \cap \overline{\iota y} \in z$ (2)

†619 $x \in \mathrm{S}‘z .\supset \mathrm{L}2$ (3)

*100 $[1 . 2 . 3 . 620 .\supset]\ 639$

†640. $(z)(y)(x)$ $z \in \mathrm{Nn} . y \in \bar{x} .\supset: x \in z .\equiv. x \cup \iota y \in \mathrm{S}‘z$

Proof. †344 $y \in \mathrm{V} .\supset. y \in \iota y$

*100 $\supset: y \in x .\vee. y \in \iota y$

†271 $\supset. y \in x \cup \iota y$ (1)

†639 $z \in \mathrm{Nn} . \mathrm{R}1 .\supset: (x \cup \iota y) \cap \overline{\iota y} \in z .\equiv. x \cup \iota y \in \mathrm{S}‘z$ (2)

†327 $\iota y \subset \bar{x} .\supset. x = (x \cup \iota y) \cap \overline{\iota y}$

*223 $\supset:. z \in \mathrm{Nn} . \mathrm{R}1 .\supset \mathrm{R}640\ [: \equiv 2]$ (3)

†342 $y \in \bar{x} .\equiv. y \in \mathrm{V} . \iota y \subset \bar{x}$ (4)

*100 $[1 . 3 . 4 .\supset] 640$

If x and y are finite classes having the same number of members, in other words if x and y belong to the same natural number, then the following theorem shows that x coincides with y if included in y.

†641. $(z)(y)(x)$ $z \in \mathrm{Nn} .\supset: x, y \in z . x \subset y .\supset. x = y$

Proof. †342 $w \in x .\supset. w \in \mathrm{V} . \iota w \subset x$

*100 $\supset. \iota w \subset x$

†326 $\supset. x = (x \cap \overline{\iota w}) \cup \iota w$ (1)

*223 $x \cap \overline{\iota w} = y \cap \overline{\iota w} .\supset:. [1 \equiv :]\ w \in x .\supset. x = (y \cap \overline{\iota w}) \cup \iota w$ (2)

*103 $[(x)1 \supset :]\ w \in y .\supset. y = (y \cap \overline{\iota w}) \cup \iota w$

*223 $\supset:: \mathrm{L}2 \supset: w \in x .\supset. x = y\ [.: \equiv 2]$ (3)

†329 $x \subset y .\supset. x \cap \overline{\iota w} \subset y \cap \overline{\iota w}$ (4)

*233

$(x)(y)\mathrm{R641} \supset: x \cap \overline{\iota w}, y \cap \overline{\iota w} \in z . \mathrm{R4} .\supset \mathrm{L2}$

*100 $\supset: [3.]\ x \cap \overline{\iota w}, y \cap \overline{\iota w} \in z . \mathrm{R4} . w \in x . w \in y .\supset. x = y$ (5)

†639 $z \in \mathrm{Nn} . w \in y .\supset: y \in \mathrm{S}‘z .\equiv. y \cap \overline{\iota w} \in z$ (6)

*110 (& D21) $x \subset y .\supset: w \in x .\supset. w \in y$

*100 $\supset: [5 . 6 .]\ (x)(y)\mathrm{R641} . x \cap \overline{\iota w} \in z . \mathrm{R4} .$

$w \in x . z \in \mathrm{Nn} . y \in \mathrm{S}‘z .\supset. x = y$ (7)

*100, *163 $(\exists w)([4 . 7 .]\ w \in x . x \cap \overline{\iota w} \in z) \supset:$

$(x)(y)\mathrm{R641} . z \in \mathrm{Nn} . y \in \mathrm{S}‘z . x \subset y .\supset. x = y$ (8)

†619 $x \in \mathrm{S}‘z .\supset \mathrm{L8}$ (9)

*100, *117

$z \in \mathrm{Nn} . (x)(y)\mathrm{R641} .\supset (x)(y)([8 . 9 .]\ x, y \in \mathrm{S}‘z . x \subset y .\supset. x = y)$ (10)

*100 $x, y \in 0 . x \subset y .\supset. x, y \in 0$

†356 (& D36) $\supset. x = y$ (11)

*637 $[(z)10 . (x)(y)11 .]\ z \in \mathrm{Nn} .\supset (x)(y)\mathrm{R641}$

*114 $\supset \mathrm{R641}$

§ 45. *Counter Sets*

WHEN WE augment a class x by throwing in the class itself as additional member, we have $x \cup \iota x$. Thus where x is the class of the twelve Apostles, $x \cup \iota x$ has thirteen members: the twelve Apostles and the class of them. Where y is this class $x \cup \iota x$ of thirteen, $y \cup \iota y$ in turn has fourteen members: the thirteen members of y and y itself. In this fashion we can generate an unending series of classes, each having one more member than its predecessor.

$x \cup \iota x$, which will be called the *self-augment* of x, is an element whenever x is. This cannot be established by direct appeal to *200, because the stratification condition is not fulfilled; but it can be established indirectly, as in the first line of the following proof. The converse also holds, as the rest of the proof shows.[1]

[1] With help of †363 it is not difficult to prove, more generally, that

$(y)(x) \qquad x \in \mathrm{V} .\equiv. x \cup \iota y \in \mathrm{V}.$

†650. (x) $x \in V . \equiv . x \cup \iota x \in V$

Proof. †273 $x \in V [. 359] . \supset . x \cup \iota x \in V$ (1)

*223 $\iota x = \Lambda . \supset : x \cup \iota x = x [. \equiv 297]$

*223 $\supset 650$ (2)

*100 $[1 . 2 . 343 . \supset] 650$

The self-augment function, whose value for any element x as argument is $x \cup \iota x$, will be called Sa.

D41. 'Sa' *for* '$\lambda_x(x \cup \iota x)$'.

In view of †650 we have the following theorems.

†651. (x) $x \in V . \supset . \mathrm{Sa}`x = x \cup \iota x$

†652. $\mathrm{r}\,\mathrm{Sa} = V$

Proofs like those of †614 and †617, using †650 instead of †610.

If x is the class of Apostles, then $\mathrm{Sa}`x$, $\mathrm{Sa}`(\mathrm{Sa}`x)$, etc. are the classes of thirteen members, fourteen members, etc. which were considered above. But if instead of starting with that twelve-member class we take $x$ as the class of zero members, viz. $\Lambda$, then $\mathrm{Sa}`x$ is the class $\Lambda \cup \iota\Lambda$, i.e. $\iota\Lambda$, having the one member Λ; $\mathrm{Sa}`(\mathrm{Sa}`x)$ is $\iota\Lambda \cup \iota\iota\Lambda$, having the two members Λ and $\iota\Lambda$; $\mathrm{Sa}`(\mathrm{Sa}`(\mathrm{Sa}`x))$ has the three members $\Lambda$, $\iota\Lambda$, and $\iota\Lambda \cup \iota\iota\Lambda$; and so on. The classes making up this series will be called *counter sets*.[1] Just as the class Nn of natural numbers comprises 0, $\mathrm{S}`0$, $\mathrm{S}`(\mathrm{S}`0)$, etc., so the class Cs of counter sets comprises Λ, $\mathrm{Sa}`\Lambda$, $\mathrm{Sa}`(\mathrm{Sa}`\Lambda)$, etc.; and accordingly, just as Nn was definable as $*\mathrm{S}``\iota 0$, so Cs is definable as $*\mathrm{Sa}``\iota\Lambda$.

D42. 'Cs' *for* '$(*\mathrm{Sa}``\iota\Lambda)$'.

The interesting feature of Cs is that it contains one member from each of the natural numbers. The counter set Λ is a member of 0, the counter set $\mathrm{Sa}`\Lambda$ is a member of $\mathrm{S}`0$, the counter set $\mathrm{Sa}`(\mathrm{Sa}`\Lambda)$ is a member of $\mathrm{S}`(\mathrm{S}`0)$, and so on (cf. †663). In the counter sets we have a guarantee that for every natural number z, no matter how high, there is a class of z members. This conclusion

[1] So-called because they are to serve as *counters* in certain formal developments rather analogous to the process of counting. The term was suggested, in a different context, by Ravven. These sets are von Neumann's natural numbers.

will be drawn formally in †664; and from this basis it will be found possible (§ 46) to proceed to a theorem establishing the infinitude of V.

The following six theorems and metatheorems are exactly analogous to six in § 44.

†653. Sa"Cs ⊂ Cs

†654. (x) $x \in$ Cs .⊃. Sa'$x \in$ Cs

†655. $\Lambda \in$ Cs

†656. (x) Sa"$x \subset x$. $\Lambda \in x$.⊃. Cs $\subset x$

***657.** *If* ψ, ϕ', *and* ϕ_0 *are like* ϕ *except for containing free occurrences respectively of* ζ, ⌜Sa'α⌝, *and* 'Λ' *wherever* ϕ *contains free occurrences of* α, *then*

⊢ ⌜$(\alpha)(\phi \supset \phi')$. ϕ_0 . $\zeta \in$ Cs .⊃ ψ⌝.

***658.** *If* [etc. as in *657] *then*

⊢ ⌜$(\alpha)(\alpha \in$ Cs . ϕ .⊃ $\phi')$. ϕ_0 . $\zeta \in$ Cs .⊃ ψ⌝.

Proofs like those of †630–†632, †635–*637, using †652, †653, and †211 instead of †617, †630, and †611.

Є is the relation which an element bears to any element of which it is a member (cf. § 42). Hence its proper ancestral ∗Є | Є is the relation of an element x to an element y such that x is a member of y or a member of a member of y or etc. (cf. end of § 39). Now the following theorem shows that no counter set bears ∗Є | Є to itself; i.e., that no counter set is a member of a member of . . . a member of itself.

†659. (x) $x \in$ Cs .⊃ ∼ (∗Є | Є)(x,x)

Proof. †562 Є ∪ I ⊂ ∗Є (1)

†422 [1 ⊃.] (Є ∪ I)(y,x) ⊃ ∗Є(y,x) (2)

*100 [650 . 344 :] $y \in x$ .v. $y \in \iota x$: R650 :⊃:

$y \in x$. $x \in$ V .v. $y \in \iota x$: $x \in x$ .v. $x \in \iota x$: R650 (3)

†271, *123

$y \in x \cup \iota x$. R650 .⊃: $y \in x$. $x \in$ V .v. $y \in \iota x$: $x \in x \cup \iota x$. R650[.:≡3](4)

†564, *123 Є$(y, x \cup \iota x)$⊃: Є(y,x) v. $y \in \iota x$: Є$(x, x \cup \iota x)$[.:≡ 4]

†554, *123 $\supset . \mathfrak{E}(y,x) \vee I(y,x) . \mathfrak{E}(x, x \cup \iota x)$
†424, *123 $\supset . (\mathfrak{E} \cup I)(y,x) . \mathfrak{E}(x, x \cup \iota x)$ (5)
†487 $*\mathfrak{E}(x \cup \iota x, y) . *\mathfrak{E}(y,x) . \supset (*\mathfrak{E} \mid *\mathfrak{E})(x \cup \iota x, x)$
†517, *224 $\supset *\mathfrak{E}(x \cup \iota x, x)$ (6)
*100, *163
$(\exists y)([6 . 5 . 2 .] *\mathfrak{E}(x \cup \iota x, y) . \mathfrak{E}(y, x \cup \iota x)) \supset . RR5 . R6$ (7)
†486, *123 $(*\mathfrak{E} \mid \mathfrak{E})(x \cup \iota x, x \cup \iota x) \supset . RR5 . R6 [:\equiv 7]$
†487 $\supset (\mathfrak{E} \mid *\mathfrak{E})(x,x)$
†524, *224 $\supset (*\mathfrak{E} \mid \mathfrak{E})(x,x)$ (8)
*223 $Sa`x = x \cup \iota x . \supset : (*\mathfrak{E} \mid \mathfrak{E})(Sa`x, Sa`x) \supset R8 [.\equiv 8]$ (9)
†190 $x \in Cs . \supset . x \in V$ (10)
*100 $[10 . 651 . 9 .] x \in Cs . R659 . \supset \sim LR9$ (11)
†564 $\mathfrak{E}(x,\Lambda) \equiv . x \in \Lambda [. 211]$ (12)
*100 $[12 . 192 . \supset] \sim (*\mathfrak{E}(\Lambda,x) . \mathfrak{E}(x,\Lambda))$ (13)
*131 $\sim(\exists x)(*\mathfrak{E}(\Lambda,x) . \mathfrak{E}(x,\Lambda)) [\equiv (x)13]$ (14)
†486, *123 $\sim(*\mathfrak{E} \mid \mathfrak{E})(\Lambda,\Lambda) [\equiv 14]$ (15)
*658 $[(x)11 . 15 .] x \in Cs . \supset R659$

It follows that no counter set is a member of itself.

†660. (x) $x \in Cs . \supset . x \bar{\in} x$

Proof. †512 $\mathfrak{E} \subset *\mathfrak{E} \mid \mathfrak{E}$ (1)
†422 $[1 \supset .] \mathfrak{E}(x,x) \supset (*\mathfrak{E} \mid \mathfrak{E})(x,x)$ (2)
*100 $[2 . 659 . 565 . \supset] 660$

I.e., Cs is included in Russell's class.

†661. $Cs \subset \hat{x}(x \bar{\in} x)$

Proof. *313 $[(x)660 \equiv] 661$

Thus V is not a counter set.

†662. $V \bar{\in} Cs$

Proof. †660 $V \in Cs . \supset . V \bar{\in} V$ (1)
*100 $[1 . 210 . \supset] 662$

It is not always true that $x \cup \iota x$ has one more member than x. If x is a member of itself, $x \cup \iota x$ is simply x; for example, $V \cup \iota V = V$. If x is not an element, $x \cup \iota x$ is again simply x (cf. †343). When x is a counter set, however, we may be sure that its self-augment $x \cup \iota x$ (or Sa'x) will have one more member than x. This is established in the following theorem.

†663. $(y)(x)$ $x \in \mathrm{Cs} \,.\, y \in \mathrm{Nn} \,.\!\supset\!: x \in y \,.\!\equiv\!.\, \mathrm{Sa}`x \in \mathrm{S}`y$

Proof. †190 $x \in \mathrm{Cs} \,.\!\supset\!.\, x \in \mathrm{V}$ (1)

*100 $[1 \,.\, 660 \,.\, 651 \,.]\, x \in \mathrm{Cs} \,.\!\supset\!.\, x \in \mathrm{V} \,.\, x \bar{\in} x \,.\, \mathrm{R}651$

*235 (& D20), *123 $\supset\!.\, x \in \bar{x} \,.\, \mathrm{R}651$ (2)

†640 $y \in \mathrm{Nn} \,.\, x \in \bar{x} \,.\!\supset\!: x \in y \,.\!\equiv\!.\, x \cup \iota x \in \mathrm{S}`y$ (3)

*223 $\mathrm{R}651 \supset\!:.\, y \in \mathrm{Nn} \,.\, x \in \bar{x} \,.\!\supset \mathrm{R}663\,[: \equiv 3]$ (4)

*100 $[2 \,.\, 4 \,.\!\supset]\, 663$

It can now be proved that every natural number has a counter set as member.

†664. (y) $y \in \mathrm{Nn} \,.\!\supset (\exists x)(x \in \mathrm{Cs} \,.\, x \in y)$

Proof. *232 $[655 \,.\, 622 \,.\!\supset]\, (\exists x)(x \in \mathrm{Cs} \,.\, x \in 0)$ (1)

*100 $[654 \,.\, 663 \,.]\, \mathrm{L}663 \,.\, x \in y \,.\!\supset\!.\, \mathrm{R}654 \,.\, \mathrm{Sa}`x \in \mathrm{S}`y$

*232 $\supset (\exists x)(x \in \mathrm{Cs} \,.\, x \in \mathrm{S}`y)$ (2)

*100, *163 $(\exists x)([2 \,.]\, x \in \mathrm{Cs} \,.\, x \in y) \supset\!: y \in \mathrm{Nn} \,.\!\supset \mathrm{R}2$ (3)

*100 $[3 \,.]\, y \in \mathrm{Nn} \,.\, \mathrm{R}664 \,.\!\supset \mathrm{R}2$ (4)

*637 $[(y)4 \,.\, 1 \,.]\, y \in \mathrm{Nn} \,.\!\supset \mathrm{R}664$

§ 46. *Finite and Infinite*

THE FOREGOING theorem shows that for every natural number y, no matter how high, there is a class (in fact a counter set) having y members. No natural number is empty.

†670. $\Lambda \bar{\in} \mathrm{Nn}$

Proof. *100 $[192 \supset] \sim (x \in \mathrm{Cs} \,.\, x \in \Lambda)$ (1)

†664 (& D8) $\Lambda \in \mathrm{Nn} \,.\!\supset \sim (x)1$ (2)

*100 $[2 \,.\, (x)1 \,.\!\supset]\, 670$

The classes belonging to the successive natural numbers run larger and larger without limit, but each is finite. On the other hand an infinite class belongs to no natural number whatever; where x is an infinite class there is no natural number y such that

x has y members. The members of an infinite class are in this sense literally *innumerable*.

If a class x were finite and yet not an element, then membership of x in a natural number would be impossible despite the finitude of x; for a non-element is incapable of membership. However, we know there are no finite non-elements (see end of § 35). Membership in a natural number thus provides a simple and adequate formal definition of finitude. The class Fin of all finite classes is the class of all members of natural numbers, i.e. $\mathfrak{E}``\mathrm{Nn}$ (cf. †567).

D43. 'Fin' *for* '$(\mathfrak{E}``\mathrm{Nn})$'.

A class is finite if and only if it belongs to a natural number.

†671. (x) $\quad x \in \mathrm{Fin} \,.\!\equiv (\exists y)(x \in y \,.\, y \in \mathrm{Nn})$

Proof.	†567, *224 (& D43)	$x \in \mathrm{Fin} \,.\!\equiv\!.\, x \in \hat{x}\,\mathrm{R671}$
	*230	$\equiv\!.\, x \in \mathrm{V} \,.\, \mathrm{R671}$
	*158	$\equiv (\exists y)(\mathrm{R191} \,.\, y \in \mathrm{Nn})$
	†191, *123	$\equiv \mathrm{R671}$

A class (element or non-element) is infinite if and only if it does not belong to Fin; i.e., if and only if it belongs to no natural number. One such class is V.

†672. (x) $\quad \sim(\mathrm{V} \in x \,.\, x \in \mathrm{Nn})$

Proof.	†333	$y \subset \mathrm{V}$	(1)
	†641	$x \in \mathrm{Nn} \,.\!\supset\!: y, \mathrm{V} \in x\, [.\, 1] \,.\!\supset\!.\, y = \mathrm{V}$	(2)
	*223	$y = \mathrm{V} \,.\!\supset\!: y\, \bar{\in}\, \mathrm{Cs}\, [.\!\equiv 662]$	(3)
	*100, *163	$(\exists y)([2 \,.\, 3 \,.]\, y \in \mathrm{Cs} \,.\, y \in x) \supset 672$	(4)
	†664	$x \in \mathrm{Nn} \,.\!\supset \mathrm{L4}$	(5)
	*100	$[4 \,.\, 5 \,.\!\supset]\, 672$	

†673. $\mathrm{V}\, \bar{\in}\, \mathrm{Fin}$

Proof.	†671 (& D8)	$\mathrm{V} \in \mathrm{Fin} \,.\!\equiv\, \sim(x)672$	(1)
	*100	$[1 \,.\, (x)672 \,.\!\supset]\, 673$	

The class V of all elements is thus infinite; in other words, there are infinitely many elements.

Whereas all finite classes are elements, it is not true conversely that all elements are finite; for V itself is an element (†210) and yet infinite. The class of *infinite elements* is $\overline{\text{Fin}}$; and V belongs to it.

†674. $\mathrm{V} \in \overline{\text{Fin}}$

Proof. *235 (& D20) 674 [≡. 210 . 673]

From †673 we see that for every finite class x there is an element not belonging to x; i.e., there is something belonging to $\bar{x}$.

†675. (x) $x \in \text{Fin} \,.\!\supset (\exists y)(y \in \bar{x})$

Proof. *223	$x = \mathrm{V} \,.\!\supset\!: x \,\bar{\in}\, \text{Fin} \,[.\!\equiv 673]$	(1)
*100	$[1 \,.\, 335 \,.] \, x \in \text{Fin} \,.\!\supset \sim (\mathrm{V} \subset x)$	
*100, *123 (& D21, 8)	$\supset (\exists y)(y \in \mathrm{V} \,.\, y \,\bar{\in}\, x)$	
*230 (& D20), *123	$\supset (\exists y)(y \in \bar{x})$	

An important property of the successor function which was not established in §§ 43–44 is as follows: if x and y are natural numbers having the same successor then $x = y$. The successor of a natural number x determines x uniquely. In fact, if w is the successor of a natural number x then x is fixed in terms cf w as

$$\hat{y}(\exists z)(z \in \bar{y} \,.\, y \cup \iota z \in w).$$

This is established in the following theorem.

†676. (x) $x \in \text{Nn} \,.\!\supset\!.\, x = \hat{y}(\exists z)(z \in \bar{y} \,.\, y \cup \iota z \in \mathrm{S}`x)$

Proof. *135	$y \in x \,.\, x \in \text{Nn} \,.\!\supset (\exists x)(y \in x \,.\, x \in \text{Nn})$	
†671	$\supset\!.\, y \in \text{Fin}$	
†675	$\supset (\exists z)(z \in \bar{y})$	(1)
*100, *117	$x \in \text{Nn} \,.\!\supset (y)([1 \supset\!:]\, y \in x \,.\!\equiv\!.\, \mathrm{R}1 \,.\, y \in x)$	
*121	$\supset\!: [189 \equiv\!.]\, x = \hat{y}(\mathrm{R}1 \,.\, y \in x)$	
*158, *123	$\supset\!.\, x = \hat{y}(\exists z)(z \in \bar{y} \,.\, y \in x)$	(2)
†640	$x \in \text{Nn} \,.\, z \in \bar{y} \,.\!\supset\!: y \in x \,.\!\equiv\!.\, y \cup \iota z \in \mathrm{S}`x$	(3)
*100, *117	$x \in \text{Nn} \,.\!\supset (y)(z)([3 \supset\!:]\, z \in \bar{y} \,.\, y \in x \,.\!\equiv\!.\, z \in \bar{y} \,.\, \mathrm{RR}3)$	
*121	$\supset\!.\, [2 \equiv]\, 676$	(4)
*100	$[4 \supset]\, 676$	

It is now easy to prove that if x and y are natural numbers having the same successor then $x = y$.

†677. $(y)(x)$ $x, y \in \text{Nn} \,.\, \mathrm{S}`x = \mathrm{S}`y \,.\!\supset\!.\, x = y$

Proof. †676 $x \in \text{Nn} \,.\!\supset\!.\, x = \hat{w}(\exists z)(z \in \bar{w} \,.\, w \cup \iota z \in \mathrm{S}`x)$ (1)

†676	$y \in \mathrm{Nn} \,.\supset.\, y = \hat{w}(\exists z)(z \in \overline{w} \,.\, w \cup \iota z \in \mathrm{S}`y)$	(2)
*228	$\mathrm{S}`x = \mathrm{S}`y \,.\, \mathrm{R1} \,.\, \mathrm{R2} \,.\equiv.\, \mathrm{S}`x = \mathrm{S}`y \,.\, x = \mathrm{RR2} \,.\, \mathrm{R2}$	
*228	$\equiv.\, \mathrm{S}`x = \mathrm{S}`y \,.\, x = y \,.\, \mathrm{R2}$	(3)
*100	$[1 \,.\, 2 \,.\, 3 \,.\supset]$ 677.	

This theorem, despite its elementary appearance, depends for its truth upon the infinitude of V. For, suppose V had only a finite number of members — say 93. Suppose, in other words, there were only 93 elements. Then no class could have 94 or 95 members; i.e., there would be no classes belonging to 94 or 95. Hence 94 and 95 would both be null; i.e.,

$$\mathrm{S}`93 = \mathrm{S}`94 = \Lambda.$$

Despite the identity of S'93 and S'94, however, 93 and 94 would remain distinct; for 93 would have V as member while 94 would have no members. Thus †677 would be violated.

Actually, however, V is infinite (†673) and no natural number reduces to Λ (†670). Proof of †677 had to be deferred till after proving these things. Note that the proof of †677 uses †676, whose proof in turn uses †675, which is a corollary of †673.

§ 47. *Powers of Relations*

The ancestral $*x$ is, we have seen, the relation of any element z to any element w such that z is w or bears x to w or bears x to something which bears x to w or etc. The pairs composing the relation $*x$ are hence classifiable as follows. First, there are the pairs $z;w$ such that z and w are the same element. The relation comprising these pairs is I. Second, there are the pairs $z;w$ such that z bears x to w. The relation comprising these pairs is $\dot{x}$ (x itself, if x is a relation). Third, there are the pairs $z;w$ such that z bears x to something which bears x to w. The relation comprising these is the relative product $x \mid x$. Fourth, there are the pairs $z;w$ such that z bears x to something which bears x to something which bears x to w. The relation comprising these is $(x \mid x) \mid x$. The fifth relation of the series is $((x \mid x) \mid x) \mid x$; and so on.[1]

[1] The position of the parentheses is of course immaterial; cf. †491.

These relations are known as the *relative powers* of x, symbolically x^0, x^1, x^2, x^3, etc. If x is the parent relation, then x^0 is I (as always); x^1 is the parent relation again; x^2 is the grandparent relation; x^3 is great-grandparent; and so on. Relative power is not to be confused with the familiar notion of power which occurs in arithmetic; the relative power stands rather to the relative product as the arithmetical power stands to the arithmetical product. The arithmetical notions of sum, product, and power will be taken up in § 48; but the present notion of relative power also has an arithmetical ingredient, in that the y of x^y is ordinarily a number.

Where y is any natural number, x^y is the relation borne by the first element z_0 to the last element z_y of any sequence $z_0, z_1, \ldots, z_y$ such that $x(z_0, z_1), x(z_1, z_2), \ldots, x(z_{y-1}, z_y)$. Where x is the father relation, e.g.,

$$x^4(\text{Kenan, Methuselah});$$

for it is written that x(Kenan, Mahalaleel) and x(Mahalaleel, Jared) and x(Jared, Enoch) and x(Enoch, Methuselah). But the sequence $z_0, z_1, \ldots, z_y$ need not always be composed, like

Kenan, Mahalaleel, Jared, Enoch, Methuselah,

of distinct elements; where x is the brother relation, e.g., Romulus will bear x^5 to Remus in view merely of the repetitive sequence

Romulus, Remus, Romulus, Remus, Romulus, Remus.

The intuitive description of x^y at the head of the foregoing paragraph turns on numerical subscripts, which must now be eliminated if we are to devise a formal definition of x^y. The business of *counting off*, accomplished informally by the subscripts, can be accomplished more formally by pairing the elements $z_0, z_1, \ldots, z_y$ with the successive natural numbers from 0 to y in the fashion $z_0;0, z_1;1, \ldots, z_y;y$. The relation which any such pair $z_i;i$ bears to the next following pair $z_{i+1}\,; i + 1$ will be called $\mathfrak{d}x$. This is a relation of pairs, and hence a class of pairs of pairs; it is the relation of any pair $m;n$ to any pair $m';(\text{S}`n)$ such that n is a natural number and $x(m,m')$.

To say that z bears x^y to an element w, now, is to say that $z;0$

is connected with $w;y$ by a $\mathfrak{d}x$-series of pairs; in other words, that $z;0$ either is $w;y$ (in which case $y = 0$) or else bears $\mathfrak{d}x$ to $w;y$ (in which case $y = 1$) or else bears $\mathfrak{d}x$ to something which bears $\mathfrak{d}x$ to $w;y$ (in which case $y = 2$) or else . . .. In short, the power x^y is the relation of any element z to any element w such that $z;0$ bears the ancestral $*\mathfrak{d}x$ to $w;y$.

The auxiliary relation $\mathfrak{d}x$ admits readily of formal definition.

D44. $\ulcorner \mathfrak{d}\zeta \urcorner$ *for*

$$\ulcorner \hat{\alpha}\hat{\beta}(\exists\alpha')(\exists\beta')(\exists\gamma)(\alpha = \alpha';\gamma \,.\, \beta = \beta';(\text{S}`\gamma) \,.\, \gamma \in \text{Nn} \,.\, \zeta(\alpha',\beta'))\urcorner.$$

Finally the power x^y is $\hat{z}\hat{w}(*\mathfrak{d}x(z;0,\ w;y))$, as remarked.

D45. $\ulcorner(\zeta^\eta)\urcorner$ *for* $\ulcorner\hat{\alpha}\hat{\beta}(*\mathfrak{d}\zeta(\alpha;0,\ \beta;\eta))\urcorner$.

Suppose, e.g., that x is the father relation. Then any father paired with any natural number will bear $\mathfrak{d}x$ to his child paired with the next natural number. Thus in particular

$\mathfrak{d}x$ (Kenan;0, Mahalaleel;1),
$\mathfrak{d}x$ (Mahalaleel;1, Jared;2),
$\mathfrak{d}x$ (Jared;2, Enoch;3),
$\mathfrak{d}x$ (Enoch;3, Methuselah;4).

Accordingly

$*\mathfrak{d}x$ (Kenan;0, Methuselah;4)

and hence

x^4(Kenan, Methuselah).

Only the first two theorems make explicit mention of $\mathfrak{d}x$.

†680. $(z)(y)(x)(w)$

$$w \in \text{Nn} \,.\!\supset\!.\, \mathfrak{d}x(y,\ z;(\text{S}`w)) \equiv (\exists y')(y = y';w \,.\, x(y',\ z))$$

Proof. †536 $\text{S}`w \in \text{V}$ (1)

†536 $\text{S}`w' \in \text{V}$ (2)

†411 $y';w \in \text{V}$ (3)

†411 $z;(\text{S}`w) \in \text{V}$ (4)

*223 $w = w' \,.\!\supset\!:\, w \in \text{Nn} \,.\!\equiv\!.\, w' \in \text{Nn}$ (5)

*226 $w = w' \,.\!\supset\!.\, \text{S}`w = \text{S}`w'$ (6)

†677 $w,\ w' \in \text{Nn} \,.\, \text{S}`w = \text{S}`w' \,.\!\supset\!.\, w = w'$ (7)

*100 [5 . 6 . 7 .] $w \in \text{Nn} \,.\!\supset\!:\, w = w' \,.\!\equiv\!.\, \text{S}`w = \text{S}`w' \,.\, w' \in \text{Nn}$ (8)

†417 $z' \in \text{V}$[. 2 . 1] $.\!\supset\!:\, z;(\text{S}`w) = z';(\text{S}`w') \,.\!\equiv\!.\, z = z' \,.\, \text{S}`w = \text{S}`w'$ (9)

*434 (& D44) LR680 ≡. y ε V [. 4] . (∃y′)(∃z′)(∃w′)(y = y′;w′ .
LR9 . w′ ε Nn . x(y′,z′)) (10)

*234b

y = y′;w . x(y′,z) .≡ (∃z′)(y = y′;w . x(y′,z′) . z = z′)

*234b, *123 ≡ (∃z′)(∃w′)(y = y′;w′ . x(y′,z′) . z = z′ . w = w′) (11)

*121 [(y′)11 ⊃:] y ε V . (∃y′)R11 .≡. y ε V . (∃y′)L11

*158 ≡ (∃y′)(y ε V . L11)

*228, *123 ≡ (∃y′)([3 .] L11) (12)

*100 (& D24), *117

w ε Nn .⊃ (y′)(z′)(w′)([8 . 9 .⊃:]
LR9 . w′ ε Nn . x(y′, z′) .≡. x(y′, z′) . z = z′ . w = w′)

*121 ⊃:. [10 ≡:] LR680 ≡. y ε V . (∃y′) R11

*122 ⊃. R680 [≡ 12]

†681. (y)(x) yεV. ⊃. x^y = ẑŵ(z = w . 0 = y .∨ (*𝔡x | 𝔡x)(z;0, w;y))

Proof. †411 z;0 ε V (1)

†563,*224 *𝔡x(z;0, w;y) ≡ (I ∪ (𝔡x | *𝔡x))(z;0, w;y)

†524,*224 ≡ (I ∪ (*𝔡x | 𝔡x))(z;0, w;y)

†424 ≡. I(z;0, w;y) ∨ (*𝔡x | 𝔡x)(z;0, w;y)

†552,*123 ≡: [1 .] z;0 = w;y .∨ (*𝔡x | 𝔡x)(z;0, w;y) (2)

†417 y, z ε V [. 611] .⊃: z;0 = w;y .≡. z = w . 0 = y (3)

*100,*117

y ε V .⊃ (z)(w)([2 . 3 .⊃:] z, w ε V . L2 .≡. z, w ε V . RR3 ∨ RR2)

*433 (& D45), *123

⊃ (z)(w)(x^y(z,w) ≡ RR681(z,w))

†447 ⊃ R681

The next two are the main theorems on relative powers.

†682. (x) x^0 = I

Proof. †624 0 ≠ S‘y′ (1)

†536 S‘y′ ε V (2)

†416 z ε V [. 611 . 2] . z;0 = z′;(S‘y′) .⊃. 0 = S‘y′ (3)

*100, *163 (∃y)(∃z′)(∃y′)([1 . 3 .] w = y;y′ . z;0 = z′;(S‘y′) .
y′ ε Nn . x(y,z′)) ⊃. z ε̄ V (4)

*434 (& D44) 𝔡x(w, z;0) ≡. w, z;0 ε V . L4 (5)

*100, *163 (∃w)([5 . 4 .] *𝔡x (y;0, w) . L5) ⊃. z ε̄ V (6)

†486, *123 (*𝔡x | 𝔡x)(y;0, z;0) ⊃. z ε̄ V [:≡ 6] (7)

†182 0 = 0 (8)

†681 [611 $\supset$.] $x^0 = \hat{y}\hat{z}(y = z\,[.\,8]\,.\!\vee L7)$

*100 (& D25), *188 $= \hat{y}\hat{z}(z \in \mathrm{V} : y = z\,.\!\vee L7)$

*100, *188 $= \hat{y}\hat{z}([7 \supset .]\; z \in \mathrm{V} \,.\, y = z)$

†553, *188 (& D26) $= \dot{I}$

*437 $= I$

†683. $(y)(x)$ $\quad y \in \mathrm{Nn} \,.\!\supset\!.\; x^{\mathrm{S}\text{‘}y} = x^y \mid x$

Proof. †624 $\quad 0 \neq \mathrm{S}\text{‘}y$ (1)

†536 $\quad \mathrm{S}\text{‘}y \in \mathrm{V}$ (2)

*100 $\quad [1 \supset] \sim (z = w \,.\, 0 = \mathrm{S}\text{‘}y)$ (3)

†681 $\quad [2 \supset .]\; x^{\mathrm{S}\text{‘}y} = \hat{z}\hat{w}(z = w \,.\, 0 = \mathrm{S}\text{‘}y \,.\!\vee (*\eth x \mid \eth x)(z;0,\, w;(\mathrm{S}\text{‘}y)))$

*100, *188 $\quad = \hat{z}\hat{w}([3 \supset]\; (*\eth x \mid \eth x)(z;0,\, w;(\mathrm{S}\text{‘}y)))$

†486, *188 $\quad = \hat{z}\hat{w}(\exists y')(*\eth x(z;0,\, y') \,.\, \eth x(y',\, w;(\mathrm{S}\text{‘}y)))$ (4)

†680, *117

$y \in \mathrm{Nn} \,.\!\supset (z)(w)(y')(\eth x(y',\, w;(\mathrm{S}\text{‘}y)) \equiv (\exists z')(x(z',w) \,.\, y' = z';y))$

*121 $\quad \supset : [4 \equiv .]$

$x^{\mathrm{S}\text{‘}y} = \hat{z}\hat{w}(\exists y')(*\eth x(z;0,\, y') \,.\, (\exists z')(x(z',w) \,.\, y' = z';y))$

*158, *123 $\quad \supset .\; x^{\mathrm{S}\text{‘}y} = \hat{z}\hat{w}(\exists y')(\exists z')(*\eth x(z;0,\, y') \,.\, x(z',w) \,.\, y' = z';y)$

*138, *123 $\quad \supset .\; x^{\mathrm{S}\text{‘}y} = \hat{z}\hat{w}(\exists z')(\exists y')($ " " " $)$

*234b, *123 $\quad \supset .\; x^{\mathrm{S}\text{‘}y} = \hat{z}\hat{w}(\exists z')(*\eth x(z;0,\, z';y) \,.\, x(z',\, w))$

*100 (& D25), *123

$\supset .\; x^{\mathrm{S}\text{‘}y} = \hat{z}\hat{w}(z \in \mathrm{V} \,.\, (\exists z')(*\eth x(z;0,\, z';y) \,.\, x(z',w)))$

*158, *123 $\quad \supset .\; x^{\mathrm{S}\text{‘}y} = \hat{z}\hat{w}(\exists z')(z \in \mathrm{V} \,.\, *\eth x(z;0,\, z';y) \,.\, x(z',w))$

*100 (& D24), *123

$\supset .\; x^{\mathrm{S}\text{‘}y} = \hat{z}\hat{w}(\exists z')(z,\, z' \in \mathrm{V} \,.\, *\eth x(z;0,\, z';y) \,.\, x(z',\, w))$

*433, *123 (& D45, 29)

$\supset$ R683

For each natural number y, the above two theorems explain x^y step by step in terms finally of I, x, and relative product.

$$x^0 = I,$$
$$x^1 = x^0 \mid x = I \mid x \qquad (= \dot{x};\ \text{cf. †559}),$$
$$x^2 = x^1 \mid x = (I \mid x) \mid x \qquad (= x \mid x),$$
$$x^3 = x^2 \mid x = ((I \mid x) \mid x) \mid x \qquad (= x \mid x \mid x),$$

etc. Where y is not a natural number, x^y is uninteresting—though of course still defined by D45.

It is next shown, by mathematical induction on the basis of †682 and †683, that the order of the relative product in †683 is inessential.

†684.	$(y)(x)$	$y \epsilon \mathrm{Nn} \mathbin{.\supset.} x \mid (x^y) = x^y \mid x$	
Proof. *226		$\mathrm{R684} \supset. (x \mid (x^y)) \mid x = (x^y \mid x) \mid x$	
†491, *224		$\supset. x \mid (x^y \mid x) = (x^y \mid x) \mid x$	(1)
*223		$\mathrm{R683} \supset:. \mathrm{R684} \supset. x \mid (x^{\mathrm{S}‘y}) = x^{\mathrm{S}‘y} \mid x\ [:\equiv 1]$	(2)
*100		$[683 \,.\, 2 \,.]\ y \epsilon \mathrm{Nn} \,.\, \mathrm{R684} \mathbin{.\supset} \mathrm{RR2}$	(3)
†559, *224		$x \mid I = I \mid x\ [.\equiv 560]$	(4)
†682, *224		$x \mid (x^0) = x^0 \mid x\ [.\equiv 4]$	(5)
*637		$[(y)3 \,.\, 5 \,.]\ y \epsilon \mathrm{Nn} \mathbin{.\supset} \mathrm{R684}$	

†685.	$(y)(x)$	$y \epsilon \mathrm{Nn} \mathbin{.\supset.} x^{\mathrm{S}‘y} = x \mid (x^y)$	
Proof. *223		$\mathrm{R684} \supset. 685\ [\equiv 683]$	(1)
*100		$[684 \,.\, 1 \mathbin{.\supset}]\ 685$	

The concluding theorem shows that if the range of a function x embraces all elements, so do the ranges of the relative powers of x.

†686.	$(y)(x)$	$y \epsilon \mathrm{Nn} \,.\, \mathfrak{r}x = \mathrm{V} \mathbin{.\supset.} \mathfrak{r}(x^y) = \mathrm{V}$	
Proof. †539		$\mathfrak{r}x = \mathrm{V} \mathbin{.\supset.} \mathfrak{r}(x^y) \subset \mathfrak{r}(x \mid (x^y))$	(1)
*223		$\mathfrak{r}(x^y) = \mathrm{V} \mathbin{.\supset:.} [1 \equiv :]\ \mathrm{L1} \supset. \mathrm{V} \subset \mathfrak{r}(x \mid (x^y))$	
†335, *123		$\supset: \mathrm{L1} \supset. \mathfrak{r}(x \mid (x^y)) = \mathrm{V}$	(2)
*223		$\mathrm{R685} \supset:: \mathfrak{r}(x^y) = \mathrm{V} \mathbin{.\supset:} \mathrm{L1} \supset. \mathfrak{r}(x^{\mathrm{S}‘y}) = \mathrm{V}\ [.:\equiv 2]$	(3)
*100, *117		$\mathrm{L1} \supset (y)([685 \,.\, 3 \,.]\ y \epsilon \mathrm{Nn} \,.\, \mathfrak{r}(x^y) = \mathrm{V} \mathbin{.\supset.} \mathfrak{r}(x^{\mathrm{S}‘y}) = \mathrm{V})$	(4)
†682, *224		$\mathfrak{r}(x^0) = \mathrm{V}\ [.\equiv 551]$	(5)
*637		$\mathrm{R4}\ [.\ 5] \,.\, y \epsilon \mathrm{Nn} \mathbin{.\supset.} \mathfrak{r}(x^y) = \mathrm{V}$	(6)
*100		$[4 \,.\, 6 \mathbin{.\supset}]\ 686$	

Whitehead and Russell (vol. 3, *301) define the relative power by a different and more complicated method. Between my version of relative power and theirs there is incidentally a minor divergence analogous to that noted in connection with the ancestral. This divergence shows itself only in the case of x^0, which is not I for Whitehead and Russell but rather

$$\hat{y}\hat{z}(y = z \,.\, (\exists w)(x(y, w) \vee x(w, y))).$$

See the end of § 49 for two more theorems on relative powers.

§ 48. *Arithmetical Sum, Product, Power*

WHERE x and y are any natural numbers, their *sum* $x + y$ is the number obtained from x by y additions of 1. In other words, $x + y$ is the successor of the successor . . . (y times) of x. Thus $x + y$ is definable as that which bears the yth relative power of the successor relation to x; symbolically $\mathrm{S}^y \text{‘} x$.

D46. $\ulcorner(\zeta + \eta)\urcorner$ *for* $\ulcorner(\mathrm{S}^\eta \text{‘}\zeta)\urcorner$.

The *product* $x \times y$ of natural numbers x and y, again, is the number obtained from 0 by y additions of x. Thus $x \times y$ is definable as that which bears the yth relative power of $\lambda_z(x + z)$ to 0; symbolically $\lambda_z(x + z)^y \text{‘}0$.

D47. $\ulcorner(\zeta \times \eta)\urcorner$ *for* $\ulcorner(\lambda_\alpha(\zeta + \alpha)^\eta \text{‘}0)\urcorner$.

The *power* $x \wedge y$, again, is the number obtained from 1 by y multiplications of x.[1] It is definable as that which bears the yth relative power of $\lambda_z(x \times z)$ to 1.

D48. $\ulcorner(\zeta \wedge \eta)\urcorner$ *for* $\ulcorner(\lambda_\alpha(\zeta \times \alpha)^\eta \text{‘}1)\urcorner$.

†683 and †682 were seen to explain x^y step by step in terms finally of I, x, and relative product, where y is any natural number (cf. § 47). Analogously the following two theorems explain $x + y$ step by step in terms finally of x and successor.

†690. (x) $\quad x \in \mathrm{Nn} \,.\!\supset.\, x + 0 = x$

Proof. †190 $\quad x \in \mathrm{Nn} \,.\!\supset.\, x \in \mathrm{V}$

†550 $\quad \supset.\, I\text{‘}x = x$

†682, *224 (& D46) $\quad \supset.\, x + 0 = x$

†691. $(y)(x)$ $\quad y \in \mathrm{Nn} \,.\!\supset.\, x + (\mathrm{S}\text{‘}y) = \mathrm{S}\text{‘}(x + y)$

[1] This notation, consisting of a radical sign inverted, is Peano's. I have adopted it in preference to the usual exponent notation for two reasons: to avoid confusion with the notion of relative power, and to prepare the way for the attachment of subscripts ‘s’ and ‘r’ in § 50. Beyond the scope of the present study these considerations cease to be relevant, and reversion to the usual exponent notation becomes possible.

Proof †190 $x \in \mathfrak{r}(\mathrm{S}^{\mathrm{S}\text{‘}y}) \,.\!\supset.\, x \in \mathrm{V}$ (1)
†538 (& D46) $x \in \mathfrak{r}(\mathrm{S}^{y}) \,.\!\supset.\, (\mathrm{S} \mid (\mathrm{S}^{y}))\text{‘}x = \mathrm{S}\text{‘}(x + y)$ (2)
†533 (& D46) $\sim \mathrm{L1} \supset.\, x + (\mathrm{S}\text{‘}y) = \Lambda$ (3)
†533 (& D46) $\sim \mathrm{L2} \supset.\, x + y = \Lambda$
*223 $\supset: \mathrm{S}\text{‘}(x + y) = \Lambda$ [.≡ 621]
*223 $\supset: \sim \mathrm{L1} \supset \mathrm{R691}$ [.≡ 3] (4)
†686 $y \in \mathrm{Nn}$ [. 617] $.\!\supset.\, \mathfrak{r}(\mathrm{S}^{y}) = \mathrm{V}$
*223 $\supset: \mathrm{L1} \supset \mathrm{L2}$ [.≡ 1] (5)
†685 $y \in \mathrm{Nn} \,.\!\supset.\, \mathrm{S}^{\mathrm{S}\text{‘}y} = \mathrm{S} \mid (\mathrm{S}^{y})$
*223 (& D46) $\supset: \mathrm{L2} \supset \mathrm{R691}$ [.≡ 2] (6)
*100 [4 . 5 . 6 .⊃] 691

Thus, where x is any natural number,

$$x + 0 = x,$$
$$x + 1 = \mathrm{S}\text{‘}(x + 0) = \mathrm{S}\text{‘}x,$$
$$x + 2 = \mathrm{S}\text{‘}(x + 1) = \mathrm{S}\text{‘}(\mathrm{S}\text{‘}x),$$
$$x + 3 = \mathrm{S}\text{‘}(x + 2) = \mathrm{S}\text{‘}(\mathrm{S}\text{‘}(\mathrm{S}\text{‘}x)), \text{ etc.}$$

Analogously the following two theorems explain $x \times y$ step by step in terms finally of 0, x, and sum.

†692. (x) $x \times 0 = 0$

Proof. †550 [611 ≡.] $I\text{‘}0 = 0$ (1)
†682, *224 (& D47) 692 [≡ 1]

†693. $(y)(x)$ $y \in \mathrm{Nn} \,.\!\supset.\, x \times (\mathrm{S}\text{‘}y) = x + (x \times y)$

Proof. †536 (& D46) $x + z \in \mathrm{V}$ (1)
†536 (& D46) $x + (x \times y) \in \mathrm{V}$ (2)
†536 (& D47) $x \times y \in \mathrm{V}$ (3)
*542 [3 . 2 .⊃.] $\lambda_z(x + z)\text{‘}(x \times y) = x + (x \times y)$ (4)
*193 [(z)1 ⊃.] $\mathrm{V} = \hat{z}\,1$
*543 $= \mathfrak{r}\,\lambda_z(x + z)$ (5)
†686 $y \in \mathrm{Nn}$ [. 5] $.\!\supset.\, \mathfrak{r}(\lambda_z(x + z)^{y}) = \mathrm{V}$
*223 $\supset: 0 \in \mathfrak{r}(\lambda_z(x + z)^{y})$ [.≡ 611]
†538 (& D47) $\supset.\, (\lambda_z(x + z) \mid (\lambda_z(x + z)^{y}))\text{‘}0 = \mathrm{L4}$
*223 $\supset:$ " " " " $= \mathrm{R4}$ [.≡ 4] (6)
†685 $y \in \mathrm{Nn} \,.\!\supset.\, \lambda_z(x + z)^{\mathrm{S}\text{‘}y} = \lambda_z(x + z) \mid (\lambda_z(x + z)^{y})$
*223 (& D47) $\supset.$ 693 [≡ 6] (7)
*100 [7 ⊃] 693

Thus, where x is any natural number,

$$x \times 0 = 0,$$
$$x \times 1 = x + (x \times 0) = x + 0 = x,$$
$$x \times 2 = x + (x \times 1) = x + x,$$
$$x \times 3 = x + (x \times 2) = x + (x + x), \text{ etc.}$$

Analogously the following two theorems explain $x \wedge y$ step by step in terms finally of 1, x, and product.

†694. (x) $x \wedge 0 = 1$

†695. $(y)(x)$ $y \in \mathrm{Nn} .\supset. x \wedge (\mathrm{S}`y) = x \times (x \wedge y)$

Proofs analogous to those of †692 and †693.

Thus, where x is any natural number,

$$x \wedge 0 = 1,$$
$$x \wedge 1 = x \times (x \wedge 0) = x \times 1 = x,$$
$$x \wedge 2 = x \times (x \wedge 1) = x \times x,$$
$$x \wedge 3 = x \times (x \wedge 2) = x \times (x \times x),$$

etc.

Where x and y are not natural numbers, $x + y$, $x \times y$, and $x \wedge y$ are uninteresting — though of course still defined by D46–48. So long as x and y are natural numbers, $x + y$, $x \times y$, and $x \wedge y$ are natural numbers as well; this is established in the following three theorems. These theorems, and subsequent ones likewise, are proved from the above six with help of mathematical induction.

†696. $(y)(x)$ $x, y \in \mathrm{Nn} .\supset. x + y \in \mathrm{Nn}$

Proof. †631 R696 $\supset$. $\mathrm{S}`(x + y) \in \mathrm{Nn}$ (1)
*223 R691 $\supset$:. R696 $\supset$. $x + (\mathrm{S}`y) \in \mathrm{Nn}$ [:≡ 1] (2)
*100 [691 . 2 .] $y \in \mathrm{Nn} : x \in \mathrm{Nn} .\supset$ R696 :$\supset$: $x \in \mathrm{Nn} .\supset$ RR2 (3)
*223 R690 $\supset$: $x + 0 \in \mathrm{Nn}$.≡. $x \in \mathrm{Nn}$ (4)
*100 [690 . 4 .] $x \in \mathrm{Nn} .\supset. x + 0 \in \mathrm{Nn}$ (5)
*637 [(y)3 . 5 .] $y \in \mathrm{Nn} .\supset: x \in \mathrm{Nn} .\supset$ R696 (6)
*100 [6 $\supset$] 696

†697. $(y)(x)$ $x, y \in \mathrm{Nn} .\supset. x \times y \in \mathrm{Nn}$

Proof. †696 $x, x \times y \in \mathrm{Nn} .\supset. x + (x \times y) \in \mathrm{Nn}$ (1)
*223 R693 $\supset$:. L1 $\supset$. $x \times (\mathrm{S}`y) \in \mathrm{Nn}$ [:≡ 1] (2)
*100, *117 $x \in \mathrm{Nn} .\supset (y)$([693 . 2 .] $y \in \mathrm{Nn} . x \times y \in \mathrm{Nn} .\supset$RR2) (3)

†692, *224 $x \times 0 \, \epsilon \, \text{Nn} \, [. \equiv 632]$ (4)

*637 $\text{R3} \, [. \, 4] \, . \, y \, \epsilon \, \text{Nn} \, . \supset . \, x \times y \, \epsilon \, \text{Nn}$ (5)

*100 $[3 \, . \, 5 \, . \supset] \, 697$

†698. $(y)(x)$ $x, y \, \epsilon \, \text{Nn} \, . \supset . \, x \wedge y \, \epsilon \, \text{Nn}$

Proof similar.

§ 49. *Familiar Identities of Arithmetic*

THE FOLLOWING is the associative law of addition.

†710. $(z)(y)(x)$ $x, y, z \, \epsilon \, \text{Nn} \, . \supset . \, (x + y) + z = x + (y + z)$

Proof. †690 $\text{R696} \supset . \, (x + y) + 0 = x + y$ (1)

†690 $y \, \epsilon \, \text{Nn} \, . \supset . \, y + 0 = y$

*223 $\supset :. \, \text{R696} \supset . \, (x + y) + 0 = x + (y + 0) \, [: \equiv 1]$ (2)

*100 $[696 \, . \, 2 \, .] \, \text{L696} \, . \supset . \, (x + y) + 0 = x + (y + 0)$ (3)

†691 $z \, \epsilon \, \text{Nn} \, . \supset . \, (x + y) + (\text{S}`z) = \text{S}`((x + y) + z)$ (4)

*223 $\text{R710} \supset :. \, [4 \equiv :] \, \text{L4} \supset . \, \text{LR4} = \text{S}`(x + (y + z))$ (5)

†696 $y, z \, \epsilon \, \text{Nn} \, . \supset . \, y + z \, \epsilon \, \text{Nn}$

†691 $\supset . \, x + (\text{S}`(y + z)) = \text{S}`(x + (y + z))$

*223 $\supset :: \text{R710} \supset : \text{L4} \supset . \, \text{LR4} = x + (\text{S}`(y + z)) [. : \equiv 5]$ (6)

†691 $\text{L4} \supset . \, y + (\text{S}`z) = \text{S}`(y + z)$

*223 $\supset ::. \, \text{L6} \supset :. \, \text{R710} \supset : \text{L4} \supset . \, \text{LR4} = x + (y + (\text{S}`z))$
$[:: \equiv 6]$ (7)

*100 $[7 \, .] \, \text{L4} \, . \, \text{L696} \supset \text{R710} \, . \supset : \text{L696} \supset . \, \text{LR4} = x + (y + (\text{S}`z))$ (8)

*637 $[(z) 8 \, . \, 3 \, .] \, \text{L4} \, . \supset . \, \text{L696} \supset \text{R710}$ (9)

*100 $[9 \supset] \, 710$

The above theorem carries an antecedent which limits the values of all the variables to natural numbers; and the same is true of the eight ensuing theorems. Proofs of such theorems can be shortened by agreeing simply to disregard clauses of the form $\ulcorner \zeta \, \epsilon \, \text{Nn} \urcorner$. This course is justified by the fact that any clause $\ulcorner \zeta \, \epsilon \, \text{Nn} \urcorner$ which would be required in the full proof of such a theorem is readily derivable from the antecedent of the theorem with help of †631 and †696–†698 (or perhaps provided outright by †632 or †633). Shortened in this fashion, the above proof assumes the following form.

†690 $(x + y) + 0 = x + y$
†690, *227 $= x + (y + 0)$ (1)
*226 R710 ⊃. $S`((x + y) + z) = S`(x + (y + z))$
†691, *224 ⊃. $(x + y) + (S`z) = x + (S`(y + z))$
†691, *224 ⊃. $(x + y) + (S`z) = x + (y + (S`z))$ (2)
*637 [(z)2 . 1 .⊃] R710

The proofs of the next eight theorems are rendered in the same elliptical fashion; the reader can easily restore the full details on comparing the above two versions of the proof of †710.

†711. (x) $x \in \text{Nn}$.⊃. $x \times 1 = x$

Proof. †615, *227 $x \times 1 = x \times (S`0)$
†693 $= x + (x \times 0)$
†692, *227 $= x + 0$
†690 $= x$

†712. $(y)(x)$ $x, y \in \text{Nn}$.⊃. $(S`x) + y = S`(x + y)$

Proof. †690 $(S`x) + 0 = S`x$
†690, *227 $= S`(x + 0)$ (1)
*226 R712 ⊃. $S`((S`x) + y) = S`(S`(x + y))$
†691, *224 ⊃. $(S`x) + (S`y) = S`(x + (S`y))$ (2)
*637 [(y)2 . 1 .⊃] R712

The next is the commutative law of addition.

†713. $(y)(x)$ $x, y \in \text{Nn}$.⊃. $x + y = y + x$

Proof. *223 R713 ⊃: [R712 ≡.] $(S`x) + y = S`(y + x)$
†691, *224 ⊃. $(S`x) + y = y + (S`x)$ (1)
*231 [(y)1 ⊃:] $x + 0 = 0 + x$.⊃. $(S`x) + 0 = 0 + (S`x)$ (2)
†182 $0 + 0 = 0 + 0$ (3)
*637 [(x)2 . 3 .⊃.] $y + 0 = 0 + y$ (4)
†183 [4 ≡.] $0 + y = y + 0$ (5)
*637 [(x)1 . 5 .⊃] R713

The next is the law of distributivity of multiplication into addition.

†714. $(z)(y)(x)$ $x, y, z \in \text{Nn}$.⊃. $x \times (y + z) = (x \times y) + (x \times z)$

Proof. †690, *227 $x \times (y + 0) = x \times y$
†690 $= (x \times y) + 0$

†692, *227 $= (x \times y) + (x \times 0)$ (1)

†710 $x + ((x \times y) + (x \times z)) = (x + (x \times y)) + (x \times z)$

†713, *227 $= ((x \times y) + x) + (x \times z)$

†710 $= (x \times y) + (x + (x \times z))$ (2)

*223 $\mathrm{R714} \supset : x + (x \times (y + z)) = \mathrm{R2}\ [. \equiv 2]$

†693, *224 $\supset . x \times (\mathrm{S}‘(y + z)) = (x \times y) + (x \times (\mathrm{S}‘z))$

†691, *224 $\supset . x \times (y + (\mathrm{S}‘z)) = (x \times y) + (x \times (\mathrm{S}‘z))$ (3)

*637 $[(z)3 \,.\, 1 \,. \supset]\ \mathrm{R714}$

Next come the associative and commutative laws of multiplication.

†715. $(z)(y)(x)$ $x, y, z \in \mathrm{Nn} \,. \supset .\, (x \times y) \times z = x \times (y \times z)$

Proof. †692 $(x \times y) \times 0 = 0$

†692 $= x \times 0$

†692, *227 $= x \times (y \times 0)$ (1)

†714 $(x \times y) + (x \times (y \times z)) = x \times (y + (y \times z))$ (2)

*223 $\mathrm{R715} \supset : (x \times y) + ((x \times y) \times z) = \mathrm{R2}\ [. \equiv 2]$

†693, *224 $\supset . (x \times y) \times (\mathrm{S}‘z) = x \times (y \times (\mathrm{S}‘z))$ (3)

*637 $[(z)3 \,.\, 1 \,. \supset]\ \mathrm{R715}$

†716. $(y)(x)$ $x, y \in \mathrm{Nn} \,. \supset .\, x \times y = y \times x$

Proof. †692 $(\mathrm{S}‘x) \times 0 = 0$

†692, †690 $= (x \times 0) + 0$ (1)

†712 $(\mathrm{S}‘x) + ((x \times y) + y) = \mathrm{S}‘(x + ((x \times y) + y))$

†710, *227 $= \mathrm{S}‘((x + (x \times y)) + y)$

†691 $= (x + (x \times y)) + (\mathrm{S}‘y)$ (2)

*223

$(\mathrm{S}‘x) \times y = (x \times y) + y \,. \supset : (\mathrm{S}‘x) + ((\mathrm{S}‘x) \times y) = \mathrm{R2}\ [. \equiv 2]$

†693, *224 $\supset . (\mathrm{S}‘x) \times (\mathrm{S}‘y) = (x \times (\mathrm{S}‘y)) + (\mathrm{S}‘y)$ (3)

*637 $[(y)3 \,.\, 1 \,. \supset .]\ (\mathrm{S}‘x) \times y = (x \times y) + y$ (4)

*223 $\mathrm{R716} \supset : [4 \equiv .]\ (\mathrm{S}‘x) \times y = (y \times x) + y$

†713, *224 $\supset . (\mathrm{S}‘x) \times y = y + (y \times x)$

†693, *224 $\supset . (\mathrm{S}‘x) \times y = y \times (\mathrm{S}‘x)$

Rest of proof like last five lines of proof of †713.

The next two are the familiar laws for breaking up exponents.

†717. $(z)(y)(x)$ $x, y, z \in \mathrm{Nn} \,. \supset .\, x \wedge (y + z) = (x \wedge y) \times (x \wedge z)$

Proof †690, *227 $x \wedge (y + 0) = x \wedge y$
†711 $= (x \wedge y) \times 1$
†694, *227 $= (x \wedge y) \times (x \wedge 0)$ (1)
*226 R717 $\supset . \ (x \wedge (y + z)) \times x = ((x \wedge y) \times (x \wedge z)) \times x$
†715, *224 $\supset . \ (x \wedge (y + z)) \times x = (x \wedge y) \times ((x \wedge z) \times x)$
†716, *224 $\supset . \ x \times (x \wedge (y + z)) = (x \wedge y) \times (x \times (x \wedge z))$
†695, *224 $\supset . \ x \wedge (\mathrm{S}`(y + z)) = (x \wedge y) \times (x \wedge (\mathrm{S}`z))$
†691, *224 $\supset : x \wedge (y + (\mathrm{S}`z)) = (x \wedge y) \times (x \wedge (\mathrm{S}`z))$ (2)
*637 $[(z)2 . 1 . \supset]$ R717

†718. $(z)(y)(x)$ $x, y, z \in \mathrm{Nn} . \supset . \ x \wedge (y \times z) = (x \wedge y) \wedge z$

Proof. †692, *227 $x \wedge (y \times 0) = x \wedge 0$
†694 $= 1$
†694 $= (x \wedge y) \wedge 0$ (1)
†693, *227 $x \wedge (y \times (\mathrm{S}`z)) = x \wedge (y + (y \times z))$
†717 $= (x \wedge y) \times (x \wedge (y \times z))$ (2)
*223 R718 $\supset : [2 \equiv .]$ L2 $= (x \wedge y) \times ((x \wedge y) \wedge z)$
†695, *224 $\supset .$ L2 $= (x \wedge y) \wedge (\mathrm{S}`z)$ (3)
*637 $[(z)3 . 1 . \supset]$ R718

Note that laws analogous to the above hold also for relative powers.

†719. $(z)(y)(x)$ $y, z \in \mathrm{Nn} . \supset . \ x^{y+z} = (x^y)|(x^z)$

†720. $(z)(y)(x)$ $y, z \in \mathrm{Nn} . \supset . \ x^{y \times z} = (x^y)^z$

Proofs similar, using †560, †682, †491, †684, †685, and †719 instead respectively of †711, †694, †715, †716, †695, and †717.

Further arithmetical notions which are readily defined are the notions of greater and less.

$\ulcorner(\zeta \geqslant \eta)\urcorner$ *for* $\ulcorner(\eta \in \mathrm{Nn} . *\mathrm{S}(\zeta, \eta))\urcorner$.
$\ulcorner(\zeta > \eta)\urcorner$ *for* $\ulcorner(\zeta \geqslant \eta . \zeta \neq \eta)\urcorner$.
$\ulcorner(\zeta \leqslant \eta)\urcorner$ *for* $\ulcorner(\eta \geqslant \zeta)\urcorner$.
$\ulcorner(\zeta < \eta)\urcorner$ *for* $\ulcorner(\eta > \zeta)\urcorner$.

If we want to set up the general notation of Arabic numerals in systematic fashion, we first introduce the digits '0', . . . , '9' in the manner begun in D36–38; then we define *xy* or *x*ˆ*y* (not to be confused with $x \times y$) in such a way that, where *x* and *y* are any natural numbers, $x \hat{} y$ proves to be the number whose designating numeral under the usual Arabic notation consists of that of x followed by that of y. A definition to this purpose is as follows:

$$\ulcorner(\zeta\ \eta)\urcorner \quad \textit{or} \quad \ulcorner(\zeta \hat{} \eta)\urcorner \quad \textit{for}$$
$$\ulcorner(\eta + (\zeta \times ((S\text{‘}9) \wedge (\imath\alpha)(\beta)((S\text{‘}9) \wedge \beta > \eta \,.\, \beta > 0 \,.\!\equiv\!.\, \beta \geqslant \alpha))))\urcorner.$$

The operation so defined is associative except where '0' appears in middle position; e.g. (58)3 and 5(83) are both 583, and (50)3 is 503, but 5(03) is rather 53 (03 being 3). But we may dispense with further derivation of theorems; the preceding developments illustrate the foundations of the arithmetic of natural numbers sufficiently for present purposes.

Proofs of †710–†718 which proceed by mathematical induction and strongly resemble the above elliptical proofs in outward appearance have been usual since Grassmann (1861).

§ 50. *Ratios*

IN THIS section and the next two the derivation of further notions of quantitative mathematics will be outlined. The presentation will be sketchy, and progressively sketchier as it proceeds. No more theorems will be proved.

Let us first take up the notion of *ratio*. One natural number is said to stand in the ratio x/y to another natural number when the one number multiplied by y equals the other multiplied by x. The number 8, e.g., stands in the ratio $2/3$ to 12; and 12 stands in the ratio $3/2$ to 8. In general, x/y is the relation of any natural number z to any natural number w such that $y \times z = x \times w$.[1]

$$\ulcorner(\zeta/\eta)\urcorner \quad \textit{for} \quad \ulcorner\hat{\alpha}\hat{\beta}(\alpha, \beta \in \mathrm{Nn} \,.\, \eta \times \alpha = \zeta \times \beta)\urcorner.$$

Like the natural numbers which they relate, ratios themselves admit of an obvious ranking in point of greater and less. One ratio is greater than another if the number which bears the one ratio to a given number is greater than the number which bears the other ratio to the given number. To determine the relative rank

[1] This version of ratios is essentially Peano's (1901, pp. 54f).

of two ratios x/y and x'/y' we first find a natural number w belonging to the converse domain of both ratios; i.e., a natural number w such that some natural number z bears x/y to w and some natural number z' bears x'/y' to w. Then we rank x/y as greater or less than x'/y' according as z is greater or less than z'.

A number w of the required kind is readily found; namely, $y \times y'$. There are natural numbers z and z' which bear the respective ratios x/y and x'/y' to $y \times y'$, namely the numbers $x \times y'$ and $x' \times y$. For, by definition, to say that $x \times y'$ bears x/y to $y \times y'$ is to say merely that

$$y \times (x \times y') = x \times (y \times y');$$

and to say that $x' \times y$ bears x'/y' to $y \times y'$ is to say merely that

$$y' \times (x' \times y) = x' \times (y \times y').$$

To judge whether x/y is greater or less than x'/y' we have therefore only to compare these particular numbers z and z', namely $x \times y'$ and $x' \times y$. The ratio x/y is greater than x'/y' if and only if $x \times y' > x' \times y$.

The analogue of '$>$' for ratios will be rendered '${}_s{>}$';[1] thus

$$(x)(y)(x')(y') \qquad x/y \;{}_s{>}\; x'/y' \;.{\equiv}.\; x \times y' > x' \times y.$$

The following is the formal definition.

$\ulcorner(\zeta \;{}_s{>}\; \eta)\urcorner$ *for*

$$\ulcorner(\exists\alpha)(\exists\beta)(\exists\gamma)(\exists\delta)(\zeta = \alpha/\beta \,.\, \eta = \gamma/\delta \,.\, \alpha \times \delta > \gamma \times \beta)\urcorner.$$

The ratios 0/1, 1/1, 2/1, 3/1, etc. will be referred to briefly as ${}_s0$, ${}_s1$, ${}_s2$, ${}_s3$, etc.; thus we define:

$$\ulcorner {}_s\zeta \urcorner \quad \textit{for} \quad \ulcorner(\zeta/1)\urcorner.$$

All ratios of this sort are functions with the range of functionality Nn. ${}_s2$ is the function "double of", ${}_s3$ is "triple of", and so on. ${}_s1$ is simply

$$\hat{x}\hat{y}(x \,\epsilon\, \mathrm{Nn} \,.\, x = y),$$

i.e., identity confined to natural numbers. ${}_s0$ is

[1] The choice of the letter 's' is arbitrary; 'r' would have been preferable, but it is reserved for use in the arithmetic of real numbers (§ 51). Whitehead and Russell afford a precedent for the use of 's', in connection with '+' and '×' (vol. 3, *305).

$$\hat{x}\hat{y}(x = 0 \,.\, y \,\epsilon\, \mathrm{Nn}),$$

i.e., the relation which 0 bears to every natural number; in other words, the "constant function" whose value is 0 for all natural numbers as arguments. All this is readily seen by checking back over the definitions and making a few obvious transformations.

The ratios ${}_s0$, ${}_s1$, ${}_s2$, etc. fall into the same order as do the natural numbers 0, 1, 2, etc.; i.e.,

$$(x)(y) \qquad {}_sx \; {}_s\!> {}_sy \,.\!\equiv\!.\, x > y.$$

But between each of these ratios ${}_sx$ and the next, viz. ${}_s(\mathrm{S}`x)$, infinitely many further ratios are packed which have no analogues among the natural numbers; between ${}_s2$ and ${}_s3$, e.g., we have 5/2, 7/3, 8/3, 9/4, 11/4, etc. without end.

It is apparent from the definition of $\ulcorner\zeta/\eta\urcorner$ that 1/0, or indeed $z/0$ for any natural number z except 0, is simply the relation $\hat{x}\hat{y}(x \,\epsilon\, \mathrm{Nn} \,.\, y = 0)$ which every natural number bears to 0; and that 0/0 is the relation $\hat{x}\hat{y}(x, y \,\epsilon\, \mathrm{Nn})$ which holds between all natural numbers. If we count these as ratios, then 1/0 will be the greatest of all ratios, whereas 0/0 will fall nowhere in the less-to-greater series; for

	$(x)(y)$	$x, y \,\epsilon\, \mathrm{Nn} \,.\, y \neq 0 \,.\!\supset\!.\, 1 \times y > x \times 0,$
i.e.	$(x)(y)$	$x, y \,\epsilon\, \mathrm{Nn} \,.\, y \neq 0 \,.\!\supset\!.\, 1/0 \;{}_s\!> x/y,$
and	$(x)(y)$	$\sim(0 \times y > x \times 0) \,.\, \sim (x \times 0 > 0 \times y),$
i.e.	$(x)(y)$	$\sim(0/0 \;{}_s\!> x/y) \,.\, \sim (x/y \;{}_s\!> 0/0).$

It is more usual, however, not to count 1/0 and 0/0 as ratios. The class Ra of ratios is therefore defined thus:

$$\text{`Ra'} \quad \textit{for} \quad \text{`}\hat{x}(\exists y)(\exists z)(y \,\epsilon\, \mathrm{Nn} \,.\, z > 0 \,.\, x = y/z)\text{'}.$$

Pressing the analogy which began with the introduction of '${}_s\!>$', '${}_s0$', '${}_s1$', etc., it is easy to define an analogue '${}_s+$' of '$+$' and an analogue '${}_s\times$' of '$\times$'. If z and z' are natural numbers bearing the respective ratios x/y and x'/y' to a natural number w, then the rational sum $(x/y) \;{}_s\!+ (x'/y')$ is explained as the ratio which the natural sum $z + z'$ bears to w. Suitable choices of z, z' and w have already been remarked; we have seen that $x \times y'$ and $x' \times y$ bear the respective ratios x/y and x'/y' to $y \times y'$. Hence

$(x/y)\ {}_s{+}\ (x'/y')$ is describable simply as the ratio of $(x \times y') + (x' \times y)$ to $y \times y'$; in short, as

$$((x \times y') + (x' \times y)) \mathbin{/} (y \times y').$$

The rational product $(x/y)\ {}_s{\times}\ (x'/y')$ is construed yet more simply, viz. as the ratio borne to $y \times y'$ by the product $x \times x'$ itself; in short, as

$$(x \times x') \mathbin{/} (y \times y').$$

The formal definitions are as follows.

$$\ulcorner(\zeta\ {}_s{+}\ \eta)\urcorner \quad \textit{for} \quad \ulcorner(\imath\gamma)(\exists\alpha)(\exists\beta)(\exists\alpha')(\exists\beta')(\zeta = \alpha/\beta \,.\, \eta = \alpha'/\beta' \,.\, \gamma = ((\alpha \times \beta') + (\alpha' \times \beta)) \mathbin{/} (\beta \times \beta'))\urcorner.$$

$$\ulcorner(\zeta\ {}_s{\times}\ \eta)\urcorner \quad \textit{for} \quad \ulcorner(\imath\gamma)(\exists\alpha)(\exists\beta)(\exists\alpha')(\exists\beta')(\zeta = \alpha/\beta \,.\, \eta = \alpha'/\beta' \,.\, \gamma = (\alpha \times \alpha') \mathbin{/} (\beta \times \beta'))\urcorner.$$

It turns out that rational addition and multiplication as defined reproduce the familiar formal laws of natural addition and multiplication, and it turns out further that they yield parallel results numerically when applied to the analogues ${}_s0$, ${}_s1$, etc. of the natural numbers. ${}_s5\ {}_s{+}\ {}_s7 = {}_s12$, e.g., and ${}_s5\ {}_s{\times}\ {}_s7 = {}_s35$. In general,

$$(x)(y) \qquad {}_sx\ {}_s{+}\ {}_sy = {}_s(x + y),$$
$$(x)(y) \qquad {}_sx\ {}_s{\times}\ {}_sy = {}_s(x \times y).$$

If we disregard the subscripts, we have in rational arithmetic simply a broader arithmetic of addition and multiplication in which the arithmetic of natural addition and multiplication is embedded as an integral part.

One feature which puts rational arithmetic in marked contrast to natural arithmetic is the efficacy of *division*. The quotient $x \div y$ could indeed have been defined in natural arithmetic, viz. as $(\imath z)(x = y \times z)$; but it would have been of little use, for there *is* no number whose product by y is x except in those relatively rare cases where x is a multiple of y. Where x and y are ratios, on the other hand, there is almost always one and only one ratio z such that $x = y\ {}_s{\times}\ z$; always, in fact, except where $y = {}_s0$. Thus it is that the definition:

$$\ulcorner(\zeta\ {}_s{\div}\ \eta)\urcorner \quad \textit{for} \quad \ulcorner(\imath\alpha)(\zeta = \eta\ {}_s{\times}\ \alpha)\urcorner$$

of rational division is useful indeed. We find, in fact, that $x\ {}_{s}\div (w/w')$ is simply $x\ {}_{s}\times\ (w'/w)$.

Along with this gain in the matter of division, however, we encounter a loss in connection with the notion of power. Let us consider how a rational analogue '${}_{s}\wedge$' of the '$\wedge$' of natural arithmetic might appropriately be construed. The suitable version of $(x/y)\ {}_{s}\wedge\ {}_{s}2$, in particular, is obvious; just as $w \wedge 2$ is $w \times w$, so we may appropriately take $(x/y)\ {}_{s}\wedge\ {}_{s}2$ as $(x/y)\ {}_{s}\times\ (x/y)$. Thus

$$(x/y)\ {}_{s}\wedge\ {}_{s}2 = (x/y)\ {}_{s}\times\ (x/y) = (x \times x)\ /\ (y \times y) = (x \wedge 2)/(y \wedge 2).$$

Similarly

$$(x/y)\ {}_{s}\wedge\ {}_{s}1 = (x \wedge 1)\ /\ (y \wedge 1) = x/y.$$

More generally, we may always explain $(x/y)\ {}_{s}\wedge\ {}_{s}x'$ as $(x \wedge x')\ /\ (y \wedge x')$. This accounts for all rational powers of the form $(x/y)\ {}_{s}\wedge\ (x'/1)$; but it remains to consider $(x/y)\ {}_{s}\wedge\ (x'/y')$ where $y' \neq 1$. Let us then examine the case $w\ {}_{s}\wedge (1/2)$, where w is any ratio. We shall want to preserve the familiar laws of exponents which hold in natural arithmetic; thus

$$(w\ {}_{s}\wedge(1/2))\ {}_{s}\times\ (w\ {}_{s}\wedge(1/2)) = w\ {}_{s}\wedge\ ((1/2)\ {}_{s}+\ (1/2)) = w\ {}_{s}\wedge\ {}_{s}1.$$

But, as just previously decided, $w\ {}_{s}\wedge\ {}_{s}1 = w$. Hence $w\ {}_{s}\wedge\ (1/2)$ must be explained as the ratio z such that $z\ {}_{s}\times\ z = w$. The difficulty is, however, that there will very commonly be no such ratio z. E.g., there is no ratio z such that $z\ {}_{s}\times\ z = {}_{s}3$; thus ${}_{s}3\ {}_{s}\wedge\ (1/2)$ is missing.

Whereas the notion of power in natural arithmetic works for any two natural numbers, the analogue for rational arithmetic fails in the observed fashion for many ratios. When the second ratio happens to be *integral*, i.e. of the form ${}_{s}x$, the power is indeed forthcoming; but if the second ratio is not integral, the power is missing more often than not. Rational arithmetic is fully as defective with respect to the notion of power as was natural arithmetic with respect to division.

§ 51. *Real Numbers*

THE MISSING powers in rational arithmetic can nevertheless be approximated by ratios, to within any preassigned interval of accuracy. Whereas e.g. there is no ratio $_s3\ _s\wedge(1/2)$, in other words no ratio z such that $z\ _s\times\ z = {_s3}$, there are nevertheless ratios z such that $z\ _s\times\ z$ falls within any preassigned neighborhood of $_s3$. Thus, consider the class of all those ratios z which are small enough so that $_s3\ _s> z\ _s\times\ z$. All these ratios are smaller e.g. than 7/4; but within the class there is no largest. By choosing larger and larger ratios z from this class we can bring $z\ _s\times\ z$ closer and closer to $_s3$, without end.

The class just now referred to has these four features. (1) Its members are ratios. (2) It does not exhaust the ratios. (3) It has no greatest member. (4) It contains all ratios except those which are too large; i.e., a ratio falls outside the class only if it exceeds all members of the class. These four features can be formulated together in the following compact fashion: the class is not the class of all ratios, and anything z belongs to the class if and only if $y\ _s> z$ for some y belonging to the class. A class of this kind is called — by caprice, we may for the moment suppose — a *real number*.[1] Thus the class Nr of real numbers is definable as follows:

$$\text{`Nr'} \quad \textit{for} \quad \text{`}\hat{x}(x \neq \text{Ra} \,.\, x = \hat{z}(\exists y)(y \in x \,.\, y\ _s> z))\text{'}.$$

The class of all ratios less than a given ratio is a real number; for it obviously has the four features noted above. Corresponding to each ratio x, thus, we have a real number $\hat{y}(x\ _s> y)$. The latter will be referred to briefly as $_rx$.

$$\ulcorner {_r\zeta} \urcorner \quad \textit{for} \quad \ulcorner \hat{\alpha}(\zeta\ _s> \alpha) \urcorner$$

[1] The fact that such classes constitute a model of the traditional real number system was pointed out by Dedekind (*Stetigkeit*). The outright identification of the real numbers with those classes was first explicitly propounded by Russell (*Principles*, Ch. XXXIII), though rather hinted previously by Peano ("Sui numeri").

But we also have real numbers which do not thus correspond to ratios; one such is the real number originally considered above, viz. $\hat{z}({}_s3 \ {}_s{>} \ z \ {}_s{\times} \ z)$. There is no ratio x such that this real number is ${}_rx$; in other words, no ratio x which is just next greater than all ratios belonging to this real number. If there were such a ratio, indeed, it would be our missing power ${}_s3 \ {}_s{\wedge} \ (1/2)$. Real numbers which correspond thus to no ratios are called *irrationals*.

$$\text{'Irrat'} \quad \textit{for} \quad \text{'}\hat{x}(x \in \text{Nr} \,.\, (y)(x \neq {}_ry))\text{'}.$$

The others may be called rational real numbers.

An irrational is a real number which, so to speak, corresponds to a missing ratio. The real number $\hat{z}({}_s4 \ {}_s{>} \ z \ {}_s{\times} \ z)$ is a rational one, for it is ${}_rx$ where $x = {}_s2$; it corresponds to the ratio ${}_s2$. On the other hand the irrational $\hat{z}({}_s3 \ {}_s{>} \ z \ {}_s{\times} \ z)$ fills in where the series of ratios exhibited only a lack: the lack of a rational power ${}_s3 \ {}_s{\wedge}(1/2)$.

A general ranking in point of greater and less is readily imposed on real numbers, in such a way that any rational reals ${}_rx$ and ${}_ry$ will follow the same order as the corresponding ratios x and y. We have merely to reflect that ${}_rx$ will include ${}_ry$ if and only if x is greater than or equal to y. Thus '${}_r{\geqslant}$' and '${}_r{>}$', the analogues of '$\geqslant$' and '$>$' for real numbers, are defined as follows:

$$\ulcorner(\zeta \ {}_r{\geqslant} \ \eta)\urcorner \quad \textit{for} \quad \ulcorner(\zeta, \eta \in \text{Nr} \,.\, \eta \subset \zeta)\urcorner,$$
$$\ulcorner(\zeta \ {}_r{>} \ \eta)\urcorner \quad \textit{for} \quad \ulcorner(\zeta \ {}_r{\geqslant} \ \eta \,.\, \zeta \neq \eta)\urcorner.$$

The series of real numbers, arranged in point of greater and less in this sense, images the series of ratios except that new numbers, the irrationals, are squeezed in everywhere. Between the series of ratios z such that ${}_s3 \ {}_s{>} \ z \ {}_s{\times} \ z$ and the series of ratios z such that $z \ {}_s{\times} \ z \ {}_s{>} \ {}_s3$, for example, there is no further ratio; but between the real numbers corresponding to the former ratios and the real numbers corresponding to the latter ratios there *is* an intermediate real number, viz. the irrational $\hat{z}({}_s3 \ {}_s{>} \ z \ {}_s{\times} \ z)$.

A ratio is called an *upper bound* of a given class of ratios if no member of the class exceeds it. A real number is called an upper bound of a class of real numbers under similar circumstances. Now a class of ratios may have an upper bound without having a least upper bound. This is true e.g. of the class $\hat{z}({}_s3 \ {}_s{>} \ z \ {}_s{\times} \ z)$ hitherto discussed; 9/5 is one ratio which no member of the class exceeds, but 7/4 is a smaller ratio having the same property, and there are smaller and smaller ones without end. On the other hand *every* class of real numbers which has an upper bound has a least one. Consider e.g. the class analogous to the above class of ratios, viz. the class of all real numbers z such that ${}_{r}3\ {}_{r}\!> z\ {}_{r}\!\times z$; this class has the irrational number $\hat{z}({}_{s}3\ {}_{s}\!> z\ {}_{s}\!\times z)$ itself as least upper bound. In general, indeed, it is readily verified that if x is a class of real numbers which has any upper bound at all then its least upper bound is simply the class $\mathfrak{E}``x$ of all members of members of x.

The fact that every bounded class of reals has a least bound is the basic formal difference between the reals and the ratios. It is in this sense that the series of real numbers is said to be *continuous* while the series of ratios is not. On the other hand both series are *dense*, in the sense that there is a third number between any two.

If a bounded class x of ratios lacks a least bound, then the class x' of those real numbers ${}_{r}z$ which correspond to the ratios z in x has an irrational least bound. Conversely, also, every irrational is the least bound of a class x' of the above kind. This is the strict meaning of the earlier loose remark to the effect that an irrational is a real number which corresponds to a missing ratio.

The irrationals do far more than supply missing powers. One familiar irrational of another sort is π. The irrationals exist in such variety, indeed, that no notation whatever is capable of providing a separate name for each of them.[1] Note in contrast that

[1] The individual signs available in any given notation are finite in number — say $\mu_1, \mu_2, \ldots, \mu_k$. (In the present notation we have indeed an infinity of variables, but they are made up of the five signs 'w', 'x', 'y', 'z', '′'.) The infinite totality of expressions which can be constructed from $\mu_1, \ldots, \mu_k$ can be ordered thus: $\mu_1, \ldots, \mu_k, \ulcorner\mu_1\mu_1\urcorner, \ldots, \ulcorner\mu_1\mu_k\urcorner, \ulcorner\mu_2\mu_1\urcorner, \ldots, \ulcorner\mu_k\mu_k\urcorner, \ulcorner\mu_1\mu_1\mu_1\urcorner, \ldots, \ulcorner\mu_1\mu_1\mu_k\urcorner, \ulcorner\mu_1\mu_2\mu_1\urcorner, \ldots$. Let us refer to them briefly as $\mu_1, \ldots, \mu_k, \mu_{k+1}, \ldots$. Consider next the unending decimal expansion $\langle x\rangle$ of a positive real number $x \leqq 1$. Because unending, $\langle x\rangle$ must be thought of not as an expression but as a function, such that $\langle x\rangle$'n is the number $\leqq 9$ which turns up in nth place on expanding x. To say that $\langle x\rangle$ is unending is to say that for every number there is a larger, n, such that $\langle x\rangle$'$n \neq 0$. We know that every positive real number $\leqq 1$ has a unique unending expansion $\langle x\rangle$ (as well as sometimes an ending one; e.g., $.43 = .42999\ldots$). Conversely, also, each such expansion determines a positive real number $\leqq 1$. Then

any ratio can be named by putting '/' between numerals.

We turn now to the problem of defining multiplication and addition for real numbers. Our object will be to maintain the laws:

$$(1) \qquad {}_{r}x \; {}_{r}\times \; {}_{r}y = {}_{r}(x \; {}_{s}\times \; y), \qquad {}_{r}x \; {}_{r}+ \; {}_{r}y = {}_{r}(x \; {}_{s}+ \; y)$$

for all ratios x and y, just as in defining '${}_{r}>$' our object was to maintain the law:

$${}_{r}x \; {}_{r}> \; {}_{r}y \;.\equiv.\; x \; {}_{s}> \; y.$$

Now clearly the following is true for all ratios (${}_{s}0$ and greater):

$$(2) \quad x \; {}_{s}\times \; y \; {}_{s}> \; z \;.\equiv\; (\exists x')(\exists y')(x \; {}_{s}> \; x' \;.\; y \; {}_{s}> \; y' \;.\; z = x' \; {}_{s}\times \; y').$$

I.e., since $w \; {}_{s}> \; z \;.\equiv.\; z \in {}_{r}w$,

$$z \in {}_{r}(x \; {}_{s}\times \; y) \;.\equiv\; (\exists x')(\exists y')(x' \in {}_{r}x \;.\; y' \in {}_{r}y \;.\; z = x' \; {}_{s}\times \; y').$$

In view of (1), then, we are to identify ${}_{r}x \; {}_{r}\times \; {}_{r}y$ with

$$\hat{z}(\exists x')(\exists y')(x' \in {}_{r}x \;.\; y' \in {}_{r}y \;.\; z = x' \; {}_{s}\times \; y').$$

So, if we treat irrationals on a par with ${}_{r}x$ and ${}_{r}y$,

$$\ulcorner(\zeta \; {}_{r}\times \; \eta)\urcorner \textit{ for } \ulcorner\hat{\alpha}(\exists\beta)(\exists\gamma)(\beta \in \zeta \;.\; \gamma \in \eta \;.\; \alpha = \beta \; {}_{s}\times \; \gamma)\urcorner.$$

Addition is more complex, because the analogue of (2) for addition can fail when x or y is ${}_{s}0$. What holds in lieu of (2) is this:

$$x \; {}_{s}+ \; y \; {}_{s}> \; z \;.\equiv:$$
$$x \; {}_{s}> \; z \;.\vee.\; y \; {}_{s}> \; z \;.\vee\; (\exists x')(\exists y')(x \; {}_{s}> \; x' \;.\; y \; {}_{s}> \; y' \;.\; z = x' \; {}_{s}+ \; y').$$

I.e., $z \in {}_{r}(x \; {}_{s}+ \; y) \;.\equiv:$

$$z \in {}_{r}x \;.\vee.\; z \in {}_{r}y \;.\vee\; (\exists x')(\exists y')(x' \in {}_{r}x \;.\; y' \in {}_{r}y \;.\; z = x' \; {}_{s}+ \; y').$$

Recalling (1), then, we are led to a definition of '${}_{r}x \; {}_{r}+ \; {}_{r}y$' which, if extended beyond ${}_{r}x$ and ${}_{r}y$ to real numbers generally, runs

consider the real number k which is determined as follows: for each n, $\langle k\rangle$'n is 1 or 2 according as μ_n does or does not name a positive real number $x \leqq 1$ such that $\langle x\rangle$'$n = 2$. Now k has no name μ_n; for, if μ_n named k, $\langle k\rangle$'n would have to satisfy the self-contradictory condition of being 2 if and only if other than 2. It is thus established that no notation is adequate to expressing every real number. Since each rational number is readily expressed with help of the fractional notation,it follows that no notation is adequate to expressing every irrational. (This argument is due in essential respects to Cantor.)

as follows:

$$\ulcorner(\zeta \,{}_{\mathrm{r}}\!+ \eta)\urcorner \textit{ for } \ulcorner(\zeta \cup \eta \cup \hat{\alpha}(\exists\beta)(\exists\gamma)(\beta \in \zeta \,.\, \gamma \in \eta \,.\, \alpha = \beta \,{}_{\mathrm{s}}\!+ \gamma))\urcorner.$$

The definition of power $\ulcorner(\zeta \,{}_{\mathrm{r}}\!\wedge \eta)\urcorner$ for real numbers will be passed over in this survey.

It was observed that natural arithmetic is virtually, but not actually, embedded in rational arithmetic. The whole numbers in rational arithmetic are not 0, 1, 2, etc., but rather the rational analogues ${}_{\mathrm{s}}0$, ${}_{\mathrm{s}}1$, ${}_{\mathrm{s}}2$, etc. Clearly the relation of rational to real arithmetic is similar; the rational reals are not ratios, but analogues ${}_{\mathrm{r}}x$ of ratios x. The whole real numbers, likewise, are not the natural numbers 0, 1, 2, etc.; they are rather real-number analogues ${}_{\mathrm{rs}}0$, ${}_{\mathrm{rs}}1$, ${}_{\mathrm{rs}}2$, etc. of the rational analogues ${}_{\mathrm{s}}0$, ${}_{\mathrm{s}}1$, ${}_{\mathrm{s}}2$, etc. of 0, 1, 2, etc. (Note incidentally that ${}_{\mathrm{rs}}0 = \Lambda$.) But when we are concerned with using arithmetic or investigating its higher reaches rather than examining its logical foundations, we of course find it convenient to leave these distinctions tacit and omit the subscripts.

§ 52. *Further Extensions*

FOR ALL REAL numbers x and y, there are real numbers $x \,{}_{\mathrm{r}}\!+ y$, $x \,{}_{\mathrm{r}}\!\times y$, and $x \,{}_{\mathrm{r}}\!\wedge y$; in this respect the arithmetic of real numbers is like that of natural numbers. For all real numbers x and y such that $y \neq {}_{\mathrm{rs}}0$, moreover, there is a real number $x \,{}_{\mathrm{r}}\!\div y$ whose product by y is x.

$$\ulcorner(\zeta \,{}_{\mathrm{r}}\!\div \eta)\urcorner \quad \textit{for} \quad \ulcorner(\iota\alpha)(\zeta = \eta \,{}_{\mathrm{r}}\!\times \alpha)\urcorner.$$

In this respect the arithmetic of real numbers is like that of ratios. And the arithmetic of real numbers outdoes both that of natural numbers and that of ratios in the following respect: for all real numbers x and y such that $y \neq {}_{\mathrm{rs}}0$ there is a yth *root* of x, symbolically $y \,{}_{\mathrm{r}}\!\surd\, x$; i.e., a number whose yth power is x.

$$\ulcorner(\zeta \,{}_{\mathrm{r}}\!\surd\, \eta)\urcorner \quad \textit{for} \quad \ulcorner(\iota\alpha)(\eta = \alpha \,{}_{\mathrm{r}}\!\wedge \zeta)\urcorner.$$

This notion of root, which is inverse to power just as quotient is inverse to product, admits equivalently of definition in terms of quotient and power:

$$\ulcorner(\zeta\ {}_{r}\!\surd\ \eta)\urcorner \quad \textit{for} \quad \ulcorner(\eta\ {}_{r}\!\wedge\ ({}_{rs}1\ {}_{r}\!\div\ \zeta))\urcorner.$$

Since in the arithmetic of real numbers the notions of product and power thus have inverses which are almost always applicable (always save in the zero case), we are led to look also for a generally applicable notion of *subtraction* inverse to addition; but here the arithmetic of real numbers, like that of natural numbers and that of ratios, is found wanting. It is not true in general, given any real numbers x and y, that there is a real number $x\ {}_{r}\!-\ y$ whose addition to y yields x; there is such a number only if $x\ {}_{r}\!\geqslant\ y$.

Just as the power defect in rational arithmetic was overcome by passing to real arithmetic, so the subtraction defect in real arithmetic is overcome by a further extension: introduction of negative numbers. Under this extension every real number x except ${}_{rs}0$ gives way to two numbers: a positive one ${}_{+}x$ and a negative one ${}_{-}x$. Only ${}_{rs}0$ is left unsplit; ${}_{+rs}0$ and ${}_{-rs}0$ are identified. The branches ${}_{+}x$ and ${}_{-}x$ of a real number x are construed, further, as distinct from the branches ${}_{+}y$ and ${}_{-}y$ of any other real number y.

Entities ${}_{+}x$ and ${}_{-}x$ fulfilling the above requirements come readily to hand; we can take ${}_{+}x$ as $\lambda_y(x\ {}_{r}\!+\ y)$, i.e. as the function of addition of x in the sense of the foregoing arithmetic of real numbers, and we can take ${}_{-}x$ as the converse of that function.

$$\ulcorner{}_{+}\zeta\urcorner \quad \textit{for} \quad \ulcorner\lambda_\alpha(\zeta\ {}_{r}\!+\ \alpha)\urcorner,$$
$$\ulcorner{}_{-}\zeta\urcorner \quad \textit{for} \quad \ulcorner\breve{\ }{}_{+}\zeta\urcorner.$$

These new numbers, real numbers in a broader sense, may be called *signed real numbers;* real numbers in the previous narrower sense may then be distinguished from these as *unsigned* real numbers. The class NR of signed real numbers is definable in obvious fashion:

$$\text{'NR'} \quad \textit{for} \quad \text{'}\hat{x}(\exists y)(y \in \text{Nr} : x = {}_{+}y \ .\vee.\ x = {}_{-}y)\text{'}.\text{[1]}$$

The *integers* form a subclass of NR, defined thus:

$$\text{'Int'} \quad \textit{for} \quad \text{'}\hat{x}((\exists y)(y \in \text{Nn} : x = {}_{+rs}y \ .\vee.\ x = {}_{-rs}y)\text{'}.$$

[1] This construction of the signed reals from the reals is modelled on Peano's construction (1901, pp. 48–49) of the signed integers from the natural numbers. Peano urges that the essential idea was entertained more or less clearly by MacLaurin and Cauchy.

Signed real numbers are ranked in point of greater and less in such a way that those of the form $_+x$ follow the same order as the corresponding unsigned reals, while the others follow the reverse order and are reckoned as smaller. Thus the series of signed real numbers constitutes an image of the unsigned series and a reverse image of the same laid end to end, with the zero as their point of contact. A definition which orders the signed real numbers in the described fashion is this:

$$\ulcorner(\zeta \ {}_{\mathrm{R}}\!> \eta)\urcorner \quad \textit{for} \quad \ulcorner(\zeta, \eta \in \mathrm{NR} \,.\, (\exists\alpha)(\zeta`\alpha \ {}_{\mathrm{r}}\!> \eta`\alpha))\urcorner.$$

A signed real number x is called *positive* if $x \ {}_{\mathrm{R}}\!> {}_{+\mathrm{rs}}0$; *negative* if ${}_{+\mathrm{rs}}0 \ {}_{\mathrm{R}}\!> x$. So far as ${}_{+\mathrm{rs}}0$ and the positive signed reals are concerned, this new arithmetic is to constitute an exact reproduction of the previous arithmetic of unsigned reals. Even throughout the negative signed reals, moreover, the notions of sum, product, and power are to conform to the same basic arithmetical laws. Definitions of sum and product which prove to meet these requirements are as follows:

$$\ulcorner(\zeta \ {}_{\mathrm{R}}\!+ \eta)\urcorner \quad \textit{for} \quad \ulcorner((\zeta \mid \eta) \cup (\eta \mid \zeta))\urcorner,$$

$$\ulcorner(\zeta \ {}_{\mathrm{R}}\!\times \eta)\urcorner \quad \textit{for}$$

$$\ulcorner(\imath\alpha)(\exists\beta)(\exists\gamma)(\zeta = {}_+\beta \,.\, \eta = {}_+\gamma \,.\mathrm{v}.\, \zeta = {}_-\beta \,.\, \eta = {}_-\gamma : \alpha = {}_+(\beta \ {}_{\mathrm{r}}\!\times \gamma) :$$
$$\mathrm{v} : \zeta = {}_+\beta \,.\, \eta = {}_-\gamma \,.\mathrm{v}.\, \zeta = {}_-\beta \,.\, \eta = {}_+\gamma : \alpha = {}_-(\beta \ {}_{\mathrm{r}}\!\times \gamma))\urcorner.$$

The general applicability of subtraction is achieved. For all signed real numbers x and y there proves to be a signed real number $x \ {}_{\mathrm{R}}\!- y$, in the sense:

$$\ulcorner(\zeta \ {}_{\mathrm{R}}\!- \eta)\urcorner \quad \textit{for} \quad \ulcorner(\imath\alpha)(\zeta = \eta \ {}_{\mathrm{R}}\!+ \alpha)\urcorner.$$

It turns out, indeed, that $x \ {}_{\mathrm{R}}\!- y$ is simply $x \ {}_{\mathrm{R}}\!+ {}_-z$ or $x \ {}_{\mathrm{R}}\!+ {}_+z$ according as y is ${}_+z$ or ${}_-z$. Division survives as in rational and unsigned real arithmetic; for all signed reals x and y such that $y \neq {}_{+\mathrm{rs}}0$ there is a signed real $x \ {}_{\mathrm{R}}\!\div y$ in the sense:

$$\ulcorner(\zeta \ {}_{\mathrm{R}}\!\div \eta)\urcorner \quad \textit{for} \quad \ulcorner(\imath\alpha)(\zeta = \eta \ {}_{\mathrm{R}}\!\times \alpha)\urcorner.$$

But the notion of power presents a difficulty. When we passed from natural numbers to ratios, we found that the gain with regard to division was offset by a loss with regard to the notion of power; and in signed real arithmetic the gain with regard to subtraction proves to be similarly offset. It turns out that if we define '${}_{R}/\!\!\wedge$' in the way in which we would have to define it in order to meet the requirements mentioned above, there will frequently be no such signed real number as $x\ {}_{R}/\!\!\wedge\ y$ when x is negative and y is not an integer.

This shortcoming proves to be attributable ultimately to the fact that there is no signed real number z such that $z\ {}_{R}\times\ z = {}_{-rs}1$. A still further arithmetic is thus prompted, the arithmetic of *complex* numbers, which does contain the analogue of such a number z. Complex numbers are construed simply as pairs $x;y$ of signed real numbers; and the notions of sum, product, and power are defined in such a way that the arithmetic of signed real numbers is imaged within a part of the new arithmetic. In this arithmetic of complex numbers, as in that of unsigned reals, the notions of sum, product, and power all prove applicable without restriction; also, as in the arithmetic of unsigned reals, the inverse operations of division and extraction of root prove applicable save in the zero case; and also, as in the arithmetic of signed reals, subtraction applies without restriction.

Analysis, which includes the differential and integral calculus, depends in its basic developments not upon the arithmetic of complex numbers but upon that of signed reals. The fundamental notion of analysis is that of the *limit* of a function. A (signed real) number z is said to be *the limit of a function x for arguments approaching y from below*, symbolically $\lim\limits_{y} x$, if for every positive number z' there is a number y' which is less than y and such that the value of x differs from z by less than z' for any argument between y' and y. Rendered symbolically, this characterization of z assumes the form:

$$(z')(z' > 0 \,.\supset\, (\exists y')(y > y' \,. \\ (w)(y > w \,.\, w > y' \,.\supset.\, (x\text{‘}w) + z' > z \,.\, z + z' > x\text{‘}w))).$$

Thus

$$\ulcorner \lim_{\eta} \zeta \urcorner \quad \textit{for} \quad \ulcorner (\imath\alpha)(\beta)(\beta > 0 \,.\supset\, (\exists\gamma)(\eta > \gamma \,. \\ (\delta)(\eta > \delta \,.\, \delta > \gamma \,.\supset.\, (\zeta\text{‘}\delta) + \beta > \alpha \,.\, \alpha + \beta > \zeta\text{‘}\delta)))\urcorner.$$

The signs '$>$', and '$+$', and '0' here are of course intended as '${}_{R}>$', '${}_{R}+$', and '${}_{+rs}0$'; but such subscripts are conveniently dropped in discourse which concerns only one kind of number.

The limit of x for arguments approaching y from above, symbolically $\overset{y}{\lim}\ x$, and the so-called limit of x at infinity, symbolically $\underset{\infty}{\lim}\ x$, admit of analogous definitions. For the one we change '$\eta > \gamma$' in the above definition to '$\gamma > \eta$' and '$\eta > \delta \,.\, \delta > \gamma$' to '$\gamma > \delta \,.\, \delta > \eta$'; for the other we simply drop '$\eta > \gamma$' and '$\eta > \delta$'.

By way of concluding this sketchy survey of the route from logic into quantitative mathematics, the definition of the *derivative* Dx of a function x of one argument is now set down without discussion.

$$\ulcorner D\zeta \urcorner \quad \textit{for} \quad \ulcorner \lambda_\alpha \overset{\alpha}{\lim}\ \lambda_\beta(((\zeta`\beta) - (\zeta`\alpha)) \div (\beta - \alpha))\urcorner.$$

The derivative of the function "square of", e.g., proves to be the function "double of".

$$D\lambda_x(x \wedge 2) = \lambda_x(2 \times x).$$

The derivative of any "constant function" (cf. § 41) proves to be the constant function whose uniform value is 0.

$$(y) \qquad D\lambda_x y = \lambda_x 0.$$

Abstract algebra is one branch of mathematics which lies aside from the direction of derivation sketched in the latter sections; and geometry would appear to be another. Actually the place where abstract algebra fits into the general structure of logic is farther back, namely within the general theory of relations and functions whose groundwork was studied in Chapter V. As for geometry, a method of reduction to logic is ready at hand in the simple expedient of identifying geometrical entities with those arithmetical entities with which they are correlated through analytic geometry (cf. Study, pp. 86–92). The more abstract, so-called nonmetrical phases of geometry go over into correspondingly general phases of arithmetical analysis — so general, in the extreme case of topology, as to transcend the specifically arithmetical domain and form part of the general theory of relations and functions. This view of geometry seems to be the most convenient one in an account of the application of geometry to nature. A scale of measurement, whether of distance or of temperature, etc., may be viewed as a scheme for the systematic assignment of numbers directly to observed objects or events (cf. Carnap, *Physikalische Begriffsbildung*, and Jeffreys, Ch. VI); and then an empirico-geometrical statement, e.g. to the effect that Boston, Albany, and Buffalo are in line, may be analyzed as attributing a certain arithmetical relationship to the numbers which have been assigned to given objects under the scheme of distance measurement.

CHAPTER SEVEN

SYNTAX

§53. *Formality* [1]

VARIABLES were described earlier (§ 12) as comprising just the letters '*w*', '*x*', '*y*', and '*z*' with or without accents. The *atomic logical formulæ* were described in turn (§ 23) as comprising just the results of putting such variables in the blanks of '(ϵ)'. Then the *logical formulæ* were described as comprising, first, the atomic logical formulæ; second, all results of putting such expressions in the blanks of '(↓)' or after a parenthesized variable; third, all results of putting expressions of the thus supplemented totality in the blanks of '(↓)' or after a parenthesized variable; and so on. Now all these characterizations are *formal,* in that they speak only of the typographical constitution of the expressions in question and do not refer to the meanings of those expressions. But this explanation of 'formal' is vague; we turn now to a more precise version.

Let us use 'S_1', 'S_2', . . . , 'S_9' as names of the respective signs or typographic shapes '*w*', '*x*', '*y*', '*z*', '′', '(', ')', '↓', and 'ϵ'; thus

S_1 = '*w*' = double-yu,
S_2 = '*x*' = ex,
S_3 = '*y*' = wye,
S_4 = '*z*' = zee = zed = izzard,
S_5 = '′' = accent,
S_6 = '(' = left parenthesis,
S_7 = ')' = right parenthesis,
S_8 = '↓' = down-arrow,
S_9 = 'ϵ' = epsilon.

Further, let us use the arch '⁀' to indicate *concatenation* of expressions. Thus $S_1 \frown S_5$ is the expression '*w*′' which is formed by writing

[1] This concluding chapter will be unintelligible to those readers in whom there is a lingering tendency to confuse use and mention of expressions (§ 4). I have not seen how to make the chapter less liable to misunderstanding except at the expense of a disproportionate increase in length. Inasmuch as the foregoing six chapters constitute a self-contained unit, it has seemed best to limit this seventh one to a modest appendage amputable at discretion.

S_1 and then S_5; again $S_9 {}^\frown S_2$ is '$\epsilon\, x$'; again $(S_1{}^\frown S_5){}^\frown(S_9{}^\frown S_2)$ is the result '$w' \epsilon\, x$' of writing $S_1{}^\frown S_5$ followed by $S_9{}^\frown S_2$. But parentheses are not needed in connection with '$\frown$', since concatenation is obviously associative; thus

$$S_1{}^\frown S_5{}^\frown S_9{}^\frown S_2 = \text{'}w' \epsilon\, x\text{'}.$$

The arch '$\frown$' and the sign names 'S_1', 'S_2', . . . , 'S_9' provide a notation for *spelling;* '$S_1{}^\frown S_5{}^\frown S_9{}^\frown S_2$' may be read 'the expression consisting of S_1 followed by S_5 followed by S_9 followed by S_2', in other words 'the expression consisting of double-yu followed by accent followed by epsilon followed by ex'.

Into this rudimentary notation, plus the notation of logic, it is possible to translate 'x is a variable' in the sense verbally defined above; also 'x is an atomic logical formula'; also 'x is a logical formula'. First it is convenient to introduce a string of preliminary notions. An expression x constitutes an initial segment of y, or begins y, symbolically $x\text{B}y$, if y consists either of x alone or of x followed by something; thus

$$(x)(y) \qquad x\text{B}y \;.\equiv:\; x = y \;.\lor\; (\exists z)(x{}^\frown z = y).$$

Similarly x is a final segment of y, or ends y, symbolically $x\text{E}y$, if y consists either of x alone or of x preceded by something.

$$(x)(y) \qquad x\text{E}y \;.\equiv.\; x = y \;.\lor\; (\exists z)(z{}^\frown x = y).$$

Now x is a string of accents, symbolically Ac x, if every initial segment of x ends in an accent.

$$(x) \qquad \text{Ac}\, x \equiv (y)(y\text{B}x \;.\supset.\; S_5\text{E}y)$$

The result of writing a left parenthesis followed by x followed by epsilon followed by y followed by a right parenthesis will be called $x\text{e}y$.

$$(x)(y) \qquad x\text{e}y = S_6{}^\frown x{}^\frown S_9{}^\frown y{}^\frown S_7.$$

When '$\downarrow$' is used instead of epsilon, the result is called $x\text{j}y$. The 'j' is intended to suggest joint denial.

$$(x)(y) \qquad x\text{j}y = S_6{}^\frown x{}^\frown S_8{}^\frown y{}^\frown S_7.$$

The result of putting an expression x in parenthesis and prefixing it to y is called x qu y. The letters 'qu' are intended to suggest universal quantification.

$$(x)(y) \qquad x \text{ qu } y = \mathrm{S}_6 \frown x \frown \mathrm{S}_7 \frown y.$$

Now x is a variable, symbolically Vbl x, if x is any one of S_1, S_2, S_3, or S_4 with or without a string of accents attached.

$$(x) \qquad \mathrm{Vbl}\ x \equiv (\exists y)(y = \mathrm{S}_1 \,.\mathsf{v}.\, y = \mathrm{S}_2 \,.\mathsf{v}.\, y = \mathrm{S}_3 \,.\mathsf{v}.\, y = \mathrm{S}_4 : x = y \,.\mathsf{v}\, (\exists z)(\mathrm{Ac}\ z \,.\, x = y \frown z)).$$

Further, where '$\mathrm{LFmla}_0\ x$' is read 'x is an atomic logical formula',

$$(x) \qquad \mathrm{LFmla}_0\ x \equiv (\exists y)(\exists z)(\mathrm{Vbl}\ y \,.\, \mathrm{Vbl}\ z \,.\, x = yez).$$

If x is formed by prefixing a parenthesized variable to y, x is said to be a quantification of y — symbolically $x\mathrm{Q}y$.

$$(x)(y) \qquad x\mathrm{Q}y \,.\!\equiv (\exists z)(\mathrm{Vbl}\ z \,.\, x = z \text{ qu } y).$$

Now the logical formulæ are to comprise the atomic logical formulæ together with all joint denials and quantifications thereof, and all joint denials and quantifications of the thus supplemented totality, and so on. By reasoning analogous to that which led to the definition of the ancestral (§ 39), then, we see that x is describable as a logical formula just in case it belongs to all classes y of the following kind: the atomic logical formulæ belong to y, and the joint denials and quantifications of members of y belong to y. Thus

$$(1) \quad (x) \qquad \mathrm{LFmla}\ x \equiv (y)((z)(w)(\mathrm{LFmla}_0\ z \,.\!\supset\!.\, z \in y : z, w \in y \,.\!\supset\!.\, zjw \in y : z \in y \,.\, w\ \mathrm{Q}\ z \,.\!\supset\!.\, w \in y) \supset .\, x \in y)$$

where 'LFmla x' is read 'x is a logical formula'.

The so-called formal characterizations of 'variable', 'atomic logical formula', and 'formula' which appeared in words at the beginning of the section thus admit of translation into terms of 'S_1', 'S_2', . . . , 'S_9', '$\frown$', and our logical notation. Whatever is "formal", in the sense intended by the earlier loose phrase 'speaking only of the typographical constitution of the expressions in question', admits of similar translation; though of course the sign-names 'S_1', 'S_2', . . . , 'S_9' have to be reconstrued and perhaps altered in number to suit the particular notation under discussion.

We are thus led to the following more rigorous criterion of formality: translatability into a notation containing only names of signs, a connective indicating concatenation, and the notation of logic.

Discourse which is "formal" in this sense, and hence translatable into the notation just now described, is called *metamathematics*, *formal syntax*, or briefly *syntax*. Since joint denial, quantification, and the membership notation suffice for logic, these same three devices plus names of signs plus a notation indicating concatenation suffice for syntax. (A still more economical notation to the same purpose will appear in § 54). The fact that 'Vbl', 'LFmla_0', and 'LFmla' are definable in this syntactical notation is perhaps best expressed henceforward by speaking of them as *syntactically definable;* the word 'formal', having served the purpose of providing a first vague indication of our present concerns, can thus be left unencumbered with technical meaning.

Any definition is *stated* with help of syntax, as a notational convention of abbreviation; but to say that an expression is *syntactically definable* is to say something more, namely that it is explicable as an abbreviation of an expression which is itself composed of just the notations of syntax.

The general notion of *formula* was explained in § 13 in terms of an unspecified notion of atomic formula. Where '$\mathrm{Fmla}\ x$' means 'x is a formula' and '$\mathrm{Fmla}_0\ x$' means 'x is an atomic formula',

$$(x) \qquad \mathrm{Fmla}\ x \equiv (y)((z)(w)(\mathrm{Fmla}_0\ z \,.\!\supset\!.\, z \in y : z, w \in y \,.\!\supset\!.\, z\mathrm{j}w \in y : z \in y \,.\, w \,\mathrm{Q}\, z \,.\!\supset\!.\, w \in y) \supset .\, x \in y)$$

in exact analogy to (1). Thus '$\mathrm{Fmla}\ x$' is *syntactically definable in terms of* 'Fmla_0'; i.e., it is translatable into a notation comprising $\ulcorner \mathrm{Fmla}_0\ \alpha \urcorner$ and nothing further except names of signs, the concatenation connective, and logic. As will become apparent in the sequel, moreover, the notions of matrix, tautology, and theorem introduced in foregoing chapters have the same status as the notion of formula; they are all syntactically definable in terms of $\ulcorner \mathrm{Fmla}_0\ \alpha \urcorner$. If we specify the atomic formulæ, so that $\ulcorner \mathrm{Fmla}_0\ \alpha \urcorner$ is itself syntactically defined in one way or another, then all these notions become syntactically definable in the absolute sense. If e.g. the atomic formulæ are taken simply as the atomic logical formulæ, then ⌜Fmla α⌝ reduces to ⌜LFmla α⌝, whose syntactical definability was observed in (1).

§ 54. *The Syntactical Primitive*

MUCH OF the discussion in previous chapters, including indeed all metatheorems, belongs to syntax; to logic or mathematics in the narrower sense belong rather the theorems themselves, some of which were stated and some of which were syntactically described. Our medium of syntactical discussion hitherto has been ordinary language, supplemented with the practical device of Greek letters and corners. But now we have a syntactical notation which is just as strict and systematic as the logical notation whereof it treats. It consists of the logical notation (the very notation whereof it treats) plus 'S_1, 'S_2', . . . , 'S_9', and '⌢'.[1]

Corners thus give way to a more analytic expedient, use of the concatenation sign; and Greek letters give way to the ordinary variables '*x*', '*y*', etc., adjuncts of the quantifier (cf. § 12). It is to be noted that these variables do not need to be reconstrued, for present purposes, as having the sense which formerly attached to the Greek letters; rather they are general variables, as always, and may refer to any entities whatever — expressions *included*. The quantification '$(x)(x = x)$', e.g., has always meant not only that V = V and 5 = 5 but also that $S_9 = S_9$ (i.e., epsilon = epsilon) and even that $S_2 = S_2$ (i.e., '*x*' = '*x*'); and the quantifier does not need to depart now from its usual sense. It happens indeed that the only values of the variables which *interest* us in syntactical applications are expressions, just as in § 39 the only values of the variables which interested us were relations and in § 49 natural numbers.

But our new syntactical scheme is capable of improvement. It

[1] This kind of approach, whereby the medium of discourse about a formalism receives strict formalization in turn, dates from Gödel (1931) and Tarski (1933). The arch notation is Tarski's.

is possible so to reformulate it that its primitive notation will comprise no terms beyond variables; and there is an advantage, as observed (§ 27), in so doing. The concatenation sign '$\frown$', like '+', '$\cap$', etc., combines terms to form terms; e.g. it combines the terms 'S_9' and 'S_2' (which are names of 'ϵ' and 'x') to form the term '$S_9 \frown S_2$' (which is a name of '$\epsilon\, x$'). But instead of this we may take as primitive a concatenation predicate 'C', or more accurately a primitive form of atomic matrix $\ulcorner C\alpha\beta\gamma \urcorner$, such that '$Cxyz$' means '$x$ is the expression formed by writing the expression y followed by the expression z'; then $y \frown z$ becomes definable as $(\imath x)\ Cxyz$. Again, the primitive names 'S_1' 'S_2', . . . , 'S_9' are eliminable by adopting a primitive form of atomic matrix $\ulcorner \alpha A \beta \urcorner$ such that 'xAy' means 'x is the sign which alphabetically just succeeds the sign y'; for then

$$S_1 = (\imath x)((\exists y)(yAx) \,.\, \sim (\exists y)(xAy)),$$
$$S_2 = (\imath x)\ x\ A\ S_1, \qquad S_3 = (\imath x)\ x\ A\ S_2, \qquad \ldots, \qquad S_9 = (\imath x)\ x\ A\ S_8,$$

where the alphabetic order of our signs is taken arbitrarily as $S_1, S_2, \ldots, S_9$.

But further economy is possible. We can get along with a single primitive form of atomic matrix $\ulcorner M\alpha\beta\gamma \urcorner$, if we endow '$Mxyz$' with this elaborate meaning: if x is a single sign (Case 1), x alphabetically just succeeds y (i.e., xAy); if x is a complex expression (Case 2), x is the result of writing y followed by z (i.e., $Cxyz$); and if x is not an expression at all (Case 3, uninteresting for syntax), $x = y$. The word 'sign' in this explanation is of course understood as referring exclusively to our chosen signs S_1, S_2, etc.; and 'expression' refers exclusively to such signs and finite rows of them.

What z may be is indifferent to Cases 1 and 3; hence if x is not a complex expression, the above account of '$Mxyz$' guarantees that $Mxyz \equiv Mxyx$. Hence if on the contrary

$$(1) \qquad Mxyz \,.\, \sim Mxyx$$

we may be sure that x is a complex expression. But then (1) must be interpreted according to Case 2; i.e., as meaning that x is the result of writing y followed by z, and not the result of writing y followed by x. But the latter clause imposes no restriction at all; since an expression is understood as a finite string of signs (one or more), an expression x could not be the result of writing an expression y followed by x again in its entirety. Hence (1) amounts merely to saying that x is the result of writing y followed by z; i.e., $\mathrm{C}xyz$.

We have seen that, where x is a complex expression, $(y) \sim \mathrm{M}xyx$. Where x is not an expression at all, $\mathrm{M}xxx$ by Case 3. Hence if

$$(2) \qquad \mathrm{M}xyx \,.\, \sim \mathrm{M}xxx$$

we may be sure that x is an expression but not a complex expression; therefore a sign. But then (2) must be interpreted according to Case 1; i.e., as meaning that x alphabetically succeeds y and does not succeed itself. Since the latter clause is clearly vacuous, (2) amounts merely to saying that $x\mathrm{A}y$.

We see therefore that both '$\mathrm{C}xyz$' and '$x\mathrm{A}y$' can be paraphrased in terms of 'M', viz. as (1) and (2). Accordingly $y\frown z$, which was $(\imath x)\, \mathrm{C}xyz$, becomes describable in terms of 'M' as

$$(\imath x)(\mathrm{M}xyz \,.\, \sim \mathrm{M}xyx);$$

and S_1, S_2, S_3, etc. become

$$(3) \quad (\imath x)((\exists y)(\mathrm{M}yxy \,.\, \sim \mathrm{M}yyy) \,.\, \sim (\exists y)(\mathrm{M}xyx \,.\, \sim \mathrm{M}xxx)),$$
$$(4) \quad (\imath x)(\mathrm{M}x\mathrm{S}_1x \,.\, \sim \mathrm{M}xxx),$$
$$(5) \quad (\imath x)(\mathrm{M}x\mathrm{S}_2x \,.\, \sim \mathrm{M}xxx),$$

etc.

Actually the '$\sim \mathrm{M}xxx$' in (4) is superfluous. The purpose of '$\sim \mathrm{M}xxx$' in (2) was to rule out Case 3 and thus compel x to be an expression; but this is already accomplished in (4) by the clause '$\mathrm{M}x\mathrm{S}_1x$', since under Case 3 '$\mathrm{M}x\mathrm{S}_1x$' would identify x with the sign S_1 and thus compel x to be an expression after all. Thus (4) reduces to '$(\imath x)\, \mathrm{M}x\mathrm{S}_1x$'; and the analogous is true of (5) and its suite.

In place of (3), moreover, the simpler description:

$$(6) \qquad (\imath x)(y)(z) \sim \mathrm{M}xyz$$

will serve for S_1. This is seen as follows. Where x is a complex expression, it has parts y and z such that x consists of y followed by z; therefore '$(y)(z) \sim \mathrm{M}xyz$' rules out Case 2. It also rules out Case 3, since '$\sim \mathrm{M}xxx$' was seen to do so. Therefore x must be a sign if $(y)(z) \sim \mathrm{M}xyz$. But then '$(y)(z) \sim \mathrm{M}xyz$' says, according to

Case 1, that the sign x is not an alphabetic successor; in other words, that it is S_1.

Our primitive syntactical notation now consists merely of 'M' and the notation of logic — ultimately 'M' and the notations of joint denial, quantification, and membership, since these latter are adequate to the logical matters. Just as the logical formulæ comprised atomic formulæ of the form $\ulcorner(\alpha \epsilon \beta)\urcorner$ and all the expressions thence constructible by joint denial and quantification, so the syntactical formulæ comprise atomic formulæ of the forms $\ulcorner(\alpha \epsilon \beta)\urcorner$ and $\ulcorner M\alpha\beta\gamma\urcorner$ and all the expressions thence constructible by joint denial and quantification.

When this syntactical notation is used for the discussion of expressions composed of signs other than the particular nine with which we have been identifying $S_1, S_2, \ldots, S_9$, the meaning intended for '$Mxyz$' will of course undergo a corresponding change in systematic fashion. The change takes place merely in the notion of alphabetic succession which is used under Case 1 in the above explanation; the relation of alphabetic succession varies with the alphabet.

It should be remarked that 'M' and the abbreviations '$\frown$', 'S_1', etc. which are defined in terms of 'M' are not a practical substitute for the former devices of quotation marks, Greek letters, and corners. The old devices are both more graphic and more conducive to brevity. And the new syntactical notation is not convenient at all for stating definitions, where signs enter which were not initially taken into account. The new notation has its value rather as a subject matter, when we come to talk *about* syntax rather than merely using syntax to talk about logic. As a medium of talking about this refined syntactical notation, in turn, the old practical devices of quotes, Greek letters, and corners recur; they have appeared to some extent in the present section and they will continue to appear in the sequel.

§ 55. *Protosyntax*

GIVEN ANY expression, we can decide once and for all by a systematic procedure of inspection whether or not it is a logical formula. Similarly, as soon as all the desired non-logical forms of atomic formulæ have been listed, we can decide by a systematic procedure whether or not a given expression is a formula at all, logical or otherwise. The systematic recognizability which thus attaches to the notions of logical formula and formula attaches in similar fashion to the notions of matrix, tautology, axiom of quantification, and axiom of membership (cf. §§ 16, 29). The notion of theorem lacks this feature, though retaining a partial sort of recognizability; appropriate to each formula which is a theorem there is a device (a so-called proof), discoverable in general only by luck, which once discovered enables us to see by a finite amount of inspection that the formula is a theorem (cf. § 16). Now any syntactical property which enjoys even this modest sort of recognizability is called *constructive*; thus the property of being a theorem, as well as the quite mechanically recognizable properties of being a formula, matrix, tautology, or axiom, are constructive.

This account of constructivity, like the notion itself, is vague. In a sharp formulation it would be necessary to avoid such vague phrases as 'mechanically recognizable', 'recognizable by a finite amount of inspection ', etc.; but then there would be an essential difficulty in establishing any correspondence between this precisely defined notion and the intuitively intended notion which those vague phrases help describe. There is a striking feature, however, which the instances of constructivity cited above have in common: they are not only syntactically definable, but they are definable even in that narrower syntax which borrows only joint denial and quantification from logic and eschews membership. The syntactical definition of 'LFmla *x*' suggested in (1) of § 53 depends indeed upon the whole of logic, membership included; but an equivalent definition is possible, as we shall see (§ 56), which involves only the 'M' of syntax plus the joint denial and quantification of logic. The same is true of the other constructive notions mentioned above (cf. §§ 56–58), and presumably equally true of *all* syntactical notions which are constructive in the vague sense above suggested.

The part of syntax which omits membership will be called *protosyntax*. Just as the logical formulæ comprise atomic formulæ of the form ⌜$(\alpha \epsilon \beta)$⌝ and the expressions thence constructible by joint denial and quantification, so the protosyntactical formulæ comprise atomic formulæ of the form ⌜M$\alpha\beta\gamma$⌝ and all the expressions thence constructible by joint denial and quantification. This meager protosyntactical notation proves to constitute a sufficient language for expressing all the metatheorems of foregoing chapters — provided of course that we do away with the unspecified extra-logical forms of atomic formulæ or at least fix upon a definite list of forms desired.

Protosyntactical definability is intended not as an approximation to constructivity, but as something more inclusive. The notion of non-theorem, e.g., is protosyntactically definable, yet presumably not constructive; we know of nothing, discoverable even by luck, which would enable us to decide that a given logical statement ϕ is not a theorem. We may find that ⌜$\sim\phi$⌝ is a theorem; but this does not exclude the possibility that ϕ is also a theorem, unless we assume the consistency for which we of course hope. Even this channel is closed if it happens, as it well may, that neither ϕ nor ⌜$\sim\phi$⌝ is a theorem.

Three precise explications of the vague phrase 'mechanically recognizable' have been independently propounded in recent years, and the three have been proved equivalent to one another. One is the notion of so-called *recursiveness* (closely connected with the kind of procedure called recursive on page 86); this was first formulated in a relatively narrow sense by Gödel ("Unentscheidbare Sätze," pp. 179–180) and afterwards extended by Gödel and Herbrand (cf. Gödel "Undecidable Propositions"). The second is Church's notion of *λ-definability* ("Unsolvable Problem," pp. 346ff, 356ff), and the third is Turing's *computability*. The second was proved equivalent to the first by Kleene, and to the third by Turing. A derivative notion, *recursive enumerability*, corresponds to constructivity; see Kleene, "Recursive Predicates." But note that protosyntactical definability is broader.

In defining the notations 'S_1,' 'S_2', . . . , 'S_9', and '⁀' in terms of 'M', in § 54, the notation ⌜$(\imath\alpha)\phi$⌝ of description was used. This device was legitimate for purposes of syntactical definition, since it reduces through D1–17 to the three logical primitives; but the device is not immediately available for protosyntactical definition, because its logical definition depends finally on 'ϵ', which is lacking in protosyntax. Of the logical definitions presented in the foregoing chapters, only D1–8 can be carried over into protosyntax; for it is only these that do not presuppose membership.

We find, however, that the logical notion of description, and likewise that of identity, admit of *other* definitions within the framework of protosyntax. It will first be shown that the appropriate sense of identity is obtainable by explaining '$x = y$' as short for:

$$(1) \qquad (z)(\mathrm{M}zxx \equiv \mathrm{M}zyy).$$

Suppose, to begin with, that x is not an expression. Then Mxxx, by Case 3 of § 54; hence, if (1) is true, it follows that Mxyy; but, by Case 3 of § 54, this identifies x with y. We see therefore that '$x = y$' in the intended sense follows from (1) when x is not an expression. Next suppose that x is an expression, and let w be the repetitive complex expression formed by writing x followed by x. Then, by Case 2 of § 54, Mwxx; hence, if (1) is true, Mwyy; but, by Case 2 of § 54, this makes y the first and last half of the repetitive expression w, thus identifying y with x. Regardless of whether x is an expression or not, therefore, '$x = y$' in the intended sense follows from (1). Conversely, moreover, (1) obviously follows from '$x = y$'. The following definition therefore serves our purpose:

Δ9. $\ulcorner(\alpha = \beta)\urcorner$ *for* $\ulcorner(\gamma)(\mathrm{M}\gamma\alpha\alpha \equiv \mathrm{M}\gamma\beta\beta)\urcorner$.

The protosyntactical definitions will be distinguished thus from the logical ones by using 'Δ' instead of 'D'. Δ1–8 are understood as D1–8.

Definition of description in protosyntax must proceed contextually, like that of abstraction in logic (§§ 24, 26). First let us deal with quasi-atomic contexts $\ulcorner\mathrm{M}\ (\imath\alpha)\phi\ \beta\ \gamma\urcorner$ having a description in the first place and variables in the other places. If w is the one and only entity x such that . . . x . . . , i.e. if

$$(x)(x = w \;.{\equiv}.\; \ldots x \ldots)$$

(cf. § 27), then we want the quasi-atomic formula:

(2) $$\mathrm{M}\,(\imath x)(\ldots x \ldots)\;y\;z$$

to be true for any objects y and z if and only if Mwyz; if on the other hand there is no such unique entity x, then we lose interest and become content to have (2) turn out in any way that proves convenient — say as false. All this is obviously accomplished by explaining (2) as an abbreviation of:

$$(\exists w)(\mathrm{M}wyz\,.\,(x)(x = w\,.\!\equiv \ldots x \ldots)).$$

Thus

Δ10. $\ulcorner \mathrm{M}\,(\imath\alpha)\phi\;\beta\;\gamma \urcorner$ *for* $\ulcorner(\exists\delta)(\mathrm{M}\delta\beta\gamma\,.\,(\alpha)(\alpha = \delta\,.\!\equiv \phi))\urcorner$.

Let us understand *terms*, for purposes of protosyntax, as comprising variables together with descriptions $\ulcorner(\imath\alpha)\phi\urcorner$ (where ϕ is a formula of protosyntax). Let us use 'ζ', 'η', and 'θ' hereafter to refer to terms in this sense. Then, repeating precisely the method used in Δ10, we can define the further quasi-atomic contexts as follows:

Δ11. $\ulcorner \mathrm{M}\;\zeta\;(\imath\alpha)\phi\;\gamma \urcorner$ *for* $\ulcorner(\exists\delta)(\mathrm{M}\zeta\delta\gamma\,.\,(\alpha)(\alpha = \delta\,.\!\equiv \phi\,))\urcorner$.
Δ12. $\ulcorner \mathrm{M}\;\zeta\;\eta\;(\imath\alpha)\phi \urcorner$ *for* $\ulcorner(\exists\delta)(\mathrm{M}\zeta\eta\delta\,.\,(\alpha)(\alpha = \delta\,.\!\equiv \phi))\urcorner$.

Now $\ulcorner \mathrm{M}\zeta\eta\theta \urcorner$ is explained for all terms ζ, η, and θ, whether variables or descriptions.

Let us next extend Δ9 to apply to terms in general, rather than simply to variables.

Δ13. $\ulcorner(\zeta = \eta)\urcorner$ *for* $\ulcorner(\gamma)(\mathrm{M}\gamma\zeta\zeta \equiv \mathrm{M}\gamma\eta\eta)\urcorner$.

It is convenient also to import D11.

Δ14. $\ulcorner(\zeta \neq \eta)\urcorner$ *for* $\ulcorner\sim(\zeta = \eta)\urcorner$.

The definitions of 'S_1', 'S_2', ..., 'S_9', and $\ulcorner \zeta\frown\eta \urcorner$ suggested in § 54 can now be set down.

Δ15. 'S_1' *for* '$(\imath x)(y)(z) \sim \mathrm{M}xyz$'.
Δ16. 'S_2' *for* '$(\imath x)\;\mathrm{M}x\;\mathrm{S}_1\;x$', '$\mathrm{S}_3$' *for* '$(\imath x)\;\mathrm{M}x\;\mathrm{S}_2\;x$', *etc. to* '$\mathrm{S}_9$'.
Δ17. $\ulcorner(\zeta\frown\eta)\urcorner$ *for* $\ulcorner(\imath\alpha)(\mathrm{M}\alpha\zeta\eta\,.\sim \mathrm{M}\alpha\zeta\alpha)\urcorner$.

A definition analogous to D6 and D7 is desirable in connection with the concatenation notation.

Δ18. $\ulcorner(\zeta_1\frown\zeta_2\frown\zeta_3)\urcorner$ *for* $\ulcorner((\zeta_1\frown\zeta_2)\frown\zeta_3)\urcorner$,
$\ulcorner(\zeta_1\frown\zeta_2\frown\zeta_3\frown\zeta_4)\urcorner$ *for* $\ulcorner((\zeta_1\frown\zeta_2\frown\zeta_3)\frown\zeta_4)\urcorner$, *etc.*

Finally we carry over nine of the ten definitions suggested in § 53. In third place among these it is convenient to insert a further definition introducing the notation '$x\mathrm{P}y$', in the sense 'x is part (or all) of y'.

Δ19. $\ulcorner(\zeta \mathrm{B} \eta)\urcorner$ *for* $\ulcorner(\zeta = \eta \,.\!\vee\, (\exists\alpha)(\zeta\frown\alpha = \eta))\urcorner$.
Δ20. $\ulcorner(\zeta \mathrm{E} \eta)\urcorner$ *for* $\ulcorner(\zeta = \eta \,.\!\vee\, (\exists\alpha)(\alpha\frown\zeta = \eta))\urcorner$.
Δ21. $\ulcorner(\zeta \mathrm{P} \eta)\urcorner$ *for* $\ulcorner(\exists\alpha)(\alpha \mathrm{B} \eta \,.\, \zeta \mathrm{E} \alpha)\urcorner$.
Δ22. $\ulcorner\mathrm{Ac}\, \zeta\urcorner$ *for* $\ulcorner(\alpha)(\alpha \mathrm{B} \zeta \,.\!\supset\!.\, \mathrm{S}_5 \mathrm{E} \alpha)\urcorner$.
Δ23. $\ulcorner(\zeta \mathrm{e} \eta)\urcorner$ *for* $\ulcorner(\mathrm{S}_6\frown\zeta\frown\mathrm{S}_9\frown\eta\frown\mathrm{S}_7)\urcorner$.
Δ24. $\ulcorner(\zeta \mathrm{j} \eta)\urcorner$ *for* $\ulcorner(\mathrm{S}_6\frown\zeta\frown\mathrm{S}_8\frown\eta\frown\mathrm{S}_7)\urcorner$.
Δ25. $\ulcorner(\zeta \text{ qu } \eta)\urcorner$ *for* $\ulcorner(\mathrm{S}_6\frown\zeta\frown\mathrm{S}_7\frown\eta)\urcorner$.
Δ26. $\ulcorner\mathrm{Vbl}\, \zeta\urcorner$ *for* $\ulcorner(\exists\alpha)(\alpha = \mathrm{S}_1 \,.\!\vee\!.\, \alpha = \mathrm{S}_2 \,.\!\vee\!.\, \alpha = \mathrm{S}_3 \,.\!\vee\!.\, \alpha = \mathrm{S}_4 :$
$\zeta = \alpha \,.\!\vee\, (\exists\beta)(\mathrm{Ac}\,\beta \,.\, \zeta = \alpha\frown\beta))\urcorner$.
Δ27. $\ulcorner(\zeta \mathrm{Q} \eta)\urcorner$ *for* $\ulcorner(\exists\alpha)(\mathrm{Vbl}\,\alpha \,.\, \zeta = \alpha \text{ qu } \eta)\urcorner$.
Δ28. $\ulcorner\mathrm{LFmla}_0\, \zeta\urcorner$ *for* $\ulcorner(\exists\alpha)(\exists\beta)(\mathrm{Vbl}\,\alpha \,.\, \mathrm{Vbl}\,\beta \,.\, \zeta = \alpha \mathrm{e} \beta)\urcorner$.

§ 56. *Formula and Matrix Defined*

AS REMARKED (§ 55), the definition of $\ulcorner\mathrm{LFmla}\,\zeta\urcorner$ suggested in (1) of § 53 is not available to protosyntax; but a protosyntactical definition is possible along other lines. The point of departure, in devising this definition, is the following reflection: x is a logical formula if and only if it belongs to a series of expressions x_1, x_2, $x_3, \ldots, x_n$ such that x_i, for each i, is either an atomic logical formula or else a joint denial of earlier expressions of the series or else a quantification of an earlier expression of the series.

Instead of speaking of a series of expressions $x_1, x_2, \ldots, x_n$, the device suggests itself of speaking rather of a single long expression composed of the successive expressions $x_1, x_2, \ldots, x_n$. But we encounter a difficulty: analysis of the long expression does not reveal uniquely the intended parts $x_1, x_2, \ldots, x_n$. Given the long expression, we have no way of knowing how much of it was intended as x_1, nor how much of the remainder was intended as x_2, and so on. This difficulty, however, is easily overcome. Instead of taking our long expression as $x_1\frown x_2\frown \ldots \frown x_n$, we can take it as

$$S_6{}^\frown S_7{}^\frown x_1{}^\frown S_6{}^\frown S_7{}^\frown x_2{}^\frown S_6{}^\frown S_7{}^\frown \ldots {}^\frown x_n{}^\frown S_6{}^\frown S_7,$$

thus framing each of $x_1, x_2, \ldots, x_n$ with occurrences of the arbitrary expression $S_6{}^\frown S_7$, i.e. '()'. Given this long expression, the parts $x_1, x_2, \ldots, x_n$ are uniquely determined as the segments reaching from one occurrence of $S_6{}^\frown S_7$ to the next; for we know that the ungrammatical combination of signs $S_6{}^\frown S_7$ will not have any additional occurrences inside the formulæ $x_1, x_2, \ldots, x_n$. Thus, if the segments reaching from one occurrence of $S_6{}^\frown S_7$ to the next in an expression z be called the *framed ingredients* of z, we can explain a *logical formula* as any framed ingredient of an expression z each of whose framed ingredients is either an atomic logical formula or a joint denial of previous framed ingredients or a quantification of a previous framed ingredient. Such an expression z will be essentially like the long expression considered above; it may depart from that pattern only in a couple of inessential respects. Changing $S_6{}^\frown S_7$ to $S_6{}^\frown S_7{}^\frown S_6{}^\frown S_7$, e.g., would make no difference so far as framed ingredients are concerned; nor would annexation of further matter at either end, so long as the added matter contained no occurrences of $S_6{}^\frown S_7$.

Let us make the above definitions explicit. To say that x is a *framed ingredient* of z, symbolically x Ing z, is to say that $S_6{}^\frown S_7{}^\frown x{}^\frown S_6{}^\frown S_7$ is part (or all) of z and $S_6{}^\frown S_7$ is not part of x.

Δ29. $\ulcorner(\zeta \text{ Ing } \eta)\urcorner$ *for* $\ulcorner(S_6{}^\frown S_7{}^\frown \zeta{}^\frown S_6{}^\frown S_7 \text{ P } \eta \,.\, \sim(S_6{}^\frown S_7 \text{ P } \zeta))\urcorner$.

One framed ingredient x of z is *prior to* another one y, symbolically $x \text{ Pr}_z\, y$, if x is a framed ingredient of an initial segment of z whereof y is not a framed ingredient.

Δ30. $\ulcorner(\zeta \text{ Pr}_\eta\, \theta)\urcorner$ *for* $\ulcorner(\exists\alpha)(\alpha \text{B}\eta \,.\, \zeta \text{ Ing } \alpha \,.\, \theta \text{ Ing } \eta \,.\, \sim(\theta \text{ Ing } \alpha))\urcorner$.

The formal definition of logical formula, then, is as follows.

(1) $\ulcorner\text{LFmla}\,\zeta\urcorner$ *for* $\ulcorner(\exists\alpha)(\zeta \text{ Ing } \alpha \,.\, (\beta)(\beta \text{ Ing } \alpha \,.\supset.\, \text{LFmla}_0\,\beta \vee (\exists\gamma)(\exists\delta)(\gamma \text{Pr}_\alpha \beta \,.\, \delta \text{ Pr}_\alpha \beta : \beta = \gamma \text{j} \delta \,.\vee.\, \beta \text{Q} \gamma)))\urcorner$.

Just as Δ28 provides for atomic formulæ of the form $\ulcorner(\alpha \in \beta)\urcorner$, other analogous protosyntactical definitions will provide for each further form of atomic formula once these further forms have been chosen. In preparation for such definitions we would presumably so reconstrue $\ulcorner \text{M}\alpha\beta\gamma\urcorner$ as to allow a longer alphabet, running beyond

S_9 to include such supplementary signs as may be wanted in the non-logical atomic formulæ (cf. § 54); but even this departure is not strictly necessary, for we could simply use the novel combinations 'ϵ'', 'ϵ''', etc. in lieu of such supplementary signs and thus keep within our nine-sign alphabet. If we bring together the definitions appropriate to the several atomic forms, then, by alternation, we obtain a general definition of $\ulcorner \text{Fmla}_0\, \zeta \urcorner$. On this basis, in turn, we can define the general notion of formula in exact analogy to the definition (1) of logical formula.

$$\textbf{Δ31.}\quad \ulcorner \text{Fmla}\, \zeta \urcorner \quad \textit{for} \quad \ulcorner (\exists\alpha)(\zeta \text{ Ing } \alpha \,.\, (\beta)(\beta \text{ Ing } \alpha \,.\!\supset\!.\, \text{Fmla}_0\, \beta \vee (\exists\gamma)(\exists\delta)(\gamma \text{Pr}_\alpha\, \beta \,.\, \delta\, \text{Pr}_\alpha\, \beta : \beta = \gamma \text{j} \delta \,.\!\vee\!.\, \beta \text{Q} \gamma))) \urcorner.$$

Definitions will hereafter be set up in the manner of Δ31 rather than (1), under the fictive assumption that a general definition of $\ulcorner \text{Fmla}_0\, \zeta \urcorner$ is at hand in place of Δ28. Just as Δ31 then defines the general notion of formula, logical and otherwise, so in subsequent definitions the notions of matrix, tautology, and theorem are defined in their full generality. If in particular the atomic formulæ are construed simply as the logical atomic formulæ, so that $\ulcorner \text{Fmla}_0\, \zeta \urcorner$ comes to be defined by Δ28 itself, then Δ31 ceases to differ at all from (1); and correspondingly for the subsequent definitions.

Definition of the notion of matrix presupposes the notions of bondage and freedom, basic to which is the notion of bound occurrence. We have first of all to ask, what is an occurrence? what kind of thing e.g. is the third occurrence of an expression *x* in an expression *y*, as distinct from the mere expression or typographic shape *x* itself? An answer, artificial but convenient and adequate, is this: an occurrence of *x* in *y* is an initial segment of *y* ending in *x*. The third "occurrence" of *x* in *y*, e.g., is construed as that initial segment of *y* which remains when (as we should ordinarily say) everything after the third occurrence of *x* is lopped off. A "later occurrence" of *x* differs from an "earlier" one in being a longer initial segment of *y*.

Thus *z* may be said to be an occurrence of *x* in *y*, symbolically $z\, \text{O}_y\, x$, if $z\text{B}y$ and $x\text{E}z$. But it is better to refine the definition in such a way as to except accented occurrence; in the context

'$z \in w''$', e.g., it is desirable to recognize an occurrence of 'w''' but to deny any occurrence of the variables 'w'' and 'w'. This refinement is accomplished by supplementing '$z\mathrm{B}y$' and '$x\mathrm{E}z$' with the further clause '$\sim(z^\frown \mathrm{S}_5 \,\mathrm{B}\, y)$'.

Δ32. $\ulcorner(\zeta \,\mathrm{O}_\theta\, \eta)\urcorner$ *for* $\ulcorner(\zeta \mathrm{B} \theta \,.\, \sim(\zeta^\frown \mathrm{S}_5 \,\mathrm{B}\, \theta) \,.\, \eta \mathrm{E} \zeta)\urcorner$.

To say that x occurs in y, symbolically x In y, is to say that there is an occurrence of x in y.

Δ33. $\ulcorner(\zeta \text{ In } \eta)\urcorner$ *for* $\ulcorner(\exists\alpha)(\alpha \,\mathrm{O}_\eta\, \zeta)\urcorner$.

Obviously 'x In y' goes beyond '$x \,\mathrm{P}\, y$' only to the extent of excepting accented occurrence.

An occurrence of a variable is *bound* in a given formula when it falls within a formula which forms part (or all) of the given formula and begins with a quantifier containing the same variable. But now that an "occurrence" is construed as an initial segment, the formulation of bondage assumes this form: z is a bound occurrence of x in y, symbolically $z \,\mathrm{BO}_y\, x$, if x is a variable and z is an occurrence of x in y and there is a formula x qu w which begins somewhere in the segment z and ends at or after the end of z (but still in y). More rigorously,

$$z \,\mathrm{BO}_y\, x \,.\!\equiv.\, \mathrm{Vbl}\, x \,.\, z \,\mathrm{O}_y\, x \,.\, (\exists w)(\exists z')(\exists x')(\mathrm{Fmla}\, w \,.\, z = z'^\frown x \,.\, z'^\frown x' \,\mathrm{O}_y\, x\mathrm{qu}w \,.\, \sim(x\mathrm{qu}w \,\mathrm{E}\, x')).$$

Thus

Δ34. $\ulcorner(\zeta \,\mathrm{BO}_\theta\, \eta)\urcorner$ *for* $\ulcorner(\mathrm{Vbl}\,\eta \,.\, \zeta \,\mathrm{O}_\theta\, \eta \,.\, (\exists\alpha)(\exists\beta)(\exists\gamma)(\mathrm{Fmla}\,\alpha \,.\, \zeta = \beta^\frown\eta \,.\, \beta^\frown\gamma \,\mathrm{O}_\theta\, \eta\mathrm{qu}\alpha \,.\, \sim(\eta\mathrm{qu}\alpha \,\mathrm{E}\, \gamma)))\urcorner$.

A *free* occurrence is one which is not bound.

Δ35. $\ulcorner(\zeta \,\mathrm{FO}_\theta\, \eta)\urcorner$ *for* $\ulcorner(\mathrm{Vbl}\,\eta \,.\, \zeta \,\mathrm{O}_\theta\, \eta \,.\, \sim(\zeta \,\mathrm{BO}_\theta\, \eta))\urcorner$.

A variable x is free in y, symbolically $x\mathrm{F}y$, if it has a free occurrence in y.

Δ36. $\ulcorner(\zeta \mathrm{F} \eta)\urcorner$ *for* $\ulcorner(\exists\alpha)(\alpha \,\mathrm{FO}_\eta\, \zeta)\urcorner$.

Matrices, finally, are formulæ with free variables, and *statements* are other formulæ.

Δ37. $\ulcorner\mathrm{Mat}\,\zeta\urcorner$ *for* $\ulcorner(\mathrm{Fmla}\,\zeta \,.\, (\exists\alpha)(\alpha\, \mathrm{F}\zeta))\urcorner$.
Δ38. $\ulcorner\mathrm{Stat}\,\zeta\urcorner$ *for* $\ulcorner(\mathrm{Fmla}\,\zeta \,.\, \sim \mathrm{Mat}\,\zeta)\urcorner$.

§ 57. *Axioms of Quantification Defined*

WE TURN next to the problem of defining tautology protosyntactically. As an auxiliary we need the notion of *truth-functional component* (cf. § 10). An immediate truth-functional component of x is anything y such that x is yjz or zjy for some formula z; and w is a truth-functional component of x, symbolically wTCx, if w belongs to a sequence of expressions each of which is either x or an immediate truth-functional component of an earlier expression of the sequence. The notion of sequence involved here can be dealt with by the same device which was used in defining formula (§ 56). Thus

Δ39. $\ulcorner(\zeta \mathrm{TC} \eta)\urcorner$ *for* $\ulcorner(\exists\alpha)(\zeta \text{ Ing } \alpha \,.\, (\beta)(\beta \text{ Ing } \alpha \,.\!\supset\!:\, \beta = \eta \,.\!\vee$
$(\exists\gamma)(\exists\delta)(\gamma \text{ Pr}_\alpha\, \beta \,.\, \text{Fmla } \delta : \gamma = \beta\text{j}\delta \,.\!\vee\!.\, \gamma = \delta\text{j}\beta)))\urcorner$.

As a further auxiliary in defining tautology we need the notion of truth set (cf. § 10) — or, since this would involve membership and thus exceed the bounds of protosyntax, we may avail ourselves rather of a parallel notion of *truth sequence.* A truth sequence of x is any expression whose framed ingredients constitute a class S fulfilling (I) of page 53. Thus to say that y is a truth sequence of x, symbolically yTSx, is to say that any joint denial zjw which is a truth-functional component of x is a framed ingredient of y if and only if neither z nor w is a framed ingredient of y.

Δ40. $\ulcorner(\zeta \text{ TS } \eta)\urcorner$ *for* $\ulcorner(\alpha)(\beta)(\text{Fmla } \alpha \,.\, \alpha\text{j}\beta \text{ TC } \eta \,.\!\supset\!::$
$\alpha\text{j}\beta \text{ Ing } \zeta \,.\!\equiv\!:\, \alpha \text{ Ing } \zeta \,.\!\downarrow\!.\, \beta \text{ Ing } \zeta)\urcorner$.

To say that a formula x is tautologous, finally, symbolically Taut x, is to say that x belongs to each of its truth sets (cf. § 10); in other words, that x is a framed ingredient of each of its truth sequences.

Δ41. $\ulcorner\text{Taut } \zeta\urcorner$ *for* $\ulcorner(\text{Fmla } \zeta \,.\, (\alpha)(\alpha \text{ TS } \zeta \,.\!\supset\!.\, \zeta \text{ Ing } \alpha))\urcorner$.

In defining tautology we have made some progress towards definition of the axioms of quantification. For the latter purpose a second necessary step is definition of the notion of *closure;* and as a preliminary to this, in turn, we need the notion of alphabetic succession of variables. The alphabet of variables is the infinite alphabet:

$$w,\ x,\ y,\ z,\ w',\ x',\ y',\ z',\ w'',\ \ldots;$$

it must not be confused with the finite alphabet of signs:

$$w,\ x,\ y,\ z,\ ',\ (,\ ),\ \downarrow,\ \epsilon.$$

Thus the *alphabetic successor* of a variable x, symbolically vs x, begins with S_2 or S_3 or S_4 or $S_1{}^\frown S_5$ according as x begins with S_1 or S_2 or S_3 or S_4; and the rest of vs x, if any, is like the rest of x.

Δ42. ⌜vs ζ⌝ *for* ⌜$(\imath\alpha)(\exists\beta)(\exists\gamma)(\beta = S_1 \,.\, \gamma = S_2 \,.\text{v}.\, \beta = S_2 \,.\, \gamma = S_3 \,.\text{v}.$
$\beta = S_3 \,.\, \gamma = S_4 \,.\text{v}.\, \beta = S_4 \,.\, \gamma = S_1{}^\frown S_5 :$
$\zeta = \beta \,.\, \alpha = \gamma \,.\text{v}\ (\exists\delta)(\zeta = \beta{}^\frown\delta \,.\, \alpha = \gamma{}^\frown\delta))$⌝.

A variable x is *alphabetically prior* to a variable y, symbolically x VPr y, if y is vs x or vs vs x or vs vs vs x or etc; or, equivalently, if vs x or vs vs x or vs vs vs x or vs vs vs vs x is an initial segment of y.

Δ43. ⌜(ζ VPr η)⌝ *for* ⌜(Vbl ζ . Vbl η : vs ζ B η .v. vs vs ζ B η .v.
vs vs vs ζ B η .v. vs vs vs vs ζ B η)⌝.

If we had dispensed with 'x', 'y', and 'z' and limited our variables to 'w' with and without accents, Δ42 and Δ43 would have been greatly simplified; the successor of a variable would be the same variable accented, and one variable would be prior to another if initial to it.

As a next step towards the notion of closure we must define 'x 1 F y', in the sense 'x is alphabetically the first free variable of y'.

Δ44. ⌜(ζ 1 F η)⌝ *for* ⌜(ζ F η . $\sim(\exists\alpha)(\alpha$ F η . α VPr ζ))⌝.

Now the *closure* of x, symbolically cl x, is readily defined in conformity with § 14.

Δ45. ⌜cl ζ⌝ *for* ⌜$(\imath\alpha)$(Fmla α . ζ Eα .
(β)(Fmla β . ζEβ . βEα .⊃ (γ)(Vbl γ . γquβ E α .≡. γ 1F β)))⌝.

We can now define the axioms of quantification of kind A (cf. § 15). To say that x is one of these, symbolically $\text{AQ}_\text{A}\, x$, is to say that x is the closure of a tautologous formula.

Δ46. ⌜$\text{AQ}_{\text{A}}\,\zeta$⌝ *for* ⌜$(\exists\alpha)(\text{Taut}\ \alpha\,.\,\zeta = \text{cl}\alpha)$⌝.

Preparatory to defining the other kinds of axioms of quantification, it is convenient to adopt the following definitions.

Δ47. ⌜$\text{n}\zeta$⌝ *for* ⌜$(\zeta\text{j}\zeta)$⌝.
Δ48. ⌜$(\zeta\ \text{al}\ \eta)$⌝ *for* ⌜$\text{n}(\zeta\text{j}\eta)$⌝.
Δ49. ⌜$(\zeta\ \text{cd}\ \eta)$⌝ *for* ⌜$(\text{n}\zeta\ \text{al}\ \eta)$⌝.

Comparison of Δ47 with D1 shows that, where x is a formula, nx is the denial of x (in primitive notation, of course). Comparison of Δ48 with D3 shows similarly that x al y is the alternation formed from x and y in that order; and comparison of Δ49 with D4 shows that x cd y is the conditional whose antecedent is x and whose consequent is y.

Now the kinds B and C of axioms of quantification are readily defined.

Δ50. ⌜$\text{AQ}_{\text{B}}\,\zeta$⌝ *for* ⌜$(\exists\alpha)(\exists\beta)(\exists\gamma)(\text{Vbl}\ \alpha\,.\,\text{Fmla}\ \beta\,.\,\text{Fmla}\ \gamma\,.$
$\zeta = \text{cl}((\alpha\ \text{qu}\ (\beta\ \text{cd}\ \gamma))\ \text{cd}\ ((\alpha\ \text{qu}\ \beta)\ \text{cd}\ (\alpha\ \text{qu}\ \gamma))))$⌝.
Δ51. ⌜$\text{AQ}_{\text{C}}\,\zeta$⌝ *for* ⌜$(\exists\alpha)(\exists\beta)(\text{Vbl}\ \alpha\,.\,\text{Fmla}\ \beta\,.\,{\sim}(\alpha\text{F}\beta)\,.$
$\zeta = \text{cl}\ (\beta\ \text{cd}\ (\alpha\ \text{qu}\ \beta)))$⌝.

Let us write '$x'\ \text{SF}_y^{y'}\ x$' to mean that x' is like x except for containing free occurrences of y' wherever x contains free occurrences of y. (The 'S' and 'F' here are to suggest the words 'substitution' and 'free'.) This must be defined before we can define the quantificational axioms of kind D. Preparatory to defining '$x'\ \text{SF}_y^{y'}x$' it is convenient to define '$x'\ \text{SF1}_y^{y'}\ x$', in the sense 'x' is like x except for containing a free occurrence of y' in place of one free occurrence of y'.

Δ52. ⌜$(\zeta'\ \text{SF1}_\eta^{\eta'}\ \zeta)$⌝ *for* ⌜$(\exists\alpha)(\exists\alpha')(\alpha\ \text{FO}_\zeta\ \eta\,.\,\alpha'\ \text{FO}_{\zeta'}\ \eta'.$
$(\beta)(\alpha = \beta\frown\eta\,.{\equiv}.\,\alpha' = \beta\frown\eta' : \zeta = \alpha\frown\beta\,.{\equiv}.\,\zeta' = \alpha'\frown\beta))$⌝.

Next let us write '$x'\ \text{SF}?_y^{y'}\ x$' to mean that x' is like x except for containing free occurrences of y' at $\geqslant 0$ places where x contains free occurrences of y; in other words, that x' belongs to a sequence of expressions each of which is either x or else like an earlier expression of the sequence except for containing a free occurrence of y' in place of a free occurrence of y. The notion of sequence involved here can be dealt with as in Δ31 and Δ39.

Δ53. $\ulcorner(\zeta'\ \mathrm{SF}?^{\eta'}_{\eta}\ \zeta)\urcorner$ *for* $\ulcorner(\exists\alpha)(\zeta'\ \mathrm{Ing}\ \alpha\ .\ (\beta)(\beta\ \mathrm{Ing}\ \alpha\ .\supset:\beta=\zeta\ .\mathrm{v}$
$(\exists\gamma)(\gamma\ \mathrm{Pr}_{\alpha}\ \beta\ .\ \beta\ \mathrm{SF1}^{\eta'}_{\eta}\ \gamma)))\urcorner$.

Finally '$x'\ \mathrm{SF}^{y'}_{y}\ x$' says that $x'\ \mathrm{SF}?^{y'}_{y}\ x$ and further that no free occurrences of y are left standing in x — unless of course y and y' are the same variable.

Δ54. $\ulcorner(\zeta'\ \mathrm{SF}^{\eta'}_{\eta}\ \zeta)\urcorner$ *for* $\ulcorner(\zeta'\ \mathrm{SF}?^{\eta'}_{\eta}\ \zeta:\eta\ \mathrm{F}\ \zeta'\ .\supset.\ \eta=\eta')\urcorner$.

Definition of the axioms of quantification of kind D is now easy.

Δ55. $\ulcorner\mathrm{AQ_D}\ \zeta\urcorner$ *for* $\ulcorner(\exists\alpha)(\exists\alpha')(\exists\beta)(\exists\beta')(\mathrm{Vbl}\ \alpha\ .\ \mathrm{Vbl}\ \alpha'\ .\ \mathrm{Fmla}\ \beta\ .$
$\beta'\ \mathrm{SF}^{\alpha'}_{\alpha}\ \beta\ .\ \zeta=\mathrm{cl}\ ((\alpha\ \mathrm{qu}\ \beta)\ \mathrm{cd}\ \beta'))\urcorner$.

§ 58. *Theorem Defined*

PREPARATORY to defining the axioms of membership, we need the notion of *stratification*. The earlier explanation of stratification (§ 28) involved putting numerals for variables; by way of keeping within our protosyntactical resources, however, let us now use strings of arrows instead of numerals — n arrows instead of the nth positive numeral. Sacrifice of '0' and the negative numerals is obviously of no consequence; stratification established with help of non-positive numerals can be established equally well by use of uniformly higher numerals all of which are positive.

First we adopt three definitions closely analogous to Δ52–54.

Δ56. $\ulcorner(\zeta'\ \mathrm{S1}^{\eta'}_{\eta}\ \zeta)\urcorner$ *for* $\ulcorner(\exists\alpha)(\exists\alpha')(\alpha\ \mathrm{O}_{\zeta}\ \eta\ .\ \alpha'\ \mathrm{O}_{\zeta'}\ \eta'\ .$
$(\beta)(\alpha=\beta\frown\eta\ .\equiv.\ \alpha'=\beta\frown\eta':\zeta=\alpha\frown\beta\ .\equiv.\ \zeta'=\alpha'\frown\beta))\urcorner$.

Δ57. $\ulcorner(\zeta'\ \mathrm{S}?^{\eta'}_{\eta}\ \zeta)\urcorner$ *for* $\ulcorner(\exists\alpha)(\zeta'\ \mathrm{Ing}\ \alpha\ .\ (\beta)(\beta\ \mathrm{Ing}\ \alpha\ .\supset:\beta=\zeta\ .\mathrm{v}$
$(\exists\gamma)(\gamma\ \mathrm{Pr}_{\alpha}\ \beta\ .\ \beta\ \mathrm{S1}^{\eta'}_{\eta}\ \gamma)))\urcorner$.

Δ58. $\ulcorner(\zeta'\ \mathrm{S}^{\eta'}_{\eta}\ \zeta)\urcorner$ *for* $\ulcorner(\zeta'\ \mathrm{S}?^{\eta'}_{\eta}\ \zeta\ .\sim(\eta\ \mathrm{In}\ \zeta'))\urcorner$.

Where y and y' are any expressions, clearly '$x'\ \mathrm{S1}^{y'}_{y}\ x$' says that x' is like x except for containing y' in place of an occurrence of y. So long as x, y, and y' are not such that substitution of y' for y in x gives rise to new occurrences of y, it is clear further that '$x'\ \mathrm{S}?^{y'}_{y}\ x$' holds if and only if x' is like x except for containing y' in place of $\geqslant 0$ occurrences of y, and that '$x'\ \mathrm{S}^{y'}_{y}\ x$' holds if and only if x' is like x except for containing y' in place of all occurrences of y. In particular these versions are correct in those cases where y is a variable and y' a string of arrows, since clearly no new occurrence of a variable y can be generated by putting a string of arrows for y in any context.

'Ar x' will mean that x is a string of arrows. The definition resembles Δ22.

Δ59. ⌜Ar ζ⌝ *for* ⌜$(\alpha)(\alpha \mathrm{B}\zeta \,.\supset .\, \mathrm{S}_8\,\mathrm{E}\alpha)$⌝.

Next let us speak of x as an *arrow projection* of y, symbolically xAPy, when x is formed from y by putting strings of arrows for all occurrences of $\geqslant 0$ variables (like strings for like variables); in other words, when x is one of a sequence of expressions each of which either is y or else is like an earlier expression of the sequence except for containing some string of arrows in place of all occurrences of some variable. Thus, using our usual sequence technique,

Δ60. ⌜$(\zeta \mathrm{AP}\eta)$⌝ *for* ⌜$(\exists\alpha)(\zeta \text{ Ing } \alpha \,.\, (\beta)(\beta \text{ Ing } \alpha \,.\supset : \beta = \eta \,.\mathrm{v}$
$(\exists\gamma)(\exists\delta)(\exists\beta')(\mathrm{Ar}\,\gamma \,.\, \mathrm{Vbl}\,\delta \,.\, \beta' \,\mathrm{Pr}_\alpha\, \beta \,.\, \beta\, \mathrm{S}^{\gamma}_{\delta}\, \beta')))$⌝.

Finally, a formula x is *stratified* if it has an arrow projection wherein (i) no variables are left unreplaced by strings of arrows and (ii) a string of arrows following 'ϵ' is always longer by one than the string preceding.

Δ61. ⌜Strat ζ⌝ *for* ⌜$(\mathrm{Fmla}\,\zeta \,.\, (\exists\alpha)(\alpha \,\mathrm{AP}\, \zeta \,.\, \sim(\exists\beta)(\mathrm{Vbl}\,\beta \,.\, \beta\,\mathrm{P}\alpha)\,.$
$(\gamma)(\delta)(\mathrm{Ar}\,\gamma \,.\, \mathrm{Ar}\,\delta \,.\, \gamma\epsilon\delta\, \mathrm{P}\alpha \,.\supset .\, \delta = \gamma^\frown \mathrm{S}_8)))$⌝.

The following definitions, needed as further preparation for the axioms of membership, are related to D2, D5, and D8 as Δ47–49 were related to D1, D3, and D4.

Δ62. ⌜$(\zeta \text{ cj } \eta)$⌝ *for* ⌜$(\mathrm{n}\zeta \text{ j } \mathrm{n}\eta)$⌝.
Δ63. ⌜$(\zeta \text{ b } \eta)$⌝ *for* ⌜$((\zeta \text{ cd } \eta) \text{ cj } (\eta \text{ cd } \zeta))$⌝.
Δ64. ⌜$(\zeta \text{ qe } \eta)$⌝ *for* ⌜$\mathrm{n}(\zeta \text{ qu } \mathrm{n}\eta)$⌝.

Thus, where x and y are formulæ, x cj y is their conjunction and xby is their biconditional (expanded into primitive notation); and, where z is a variable, z qe x is the existential quantification of x with respect to z.

Similarly, where x and y are variables, their identity will be called x id y. More accurately, x id y is to be the formula (in primitive notation) whose abbreviation via D1–10 is formed by putting *x* and *y* in the respective blanks of '(=)'. The odd bound variable α of D10 may conveniently be understood always as the alphabetically earliest variable other than *x* and *y* — hence the alphabetically earliest variable foreign to $x \frown y$ — in conformity with the plan suggested earlier (§ 24). In general, let us speak of the alphabetically earliest variable foreign to *z* as f*z*.

Δ65. $\ulcorner \mathrm{f}\zeta \urcorner$ *for* $\ulcorner (\imath\alpha)({\sim}(\alpha \text{ In } \zeta) \,.\, \text{Vbl } \alpha \,.\, (\beta)(\beta \text{ VPr } \alpha \,.\!\supset\!.\, \beta \text{ In } \zeta))\urcorner$.

Then, in view of D10,

Δ66. $\ulcorner(\zeta \text{ id } \eta)\urcorner$ *for* $\ulcorner(\mathrm{f}(\zeta \frown \eta) \text{ qu } ((\mathrm{f}(\zeta \frown \eta) \text{ e } \zeta) \text{ b } (\mathrm{f}(\zeta \frown \eta) \text{ e } \eta)))\urcorner$.

When *x* and *y* are variables and *z* is a formula, *x* e *yaz* is to be the formula (in primitive notation) whose abbreviation via D1–9 is formed by putting *x* in the first blank of '($\epsilon\hat{\ }$)', *z* in the last, and *y* under the circumflex. The definition is apparent from D9. The odd variable γ of D9 is taken as the alphabetically earliest variable other than *x*, *y*, and the variables of *z* — hence as $\mathrm{f}(x \frown y \frown z)$.

Δ67. $\ulcorner(\zeta \text{ e } \eta\text{a}\theta)\urcorner$ *for*
$\ulcorner(\mathrm{f}(\zeta \frown \eta \frown \theta) \text{ qe}((\zeta \text{ e } \mathrm{f}(\zeta \frown \eta \frown \theta)) \text{ cj } (\eta \text{ qu } ((\eta \text{ e } \mathrm{f}(\zeta \frown \eta \frown \theta)) \text{ cd } \theta))))\urcorner$.

Where *x*, *y*, and *z* are as before, *x* id *yaz* is to be the formula whose abbreviation via D1–10 is formed by putting *x* in the first blank of '($=\hat{\ }$)', *z* in the last, and *y* under the circumflex. Thus, in view of D10,

Δ68. $\ulcorner(\zeta \text{ id } \eta\text{a}\theta)\urcorner$ *for* $\ulcorner(\mathrm{f}(\zeta \frown \eta \frown \theta) \text{ qu}((\mathrm{f}(\zeta \frown \eta \frown \theta) \text{ e } \zeta) \text{ b } (\mathrm{f}(\zeta \frown \eta \frown \theta) \text{ e } \eta\text{a}\theta)))\urcorner$.

Where *x* and *y* are variables and *z* and *w* are formulæ, *xaz* e *yaw* is to be the formula whose abbreviation via D1–12 is formed by putting *z* and *w* in the respective blanks and *x* and *y* under the respective circumflexes of '($\hat{\ }$ $\epsilon\hat{\ }$)'. Thus, in view of D12,

Δ69. $\ulcorner(\zeta\text{a}\eta \text{ e } \zeta'\text{a}\eta')\urcorner$ *for*
$\ulcorner(\mathrm{f}(\zeta \frown \eta \frown \zeta' \frown \eta') \text{ qe } ((\mathrm{f}(\zeta \frown \eta \frown \zeta' \frown \eta') \text{ id } \zeta\text{a}\eta) \text{ cj } (\mathrm{f}(\zeta \frown \eta \frown \zeta' \frown \eta') \text{ e } \zeta'\text{a}\eta')))\urcorner$.

Where *x* is a variable, *x* e v is to be the formula whose abbreviation via D1–15 is formed by putting *x* in the blank of '(ϵV)'; and, where *y* is a formula, *xay* e v is to be the formula whose abbreviation is formed by putting x under the circumflex and y in the blank of '(^ ϵV)'. Thus, in view of D15,

Δ70. $\ulcorner(\zeta \text{ e v})\urcorner$ *for* $\ulcorner(\zeta \text{ e } S_2\text{a}(S_2 \text{ id } S_2))\urcorner$.

Δ71. $\ulcorner(\zeta\text{a}\eta \text{ e v})\urcorner$ *for* $\ulcorner(\zeta\text{a}\eta \text{ e } S_2\text{a}(S_2 \text{ id } S_2))\urcorner$.

Let us write '$\text{R}x$' to mean that every quantification which x contains is restricted to elements (except for occurrences of quantifications within the definitional expansions of the restricting clauses themselves).

△72. $\ulcorner\text{R}\zeta\urcorner$ *for* $\ulcorner(\alpha)(\beta)(\gamma)(\text{Vbl } \alpha \,.\, \text{Fmla } \beta \,.\, \gamma \text{ O}_\zeta\, \alpha\text{qu}\beta \,.\supset$
$(\exists\delta)(\exists\delta')(\beta = (\alpha\text{ev}) \text{ cd } \delta \,.\text{v}.\, \text{Vbl } \delta \,.\, \gamma\frown\delta' \text{ O}_\zeta\, \delta\text{ev} \,.\, \sim \delta\text{ev P } \delta'))\urcorner$.

Finally let us write '$\text{AM}_\text{A}x$' to mean that x is an axiom of membership of the first kind, i.e., an axiom of elementhood. For simplicity of definition I shall construe these axioms somewhat more broadly now than in § 28; but the reader can easily satisfy himself that this enlargement of the class of axioms of elementhood is inessential, in the sense of not enlarging the class of theorems.

Δ73. $\ulcorner\text{AM}_\text{A}\, \zeta\urcorner$ *for* $\ulcorner(\exists\alpha)(\exists\beta)(\exists\gamma)(\text{Fmla } \alpha \,.\, \text{Vbl } \beta \,.\, \text{Strat } \gamma \,.\, \text{R}\gamma \,.$
$(\delta)(\delta \text{ F}\gamma \,.\, \delta \neq \beta \,.\supset \text{Taut } \alpha\text{cd}(\delta\text{ev})) \,.\, \zeta = \text{cl}(\alpha \text{ cd } (\beta\text{a}\gamma \text{ e v})))\urcorner$.

The other two kinds of axioms of membership, described in § 29, present no difficulty.

Δ74. $\ulcorner\text{AM}_\text{B}\, \zeta\urcorner$ *for* $\ulcorner(\exists\alpha)(\exists\alpha')(\exists\beta)(\exists\beta')(\text{Vbl } \alpha \,.\, \text{Vbl } \alpha' \,.$
$\text{Fmla}_0\, \beta \,.\, \beta' \text{ S1}^{\alpha'}_{\alpha}\, \beta \,.\, \zeta = \text{cl}((\alpha \text{ id } \alpha') \text{ cd } (\beta \text{ cd } \beta')))\urcorner$.

Δ75. $\ulcorner\text{AM}_\text{C}\, \zeta\urcorner$ *for* $\ulcorner(\exists\alpha)(\exists\beta)(\exists\gamma)(\text{Vbl } \alpha \,.\, \text{Vbl } \beta \,.\, \text{Fmla } \gamma \,.$
$\beta \neq \alpha \,.\, \sim (\beta\text{F}\gamma) \,.\, \zeta = \text{cl}(\beta \text{ qe } (\alpha \text{ qu } ((\alpha \text{ e } \beta) \text{ b } ((\alpha \text{ e v}) \text{ cj } \gamma)))))\urcorner$.

The general notion of axiom is now definable.

Δ76. $\ulcorner\text{Ax}\, \zeta\urcorner$ *for* $\ulcorner(\text{AQ}_\text{A}\, \zeta \text{ v } \text{AQ}_\text{B}\, \zeta \text{ v } \text{AQ}_\text{C}\, \zeta \text{ v } \text{AQ}_\text{D}\, \zeta \text{ v}$
$\text{AM}_\text{A}\, \zeta \text{ v } \text{AM}_\text{B}\, \zeta \text{ v } \text{AM}_\text{C}\, \zeta)\urcorner$.

Now x is a theorem if it belongs to a sequence of expressions each of which is an axiom or a ponential of earlier expressions of the sequence.

Δ77. $\ulcorner\text{Thm}\, \zeta\urcorner$ *for* $\ulcorner(\exists\alpha)(\zeta \text{ Ing } \alpha \,.\, (\beta)(\beta \text{ Ing } \alpha \,.\supset.\, \text{Ax } \beta \text{ v}$
$(\exists\gamma)(\gamma \text{ Pr}_\alpha\, \beta \,.\, \gamma\text{cd}\beta \text{ Pr}_\alpha\, \beta)))\urcorner$.

§ 59. *Protosyntax Self-Applied*

IN § 55–58 we have considered the protosyntax of logic; let us now consider the protosyntax of protosyntax itself. Instead of using the notation of protosyntax to discuss expressions composed of the alphabet:

$$w, \quad x, \quad y, \quad z, \quad ', \quad (, \quad), \quad \downarrow, \quad \epsilon,$$

we shall now use it to discuss expressions composed of the very alphabet:

$$w, \quad x, \quad y, \quad z, \quad ', \quad (, \quad), \quad \downarrow, \quad \mathrm{M}$$

in which protosyntax itself is couched. The explanation of '$\mathrm{M}xyz$' in § 54 accordingly needs to be reconstrued to this slight extent: the alphabet mentioned under Case 1 ceases to have the typographical shape 'ϵ' as its ninth member, and comes to have the shape 'M' instead. In other words, S_9 ceases to be 'ϵ' and becomes 'M'.

The protosyntactical definition of protosyntactical formula proceeds just like the protosyntactical definition of logical formula, except that the atomic formulæ are now described as having the form $\ulcorner \mathrm{M}\alpha\beta\gamma \urcorner$ instead of $\ulcorner(\alpha \epsilon \beta)\urcorner$. Thus the definitions Δ1–30 are taken over intact except that Δ23 and Δ28 give way to:

$\ulcorner \mathrm{m}\{\zeta, \eta, \theta\} \urcorner$ *for* $\ulcorner(\mathrm{S}_9 \frown \zeta \frown \eta \frown \theta)\urcorner$,
$\ulcorner \mathrm{PFmla}_0\, \zeta \urcorner$ *for*

$$\ulcorner(\exists\alpha)(\exists\beta)(\exists\gamma)(\mathrm{Vbl}\ \alpha\ .\ \mathrm{Vbl}\ \beta\ .\ \mathrm{Vbl}\ \gamma\ .\ \zeta = \mathrm{m}\{\alpha, \beta, \gamma\})\urcorner;$$

and Δ31 is taken over with 'Fmla_0' and 'Fmla' changed to 'PFmla_0' and 'PFmla'.

Just as the theorems of logic comprised the ponentials of ponentials of . . . axioms of quantification and axioms of membership, so the theorems of protosyntax would comprise the ponentials of ponentials of . . . axioms of quantification and various axioms appropriate to 'M'. The protosyntactical definition of protosyntactical theorem would thus proceed analogously to the protosyntactical definition of theorem hitherto constructed for logic. We would first define the axioms of quantification as in § 57; next, instead of defining axioms of membership as in § 58, we would define appropriate axioms governing 'M'; and from here we would proceed to 'PThm *x*', '*x* is a protosyntactical theorem', through definitions precisely parallel to Δ76–77.

What should we choose by way of axioms governing 'M'? An ideal choice would be such that the theorems would come to exhaust those protosyntactical formulæ which are true statements, and of course exclude all which are false. It turns out, however, that such a choice is impossible — so long as we insist on its specifiability in protosyntactical terms. It turns out that *no* protosyntactically definable notion of protosyntactical theorem can exhaust the protosyntactical truths and exclude the falsehoods. The remainder of the section will be devoted to establishing this conclusion, which has far-reaching significance. It will be shown that if ⌜PThm ζ⌝ is so construed as to be translatable into protosyntax at all then there is a protosyntactical statement which is true if and only if it is not a theorem. If the statement in question is true then it is not a theorem, so that the theorems do not exhaust the protosyntactical truths; and otherwise it is a theorem, so that the theorems do not exclude falsehoods.

In a word, then, our problem is to construct a protosyntactical statement to the effect:

(1) (1) is not a theorem.

As it stands, the statement (1) refers to itself by means of a demonstrative pronoun '(1)'; but we can gain the same effect without this device. If for purposes of a preliminary venture we allow appeal to quotation, we can reformulate the self-mentioning statement (1) thus:

(2) The result of putting the quotation of 'The result of putting the quotation of *w* for S_1 in *w* is not a theorem' for S_1 in 'The result of putting the quotation of *w* for S_1 in *w* is not a theorem' is not a theorem.

This is more intelligible than it looks. It says that a certain result is not a theorem. It says, moreover, how to construct the result in question. To construct it we are to substitute a certain quotation for S_1 (not for 'S_1' but for S_1, which is the variable 'w') in the matrix:

(3) The result of putting the quotation of w for S_1 in w is not a theorem.

Moreover, the quotation which we are told to substitute is the quotation of (3) itself — i.e., it is (3) plus an enclosing pair of quotation marks. When we perform this substitution we come out with (2) again as our result. So (2) says that (2) is not a theorem.

When we try to put (2) into the notation of protosyntax, the problem confronts us of somehow translating or eliminating the two idioms 'result of putting' and 'quotation of'. Let us deal with the former of these first. Imagining '——' to be some expression, and imagining '$\ldots w \ldots$' to be some matrix with free 'w', we observe to begin with that the result of putting '——' for S_1 (i.e. for 'w') in '$\ldots w \ldots$' is equivalent to:

$$(w)(w = \text{——} \,.\!\supset \ldots w \ldots)$$

(cf. *234a). This observation may, with help of the notations introduced in △25 and △49 (pp. 295, 301), be rephrased as follows: the result of putting an expression u for free S_1 in a matrix y is equivalent to S_1qu(z cd y) where z consists of '$w =$' followed by u. Now if in particular we take u as the quotation of some expression x, the foregoing observation takes on the following form:

(4) The result of putting the quotation of x for free S_1 in y is equivalent to S_1qu(z cd y) where z is '$w =$ ' followed by the quotation of x.

Quotation, we know, is extraneous to the notation and subject matter of protosyntax. Nevertheless it is possible in protosyntax to construct corresponding to any given expression x a matrix which as a whole is equivalent to '$w =$' followed by the quotation of x. How this can be done will be seen later; meanwhile let us suppose it done and get on. The matrix in question will be called the *identity matrix* of x, briefly im x. Thus

(5) im x is equivalent to '$w =$' followed by the quotation of x.

By (4), then,

(6) The result of putting the quotation of *x* for free S_1 in *y* is equivalent to S_1qu(im*x* cd *y*).

Accordingly the purpose of (3) will be served just as well by:

(7) ∼PThm S_1qu(im*w* cd *w*).

Then (2) becomes:

(8) The result of putting the quotation of '∼PThm S_1qu (im*w* cd *w*)' for S_1 in '∼PThm S_1qu(im*w* cd *w*)' is not a theorem.

Now if in turn we translate (8) as a whole in conformity with (6), we obtain:

(9) ∼PThm S_1qu(im'∼PThm S_1qu(im*w* cd *w*)' cd '∼PThm S_1qu(im*w* cd *w*)').

We are imagining that ⌜PThm ζ⌝ has some protosyntactical definition. Let us also suppose that we have succeeded (as in time we shall) in devising an appropriate protosyntactical definition for ⌜im ζ⌝. Then (7) can be expanded into the primitive notation of protosyntax. The expanded formula can in turn be named in protosyntax, by spelling (cf. page 284). This name, which we may abbreviate by the letter 'H', is as good as a quotation of (7); so (9) becomes:

(10) ∼PThm S_1qu(imH cd H).

Here — pending our remaining task of defining ⌜im ζ⌝ — is a protosyntactical analogue of (2) or indeed of (1). Here, under abbreviation, is a protosyntactical formula which is true if and only if it is not a protosyntactical theorem. The existence of such a formula proves that the theorems either exclude some protosyntactical truths or include some falsehoods. The argument holds for any protosyntactically definable concept of protosyntactical theorem.

We have now to fill in the one gap in the argument; namely, to devise a protosyntactical definition of ⌜im ζ⌝ conforming, for every protosyntactical expression *x*, to the requirement (5).

Just as in §§ 57–58 we adopted definitions △47–49, △62–64, and

Δ66 for convenience in referring to those formulæ which the logical definitions D1–5, D8 and D10 serve to abbreviate, let us now adopt definitions which will enable us similarly to refer to those protosyntactical formulæ which the protosyntactical definitions Δ1–5, Δ8, and Δ9 serve to abbreviate. The first six of these new definitions, introducing $\ulcorner n\zeta \urcorner$, $\ulcorner \zeta \text{ al } \eta \urcorner$, $\ulcorner \zeta \text{ cd } \eta \urcorner$, $\ulcorner \zeta \text{ cj } \eta \urcorner$, $\ulcorner \zeta \text{ b } \eta \urcorner$, and $\ulcorner \zeta \text{ qe } \eta \urcorner$, simply duplicate Δ47–49 and Δ62–64; for Δ1–5 and Δ8 are the same as D1–5 and D8. The remaining one of our new definitions is related to Δ9 as Δ66 is related to D10; Δ65 is to be understood as carried over in preparation.

$$\ulcorner(\zeta \text{ id}_P\, \eta)\urcorner \quad \textit{for} \quad \ulcorner(\text{f}(\zeta \frown \eta) \text{ qu } (\text{m}\{\text{f}(\zeta \frown \eta), \zeta, \zeta\} \text{ b m}\{\text{f}(\zeta \frown \eta), \eta, \eta\}))\urcorner.$$

Thus, where x and y are variables, $x \text{ id}_P\, y$ is the protosyntactical formula whose definitional abbreviation via Δ1–9 is '(=)' with x and y in the respective blanks.

Now let Z_1 be the protosyntactical formula '$(x)(y) \sim Mwxy$', or rather the primitive expansion:

$$(x)(y)(Mwxy \downarrow Mwxy)$$

thereof; i.e.,

$$\text{'}Z_1\text{'} \quad \textit{for} \quad \text{'}(S_2 \text{ qu } (S_3 \text{ qu n m}\{S_1, S_2, S_3\}))\text{'}.$$

Let Z_2 be the primitive expansion of the result of putting Z_1 in the blank of:

$$(11) \qquad (\exists x)((\exists w)(x = w \,.\qquad\qquad) \,.\, Mwxw);$$

i.e.,

$$\text{'}Z_2\text{'} \quad \textit{for} \quad \text{'}(S_2 \text{ qe } ((S_1 \text{ qe } ((S_2 \text{ id}_P\, S_1) \text{ cj } Z_1)) \text{ cj m}\{S_1, S_2, S_1\}))\text{'}.$$

Similarly let Z_3 be the primitive expansion of the result of putting Z_2 in the blank of (11); and so on to Z_9. The definitions of 'Z_3', . . . , 'Z_9' are just like the above definition of 'Z_2' except for containing 'Z_2', . . . , 'Z_8' in place of 'Z_1'.

We have noted (§ 54) that $(x)(y) \sim Mwxy$ if and only if $w = S_1$. Thus the formula Z_1 is equivalent to '$w = S_1$'. Accordingly, putting '$w = S_1$' in the blank of (11), we conclude that Z_2 is equivalent to:

$$(\exists x)((\exists w)(x = w \,.\, w = S_1) \,.\, Mwxw),$$

which is equivalent in turn (cf. *234b) to '$\mathrm{M}wS_1w$' and hence to '$w = S_2$' (cf. § 54). Putting the latter in the blank of (11), we conclude in similar fashion that Z_3 is equivalent to '$w = S_3$'; and so on. $Z_1, Z_2, \ldots, Z_9$ are protosyntactical formulæ (in primitive notation) which are equivalent respectively to '$w = S_1$', '$w = S_2$', ..., '$w = S_9$'. It is therefore consonant with (5) to explain im S_i as Z_i, for each i from 1 to 9. But there remains the task of formulating im x where x is longer than a single sign.

By z cid z', next, will be meant the primitive expansion of the result of putting z and z' (whatever expressions these may be) in the respective blanks of:

(12) $(\exists x)(\exists y)((\exists w)(x = w \,.\qquad) \,.\, (\exists w)(y = w \,.\qquad) \,.\, \mathrm{M}wxy \,.\, \sim \mathrm{M}wxw)$.

Thus

$\ulcorner(\zeta \text{ cid } \eta)\urcorner$ *for* $\ulcorner(S_2 \text{ qe } (S_3 \text{ qe } ((((S_1 \text{ qe } ((S_2 \text{ id}_P\, S_1) \text{ cj } \zeta)) \text{ cj } (S_1 \text{ qe } ((S_3 \text{ id}_P\, S_1) \text{ cj } \eta))) \text{ cj m}\{S_1, S_2, S_3\}) \text{ cj n m}\{S_1, S_2, S_1\})))\urcorner$.

Consider, e.g., Z_h cid Z_k (supposing h and k chosen from among 1, 2, ..., 9). Since Z_h and Z_k are equivalent respectively to '$w = S_h$' and '$w = S_k$', Z_h cid Z_k is equivalent to the result of putting '$w = S_h$' and '$w = S_k$' in the respective blanks of (12). But this reduces to '$\mathrm{M}wS_hS_k \,.\, \sim\mathrm{M}wS_hw$' (cf. *234b), which is equivalent in turn to '$w = S_h{}^\frown S_k$' (cf. § 54). Thus Z_h cid Z_k is equivalent to '$w = S_h{}^\frown S_k$'.[1] By the same reasoning, again, (Z_h cid Z_k) cid Z_m is seen to be equivalent to '$w = S_h{}^\frown S_k{}^\frown S_m$'; likewise (($Z_h$ cid Z_k) cid Z_m) cid Z_n is equivalent to '$w = S_h{}^\frown S_k{}^\frown S_m{}^\frown S_n$'; and so on. In general, (... ((Z_{q_1} cid Z_{q_2}) cid Z_{q_3}) ...) cid Z_{q_r} is a protosyntactic formula equivalent to '$w = S_{q_1}{}^\frown S_{q_2}{}^\frown \ldots {}^\frown S_{q_r}$', and hence exactly fits the characterization (5) of im x when x is $S_{q_1}{}^\frown S_{q_2}{}^\frown \ldots {}^\frown S_{q_r}$. Since each expression in protosyntax is $S_{q_1}{}^\frown S_{q_2}{}^\frown \ldots {}^\frown S_{q_r}$ for one or another choice of $q_1, q_2, \ldots, q_r$ (where $r \geqslant 1$ and $1 \leqslant q_i \leqslant 9$ for each i), we have now succeeded in identifying im x with a determinate protosyntactical expression for every determinate choice of x. We explain im S_i as Z_i for each i, and im($S_i{}^\frown S_j$) as Z_i cid Z_j for each i and j, and im($S_i{}^\frown S_j{}^\frown S_k$) as ($Z_i$ cid Z_j) cid Z_k for each i, j, and

[1] The connective 'cid' is supposed to suggest the words 'concatenation' and 'identity.'

k, and so on.

But there still remains the further task of devising a protosyntactical definition of 'im x' where x is unspecified. In seeking this objective it will be convenient to concentrate not on im x itself but on im$x^\frown x$, the concatenate of im x with x. From im$x^\frown x$ as a whole then we can single out im x afterward as *the* initial protosyntactical formula thereof; for a survey of our primitive notation reveals that a string of signs cannot have two *formulæ* as initial segments, when the full parentheses appropriate to joint denial are taken into consideration.

Let us write 'IMC w' to mean that w is im$x^\frown x$ for some x. To see how to express this in protosyntax, we reflect that 'IMC w' is to hold if and only if w is either $Z_i{}^\frown S_i$ for some i, or $(Z_i \text{ cid } Z_j)^\frown S_i{}^\frown S_j$ for some i and j, or $((Z_i \text{ cid } Z_j) \text{ cid } Z_k)^\frown S_i{}^\frown S_j{}^\frown S_k$ for some i, j, and k, or etc. In other words, IMC w if and only if w is one of a sequence of expressions each of which is either $Z_i{}^\frown S_i$ for some i or else $(z \text{ cid } Z_i)^\frown y^\frown S_i$ for some i where $z^\frown y$ is an earlier expression of the sequence (and z is a formula). Accordingly, applying our usual sequence technique,

$$\ulcorner\text{IMC } \zeta\urcorner \textit{ for } \ulcorner(\exists\alpha)(\zeta \text{ Ing } \alpha \,.\, (\beta)(\beta \text{ Ing } \alpha \,.\!\supset\!: \beta = Z_1{}^\frown S_1 \,.\!\vee\!. \ldots$$
$$.\!\vee\!.\ \beta = Z_9{}^\frown S_9 \,.\!\vee\, (\exists\gamma)(\exists\delta)(\text{PFmla } \gamma \,.\, \gamma^\frown\delta \text{ Pr}_\alpha\, \beta : \beta = (\gamma \text{ cid } Z_1)^\frown\delta^\frown S_1 \,.\!\vee\!.$$
$$\ldots .\!\vee\!.\ \beta = (\gamma \text{ cid } Z_9)^\frown\delta^\frown S_9)))\urcorner.$$ [1]

Now the definition of $\ulcorner$im $\zeta\urcorner$ is obvious.

$$\ulcorner\text{im } \zeta\urcorner \quad \textit{for} \quad \ulcorner(\iota\alpha)(\text{PFmla } \alpha \,.\, \text{IMC } \alpha^\frown\zeta)\urcorner.$$

§ 60. *Incompleteness*

WE HAVE seen that protosyntax is *protosyntactically incompletable*, in the sense that no protosyntactically definable notion of protosyntactical theorem can exhaust the protosyntactical truths and exclude the falsehoods. Protosyntactical truth is protosyntactically indefinable. Where K is the class of all those statements which

[1] Obviously the dots '. . .' here mark no breach of rigor. The reader can write in the fourteen missing clauses.

are built up of atomic matrices ⌜$M\alpha\beta\gamma$⌝ by joint denial and quantification and are in fact true under the intended interpretation of 'M', ' ↓ ', etc., no matrix '. . . *x* . . .' composed of those same materials will fulfill the condition:

(1) $(x)(\ldots x \ldots . \equiv . x \in K)$.

In the course of the present section we shall find that logic itself is likewise protosyntactically incompletable; that no protosyntactically definable notion of logical theorem can exhaust the logical truths and exclude the falsehoods. Logical truth, like protosyntactical truth, is protosyntactically indefinable.

'M' was explained (§ 54) in a fashion equivalent to the following: M*xyz* if and only if

(i) *x* and *y* are respectively '*x*' and '*w*', or '*y*' and '*x*', or . . . (through the nine-sign alphabet), or else

(ii) *y* and *z* are expressions (composed of those nine signs) and *x* consists of *y* followed by *z*, or else

(iii) *x* is not an expression and *y* is *x*.

Now obviously the truth of the statements in K is not contingent upon geometrical peculiarities of the particular typographical shapes '*w*', '*x*', '*y*', '*z*', '′', '(', etc. which were chosen for purposes of (i); any other shapes, e.g. the numerals '1', '2', . . . , '9', would serve as well. Indeed, we need not even insist upon typographical shapes. Instead of considering the shapes '1', '2', . . . , '9' and their complexes '11', '12', . . . , '99', '111', etc., we could just as well consider the corresponding natural numbers, 1, 2, . . ., 9, 11, 12, . . . , 99, 111, . . . themselves.

We are thus led to an arithmetical analogue of the protosyntactical 'M'. By way of notation for the analogue, let us use the letter 'ϵ' in place of 'M'. This sense of 'ϵ' has nothing to do with the membership sense of 'ϵ'; but there is no danger of confusion, since 'ϵ' in the new sense has contexts of the form ⌜$\epsilon\zeta\eta\theta$⌝ while 'ϵ' in the membership sense has contexts rather of the form ⌜$(\zeta \in \eta)$⌝. These new contexts are explicable in direct analogy to (i)–(iii) above: ϵxyz if and only if

(iv) *x* and *y* are respectively 2 and 1, or 3 and 2, or . . . , or 9 and 8, or else

(v) y and z are *perpositive* numbers (natural numbers in whose Arabic numerals '0' does not appear) and x is $y^\frown z$ (cf. end of § 49), or else

(vi) x is not a perpositive number and y is x.

Where K′ is the class of all those true statements which are built up of matrices $\ulcorner \epsilon\alpha\beta\gamma \urcorner$ by joint denial and quantification, the only difference between K′ and the previously described class K is that the statements of the one class contain 'ϵ' where the statements of the other class contain 'M'.

This new idiom 'ϵxyz' is readily given a logical definition. To the series of logical definitions D1–48 we first add the ones suggested at the end of § 49, together with the following definition of the class of perpositive numbers.

D61. 'Pp' *for*

$$`\hat{x}(x \epsilon \mathrm{Nn} \,.\, (y)(x \neq y^\frown 0 \,.\, (z)(x = (z^\frown 0)^\frown y \;.\supset.\; z = 0)))'.$$

Then we simply reproduce (iv)–(vi) as a formal definition.

D62. $\ulcorner \epsilon\zeta\eta\theta \urcorner$ *for*

$$\ulcorner(8 \geqslant \eta \,.\, \eta \geqslant 1 \,.\, \zeta = \mathrm{S}`\eta \;.\mathrm{v}.\; \eta, \theta \,\epsilon\, \mathrm{Pp} \,.\, \zeta = \eta^\frown\theta \;.\mathrm{v}.\; \zeta \,\bar{\epsilon}\, \mathrm{Pp} \,.\, \zeta = \eta)\urcorner.$$

Let us return now to the protosyntax of logic, begun in §§ 55–58. The definitions Δ47–49, Δ62–64, and Δ66–71 were set up in such a way as to parallel D1–5, D8–10, D12, and D15; and by continuing the series in obvious fashion we finally reach a protosyntactical definition:

$\ulcorner \mathrm{e}\{\zeta, \eta, \theta\} \urcorner$ *for* . . .

paralleling D62. Where x, y, and z are any variables, $\mathrm{e}\{x, y, z\}$ is the logical formula (built up of matrices $\ulcorner(\alpha \,\epsilon\, \beta)\urcorner$ by joint denial and quantification) which D1–D62 abbreviate as 'ϵ' followed by the successive variables x, y, and z (whichever ones they may be).

Suppose now that y is a logical formula (in primitive notation, of course), and that x is an abbreviation of y by D1–62. It may happen that x contains only primitive notation plus abbreviations of the form $\ulcorner \epsilon\alpha\beta\gamma \urcorner$ explained by D62; x may be devoid of other abbreviation notation. When all these circumstances obtain, let us call x an *epsilon abbreviate* of y; symbolically $x\mathrm{EA}y$. Where $x\mathrm{EA}y$, thus, x consists only of the alphabet $\mathrm{S}_1, \ldots, \mathrm{S}_9$, i.e. '$w$',

'x', . . . , '$\downarrow$', 'ϵ', whereof the protosyntax of logic speaks; but this does not make x a logical formula, for the mode of combination of signs which is essential to a logical formula is departed from in $\ulcorner \epsilon\alpha\beta\gamma \urcorner$. At the same time x is an abbreviation, via D1–62, of the genuine logical formula y.

To say that x is an epsilon abbreviate of y is to say that x is one of a sequence of expressions each of which either is y or else is formed from an earlier expression of the sequence by abbreviating a part $\mathrm{e}\{z, z', z''\}$ as $S_9 \frown z \frown z' \frown z''$ (z, z', and z'' being any variables). The protosyntactical definition of this notion of epsilon abbreviate is then readily accomplished, following our usual method of dealing with sequences.

$$\ulcorner(\zeta \mathrm{EA} \eta)\urcorner \textit{ for } \ulcorner(\exists\alpha)(\zeta \text{ Ing } \alpha \,.\, (\beta)(\beta \text{ Ing } \alpha \,.\!\supset:\beta = \eta \,.\!\vee (\exists\gamma)(\exists\delta)(\exists\delta')(\exists\delta'')(\mathrm{Vbl}\,\delta \,.\, \mathrm{Vbl}\,\delta' \,.\, \mathrm{Vbl}\,\delta'' \,.\, \gamma \,\mathrm{Pr}_\alpha\, \beta \,.\, \beta \,\mathrm{S1}^{S_9 \frown \delta \frown \delta' \frown \delta''}_{\mathrm{e}\{\delta,\, \delta',\, \delta''\}}\, \gamma)))\urcorner.$$

Under this definition, '$x\mathrm{EA}y$' is true also in the limiting case where x and y are the same expression; but this departure is of no consequence.

Let us call x a *quasi-theorem*, symbolically $\mathrm{Thm}'x$, if x is built up solely of expressions $\ulcorner \epsilon\alpha\beta\gamma \urcorner$ by joint denial and quantification, and is furthermore an abbreviation by D1–62 of a theorem of logic. This notion is readily defined in protosyntax, given any protosyntactical definition of theorem (e.g. that of § 58).

$$\ulcorner \mathrm{Thm}'\zeta \urcorner \textit{ for } \ulcorner((\exists\alpha)(\mathrm{Thm}\,\alpha \,.\, \zeta \mathrm{EA} \alpha) \,.\, \sim(\exists\beta)(\mathrm{Fmla}\,\beta \,.\, \beta \mathrm{P} \zeta))\urcorner.$$

Suppose now that the protosyntactical definition of $\ulcorner \mathrm{Thm}\,\zeta \urcorner$ is such as to include among the theorems all those logical formulæ which are true, and exclude all falsehoods. Then an expression x will be a quasi-theorem, in the above sense, if and only if it belongs to the class K′ described earlier. Thus if we expand '$\mathrm{Thm}'x$' into protosyntactical primitives we have a protosyntactical matrix '. . . x . . .' such that

$$(2) \qquad (x)(\ldots x \ldots .\!\equiv\!. \, x \in \mathrm{K}').$$

Whereas the protosyntactical 'M' (occurring in '. . . x . . .') has thus far been construed in the fashion appropriate to the protosyntax of logic, i.e. in such fashion that S_9 = 'ϵ', let us now reconstrue it in the fashion appropriate rather to the protosyntax of protosyntax (§ 59); i.e., in such fashion that S_9 becomes 'M'. The matrix '. . . *x* . . .' then comes to satisfy the condition (1) rather than (2), since the expressions belonging to K′ go over into those of K when 'ϵ' is changed to 'M'. But we know that no protosyntactical matrix '. . . *x* . . .' can satisfy (1). The initial assumption of the preceding paragraph is therefore false; no protosyntactical definition of theorem can include all true logical formulæ among the theorems and exclude the falsehoods. Logic, like protosyntax itself, is protosyntactically incompletable.

It follows in particular that the notion of theorem which was developed in earlier chapters, and defined protosyntactically in § 58, does not accord the status of theorem to all those logical formulæ which are true statements — or else, worse, that it accords the status of theorem to some falsehoods. And, as the foregoing argument shows, any alternative notion of theorem which we might devise will suffer a similar fate, so long as we insist on protosyntactical definability. Nor is the demand of protosyntactical definability easily waived; it seems already to be more liberal than the normal practical demand of constructivity — the demand that for each theorem there exist an at least fortuitously discoverable method of confirming its theoremhood (cf. § 55).

The protosyntactical incompletability of logic is of more immediate concern to us than was the protosyntactical incompletability of protosyntax; but the latter is more remarkable than the former, inasmuch as protosyntax has a far more limited subject matter than logic. Apprised only of the protosyntactical incompletability of logic, we should probably have blamed the difficulty on the connective 'ϵ' of membership and hence questioned the admissibility of that connective. The notion of membership is a natural object of suspicion; for it is this notion that imports the whole realm of classes of higher and higher orders of abstractness, and even calls for *ad hoc* measures such as the distinction between element and non-element for the avoidance of contradiction. Protosyntax itself, on the other hand, is wholly independent of the notion of membership and the theory of classes; it calls for no non-elements, indeed no entities whatever beyond an infinite domain of finite expressions each of which is nameable within the notation of protosyntax in systematic fashion. Alternatively, we have seen, those entities can be construed as natural numbers. When protosyntactical incompletability reasserts itself in as simple a field as this, we cease to regard such incompletability as a ground of suspicion and come rather to expect it in every fairly untrivial field.

Protosyntax, though protosyntactically incompletable, can indeed be shown to be *syntactically* completable. I.e., whereas no matrix constructed solely of atomic matrices $\ulcorner M\alpha\beta\gamma\urcorner$ by joint denial and quantification will fulfill (1) when put for the blank '. . . *x* . . .', introduction of additional atomic matrices of the form $\ulcorner(\alpha \in \beta)\urcorner$ does enable us to construct a matrix which will fulfill (1) when put for '. . . *x* . . .'. On the other hand syntax itself, embracing membership in addition to the resources of protosyntax, is not even syntactically completable. This can be shown by applying syntax to itself just as protosyntax was applied to itself in § 59. (For this purpose the alphabet implicit in the interpretation of 'M' is of course reconstrued as embracing both 'ϵ' and 'M', as S_9 and S_{10}.) Whereas protosyntactical truth is syntactically but not protosyntactically definable, syntactical truth is not even syntactically definable.

Protosyntactical definability was thought of as moderately broader than constructivity, and syntactical definability was thought of as corresponding to the much broader property of *formality* (cf. §§ 53, 55). Insofar as this way of thinking does justice to the intent of the vague terms 'constructive' and 'formal', then, we must conclude that protosyntactical truth is a nonconstructive formal notion and that syntactical truth is not formal at all. Whoever finds the latter conclusion intolerable had best explain his sense of 'formal'.

The argument from the protosyntactical incompletability of protosyntax to the protosyntactical incompletability of logic turned essentially upon the fact that a set of entities behaving like the objects of protosyntax — in short, a *model* of protosyntax — is discoverable among logically definable objects; viz., in the arithmetic of natural numbers. Now it is possible similarly, by exhibiting a logical model of syntax, to argue from the syntactical incompletability of syntax to the syntactical incompletability of logic. Logical truth, like syntactical truth, is syntactically indefinable. Logical truth may, subject to the cautions of the preceding paragraph, be said to be informal.

Thus even if we go so far as to waive the demand of protosyntactical definability in order to come by a notion of logical theorem which will cover all logical truths and exclude all falsehoods, the outlook remains dark; we should have to renounce syntactical definability as well. Indeed, a notion of theorem capable of exhausting those logical formulæ which are true and excluding those which are false will be definable only in a medium so rich and complex as not to admit of a model anywhere in the reaches of the theory of logic which is under investigation. An exhaustive formulation of logical truth which carries general recognizability with it, even of the most tenuous sort, is not to be aspired to. We must in practice rest content with one or another notion of theorem which covers an important subclass of the logical truths and does not encroach upon the falsehoods.

The fact that logic generally and the elementary theory of natural numbers in particular is incompletable by constructive means (cf. § 55) was discovered and established by Gödel (1931); and it is to him that the essential idea behind the technique of self-application (§ 59) is due. The protosyntactical incompletability of protosyntax and the syntactical incompletability of logic are cases of a more sweeping observation which is due to Tarski ("Wahrheitsbegriff," pp. 370ff; also "Undecidable Statements").

APPENDIX

Theorem versus Metatheorem

Many readers of the first edition have expressed uneasiness over the informality of the proofs of metatheorems such as *111. Some have even sensed here a circularity of reasoning. Thus, it is argued, the proof of *111 makes implicit use of the principles of mathematical induction and the substitutivity of identity; yet these principles are established only in *222 and *636, on the basis of *111 itself.

These misgivings turn on a confusion between theorem and metatheorem which a more careful reading would probably clear up. The matter has proved troublesome enough, however, to warrant some supplementary exposition at this point.

By a *formal deduction* let me be understood to mean any list of formulas such that every formula of the list is either an axiom of quantification (or of membership, after § 29) or a ponential of two earlier formulas of the list. Thus each single axiom is itself a short formal deduction, one line long. But none of the so-called proofs in the book are formal deductions; indeed the only formal deduction appearing anywhere in the book, apart from single axioms, is the sample at the middle of page 86.

Now a *theorem*, as the term is used in this book, is no more nor less than a statement that can be reached as a line of a formal deduction (cf. page 86). To show that a statement is a theorem it is sufficient, therefore, to write down a formal deduction which terminates with the statement in question.

But we also may know that a statement is a theorem without ever seeing a formal deduction of it. To know that a statement is a theorem is to know that the statement possesses a formal deduction which *could* be written down.

Suppose, e.g., that we have two formal deductions before us, and that their respective last lines have the forms:

$$(1) \qquad (y)(x)(w)(\text{- - -} \supset \ldots.), \qquad (y)(x)(\text{- - -})$$

where the matrix represented as '- - -' is the same in both cases. Now it is true, but not obvious, that any two such deductions can be combined and pieced out with a few more lines so as to yield a third formal deduction whose last line is '$(y)(w)(\ldots.)$' — supposing 'y' and 'w' to be the free variables of the matrix represented as '. . . .'. Once we have assured ourselves of the unobvious fact that such a third formal deduction is constructible in all such cases, we can thenceforward always leap directly from a knowledge of the theoremhood of two statements of type (1) to a knowledge of the theoremhood of the corresponding statement '$(y)(w)(\ldots.)$'. Time and space are thereby saved.

Now the unobvious fact spoken of above is what *111 affirms. *111 says of all formulæ ϕ and ψ that if the closures of $\ulcorner\phi \supset \psi\urcorner$ and ϕ possess formal deductions, so does the closure of ψ. Because the fact is not obvious, we have to convince ourselves of it; and this is the purpose of the so-called proof on pages 90–91. It is satisfactory in this proof to avail ourselves of anything that we know in fact to be true, such as that the closure of $\ulcorner\phi \supset \psi\urcorner$ is the closure of $\ulcorner(\alpha)(\phi \supset \psi)\urcorner$ (where α is alphabetically the earliest free variable), or such as that if 0 has some property and the number after each number having the property has the property then every positive integer has the property. It is satisfactory to avail ourselves thus of anything we know to be true because we are not seeking a formal deduction; we are merely seeking the truth about formal deductions.

The worry over a circular presupposition of *222 or *636, in the argument of *111, involves a misunderstanding of the content of *222 and *636. These metatheorems say that the closures of certain conditionals are theorems, i.e., possess formal deductions. These facts are not obvious, and no knowledge of them is presupposed in proving *111. The most that is presupposed there is that the closures of conditionals like those concerned in *222 and *636 are true; and this much is too evident to require support.

It must be emphasized that common-sense reasoning is never smuggled into the formal deductions themselves. Formal deductions proceed from axioms by means exclusively of modus ponens.

Metatheorems are informal observations *about* theoremhood — hence about the possibility of formal deductions. But it must be noted that even in the case of actual theorems, e.g. †182–†185, the "proofs" which are given in the book are on a par with the proofs of metatheorems; none are formal deductions, each is merely an argument to show that an appropriate formal deduction *could* be obtained. This is why metatheorems can quite properly be cited in the proofs of theorems. In the case of †182, as it happens, a formal deduction could also very easily be presented, consisting merely of †182 itself (expanded into primitive notation); for this is actually an axiom of quantification of kind A (see page 81). In the case of †183–†185 and most other theorems stated in the book, however, the formal deductions would run to considerable lengths.

As long as we are interested only in formalizing logic and not in formalizing our discourse about the formalization of logic, the proofs of metatheorems and theorems presented are open to no objection on the score of informality. All these proofs are informal arguments regarding the existence of formal deductions. But there is reason for interest also in formalizing our discourse about the formalization of logic; and this topic is indeed dealt with, in certain of its aspects, in Chapter VII.

LIST OF DEFINITIONS

The definitions D1–48, introduced in the course of §§ 9–48, are assembled here for convenience of reference. The numerous definitions of §§ 49–59, not having been used in formal proofs, are omitted from the list. Various definitions which were casually suggested in the course of earlier discussion (pp. 49, 141, 201, 208, 225, 227) are likewise omitted.

	PAGE
D1. $\ulcorner \sim\phi \urcorner$ *for* $\ulcorner (\phi \downarrow \phi) \urcorner$.	47
D2. $\ulcorner (\phi \,.\, \psi) \urcorner$ *for* $\ulcorner (\sim\phi \downarrow \sim\psi) \urcorner$.	47
D3. $\ulcorner (\phi \vee \psi) \urcorner$ *for* $\ulcorner \sim(\phi \downarrow \psi) \urcorner$.	47
D4. $\ulcorner (\phi \supset \psi) \urcorner$ *for* $\ulcorner (\sim\phi \vee \psi) \urcorner$.	48
D5. $\ulcorner (\phi \equiv \psi) \urcorner$ *for* $\ulcorner ((\phi \supset \psi) \,.\, (\psi \supset \phi)) \urcorner$.	48
D6. $\ulcorner (\phi \,.\, \psi \,.\, \chi) \urcorner$ *for* $\ulcorner ((\phi \,.\, \psi) \,.\, \chi) \urcorner$, *etc.*	48
D7. $\ulcorner (\phi \vee \psi \vee \chi) \urcorner$ *for* $\ulcorner ((\phi \vee \psi) \vee \chi) \urcorner$, *etc.*	48
D8. $\ulcorner (\exists\alpha) \urcorner$ *for* $\ulcorner \sim(\alpha)\sim \urcorner$.	102
D9. $\ulcorner (\beta \epsilon \hat{\alpha}\phi) \urcorner$ *for* $\ulcorner (\exists\gamma)(\beta \epsilon \gamma \,.\, (\alpha)(\alpha \epsilon \gamma \,.\!\supset \phi)) \urcorner$.	133
D10. $\ulcorner (\zeta = \eta) \urcorner$ *for* $\ulcorner (\alpha)(\alpha \epsilon \zeta \,.\!\equiv\!.\, \alpha \epsilon \eta) \urcorner$.	136
D11. $\ulcorner (\zeta \neq \eta) \urcorner$ *for* $\ulcorner \sim(\zeta = \eta) \urcorner$.	137
D12. $\ulcorner (\hat{\alpha}\phi \epsilon \zeta) \urcorner$ *for* $\ulcorner (\exists\beta)(\beta = \hat{\alpha}\phi \,.\, \beta \epsilon \zeta) \urcorner$.	140
D13. $\ulcorner (\zeta \,\bar{\epsilon}\, \eta) \urcorner$ *for* $\ulcorner \sim(\zeta \epsilon \eta) \urcorner$.	140
D14. $\ulcorner (\zeta, \eta \epsilon \theta) \urcorner$ *for* $\ulcorner (\zeta \epsilon \theta \,.\, \eta \epsilon \theta) \urcorner$, *etc.*	140
D15. 'V' *for* '$\hat{x}(x = x)$'.	144
D16. 'Λ' *for* '$\hat{x}(x \neq x)$'.	144
D17. $\ulcorner (\iota\alpha)\phi \urcorner$ *for* $\ulcorner \hat{\beta}(\exists\gamma)(\beta \epsilon \gamma \,.\, (\alpha)(\alpha = \gamma \,.\!\equiv \phi)) \urcorner$.	147
D18. $\ulcorner (\zeta \frown \eta) \urcorner$ *for* $\ulcorner \hat{\alpha}(\alpha \epsilon \zeta \,.\, \alpha \epsilon \eta) \urcorner$.	180
D19. $\ulcorner (\zeta \smile \eta) \urcorner$ *for* $\ulcorner \hat{\alpha}(\alpha \epsilon \zeta \,.\!\vee\!.\, \alpha \epsilon \eta) \urcorner$.	180
D20. $\ulcorner \bar{\zeta} \urcorner$ *for* $\ulcorner \hat{\alpha}(\alpha \,\bar{\epsilon}\, \zeta) \urcorner$.	180
D21. $\ulcorner (\zeta \subset \eta) \urcorner$ *for* $\ulcorner (\alpha)(\alpha \epsilon \zeta \,.\!\supset\!.\, \alpha \epsilon \eta) \urcorner$.	185
D22. $\ulcorner \iota\zeta \urcorner$ *for* $\ulcorner \hat{\alpha}(\alpha = \zeta) \urcorner$.	189
D23. $\ulcorner (\zeta;\eta) \urcorner$ *for* $\ulcorner (\iota\iota\zeta \smile \iota(\iota\zeta \smile \iota\eta)) \urcorner$.	198
D24. $\ulcorner (\zeta(\eta, \theta)) \urcorner$ *for* $\ulcorner (\eta, \theta \epsilon \mathrm{V} \,.\, \eta;\theta \epsilon \zeta) \urcorner$.	200
D25. $\ulcorner \hat{\alpha}\hat{\beta}\phi \urcorner$ *for* $\ulcorner \hat{\gamma}(\exists\alpha)(\exists\beta)(\alpha, \beta \epsilon \mathrm{V} \,.\, \gamma = \alpha;\beta \,.\, \phi) \urcorner$.	202
D26. $\ulcorner \dot{\zeta} \urcorner$ *or* $\ulcorner {}^{\cdot}\zeta \urcorner$ *for* $\ulcorner \hat{\alpha}\hat{\beta}(\zeta(\alpha, \beta)) \urcorner$.	205

PAGE

D27. $\ulcorner \breve{\zeta} \urcorner$ *or* $\ulcorner {}^{\smile}\zeta \urcorner$ *for* $\ulcorner \hat{\alpha}\hat{\beta}(\zeta(\beta, \alpha)) \urcorner$. 208

D28. $\ulcorner (\zeta``\eta) \urcorner$ *for* $\ulcorner \hat{\alpha}(\exists\beta)(\zeta(\alpha, \beta) \,.\, \beta \in \eta) \urcorner$. 210

D29. $\ulcorner (\zeta \mid \eta) \urcorner$ *for* $\ulcorner \hat{\alpha}\hat{\gamma}(\exists\beta)(\zeta(\alpha, \beta) \,.\, \eta(\beta, \gamma)) \urcorner$. 213

D30. $\ulcorner {*}\zeta \urcorner$ *for* $\ulcorner \hat{\alpha}\hat{\beta}(\gamma)(\zeta``\gamma \subset \gamma \,.\, \beta \in \gamma \,.\!\supset\!.\, \alpha \in \gamma) \urcorner$. 216

D31. $\ulcorner (\zeta`\eta) \urcorner$ *for* $\ulcorner (\imath\alpha)\, \zeta(\alpha, \eta) \urcorner$. 222

D32. $\ulcorner \mathrm{r}\zeta \urcorner$ *for* $\ulcorner \hat{\beta}(\exists\gamma)(\alpha)(\alpha = \gamma \,.\!\equiv \zeta(\alpha, \beta)) \urcorner$. 222

D33. $\ulcorner \lambda_{\alpha}\zeta \urcorner$ *for* $\ulcorner \hat{\beta}\hat{\alpha}(\beta = \zeta) \urcorner$. 226

D34. 'I' *for* '$\lambda_x x$'. 229

D35. 'Є' *for* '$\hat{x}\hat{y}(x \in y)$'. 232

D36. '0' *for* '$\iota\Lambda$'. 237

D37. '1' *for* '$\hat{x}(\exists y)(y \in x \,.\, x \cap \overline{\iota y} \in 0)$'. 238

D38. '2' *for* '$\hat{x}(\exists y)(y \in x \,.\, x \cap \overline{\iota y} \in 1)$'. 238

D39. 'S' *for* '$\lambda_z \hat{x}(\exists y)(y \in x \,.\, x \cap \overline{\iota y} \in z)$'. 238

D40. 'Nn' *for* '$({*}\mathrm{S}``\iota 0)$'. 242

D41. 'Sa' *for* '$\lambda_x(x \cup \iota x)$'. 247

D42. 'Cs' *for* '$({*}\mathrm{Sa}``\iota\Lambda)$'. 247

D43. 'Fin' *for* '(Є``Nn)'. 251

D44. $\ulcorner \mathfrak{d}\zeta \urcorner$ *for*
$\ulcorner \hat{\alpha}\hat{\beta}(\exists\alpha')(\exists\beta')(\exists\gamma)(\alpha = \alpha';\gamma \,.\, \beta = \beta';(\mathrm{S}`\gamma) \,.\, \gamma \in \mathrm{Nn} \,.\, \zeta(\alpha', \beta')) \urcorner$. 255

D45. $\ulcorner (\zeta^{\eta}) \urcorner$ *for* $\ulcorner \hat{\alpha}\hat{\beta}({*}\mathfrak{d}\zeta(\alpha;0, \beta;\eta)) \urcorner$. 255

D46. $\ulcorner (\zeta + \eta) \urcorner$ *for* $\ulcorner (\mathrm{S}^{\eta}\, `\zeta) \urcorner$. 259

D47. $\ulcorner (\zeta \times \eta) \urcorner$ *for* $\ulcorner (\lambda_{\alpha}(\zeta + \alpha)^{\eta}\, `0) \urcorner$. 259

D48. $\ulcorner (\zeta \wedge \eta) \urcorner$ *for* $\ulcorner (\lambda_{\alpha}(\zeta \times \alpha)^{\eta}\, `1) \urcorner$. 259

LIST OF THEOREMS AND METATHEOREMS

Various of the theorems and metatheorems are assembled here for convenience of reference. A bookmark at this point will prove useful.

The plan has been to include a theorem or metatheorem in the list just in case it is cited by number at least twice in proofs subsequent to the section in which it made its own appearance. Hence the reader should not use this list in looking up numbers which are close to the number of the theorem or metatheorem whose proof he is examining; instead he should merely glance back over neighboring pages. In cases where the theorem or metatheorem sought is missing from the list because of not having been used twice, the reader will indeed consult the list and be disappointed; but from the list he will learn the page numbers of neighboring theorems and metatheorems, thus expediting his consultation of the text.

Initial universal quantifiers have for brevity been omitted in listing theorems.

PAGE

***100.** *If* ϕ *is tautologous,* $\vdash \phi$. 88

***101.** $\vdash \ulcorner (\alpha)(\phi \supset \psi) \supset . (\alpha)\phi \supset (\alpha)\psi \urcorner$. 88

***102.** *If* α *is not free in* ϕ, $\vdash \ulcorner \phi \supset (\alpha)\phi \urcorner$. 88

***103.** *If* ϕ' *is like* ϕ *except for containing free occurrences of* α' *wherever* ϕ *contains free occurrences of* α, $\vdash \ulcorner (\alpha)\phi \supset \phi' \urcorner$. 88

***110.** $\vdash \ulcorner (\alpha)\phi \supset \phi \urcorner$. 90

***114.** $\vdash \ulcorner (\alpha_1) \ldots (\alpha_n)\phi \supset \phi \urcorner$. 93

***117.** *If* $\vdash \ulcorner \phi \supset \psi \urcorner$, *and none of* $\alpha_1, \ldots, \alpha_n$ *is free in* ϕ, *then* $\vdash \ulcorner \phi \supset (\alpha_1) \ldots (\alpha_n)\psi \urcorner$. 95

***118.** *If* α *is not free in* ϕ, $\vdash \ulcorner \phi \equiv (\alpha)\phi \urcorner$. 95

***119.** $\vdash \ulcorner (\alpha)(\beta)\phi \equiv (\beta)(\alpha)\phi \urcorner$. 95

***121.** *If* ψ' *is like* ψ *except for containing* ϕ' *at some places where* ψ *contains* ϕ, *and* $\alpha_1, \ldots, \alpha_n$ $(n \geqslant 0)$ *exhaust the variables with respect to which those occurrences of* ϕ *and* ϕ' *are bound*

PAGE

in ψ *and* ψ', *then* $\vdash\ulcorner(\alpha_1)\ldots(\alpha_n)(\phi \equiv \phi') \supset . \psi \equiv \psi'\urcorner$. 99

***122.** *If* ψ' *is like* ψ *except for containing free occurrences of* ϕ' *at some places where* ψ *contains free occurrences of* ϕ, *then* $\vdash\ulcorner\phi \equiv \phi' . \supset . \psi \equiv \psi'\urcorner$. 100

***123.** *If* $\vdash\ulcorner\phi \equiv \phi'\urcorner$, *and* ψ' *is formed from* ψ *by putting* ϕ' *for some occurrences of* ϕ, *then* $\vdash\ulcorner\psi \equiv \psi'\urcorner$. 100

***130.** $\vdash\ulcorner\sim(\alpha)\phi \equiv (\exists\alpha)\sim\phi\urcorner$. 102

***131.** $\vdash\ulcorner\sim(\exists\alpha)\phi \equiv (\alpha)\sim\phi\urcorner$. 103

***134.** *If* [etc. as in *103] *then* $\vdash\ulcorner\phi' \supset (\exists\alpha)\phi\urcorner$. 103

***135.** $\vdash\ulcorner\phi \supset (\exists\alpha)\phi\urcorner$. 103

***137.** *If* α *is not free in* ϕ, $\vdash\ulcorner\phi \equiv (\exists\alpha)\phi\urcorner$. 104

***138.** $\vdash\ulcorner(\exists\alpha)(\exists\beta)\phi \equiv (\exists\beta)(\exists\alpha)\phi\urcorner$. 104

***140.** $\vdash\ulcorner(\alpha)(\phi . \psi) \equiv . (\alpha)\phi . (\alpha)\psi\urcorner$. 105

***149.** $\vdash\ulcorner(\alpha)(\phi \supset \psi) \supset . (\exists\alpha)\phi \supset (\exists\alpha)\psi\urcorner$. 107

***156.** $\vdash\ulcorner(\exists\alpha)(\phi . \psi) \supset . (\exists\alpha)\phi . (\exists\alpha)\psi\urcorner$. 108

***158.** *If* α *is not free in* ϕ, $\vdash\ulcorner(\exists\alpha)(\phi . \psi) \equiv . \phi . (\exists\alpha)\psi\urcorner$. 108

***159.** *If* α *is not free in* ϕ, $\vdash\ulcorner(\alpha)(\phi \vee \psi) \equiv . \phi \vee (\alpha)\psi\urcorner$. 109

***161.** *If* α *is not free in* ψ, $\vdash\ulcorner(\alpha)(\phi \supset \psi) \equiv . (\exists\alpha)\phi \supset \psi\urcorner$. 109

***163.** *If* $\vdash\ulcorner\phi \supset \psi\urcorner$, *and none of* $\alpha_1, \ldots, \alpha_n$ *is free in* ψ, *then* $\vdash\ulcorner(\exists\alpha_1)\ldots(\exists\alpha_n)\phi \supset \psi\urcorner$. 109

***171.** *If* ϕ *and* ϕ' *are alphabetic variants,* $\vdash\ulcorner\phi \equiv \phi'\urcorner$. 113

†182. $x = x$ 137

†183. $x = y . \equiv . y = x$ 137

†184. $x = y . \supset : x = z . \equiv . y = z$ 137

***186.** *If* η' *is like* η *except for containing* ϕ' *at some places where* η *contains* ϕ, *and* $\alpha_1, \ldots, \alpha_n$ *exhaust the variables with respect to which those occurrences of* ϕ *and* ϕ' *are bound in* η *and* η', *then* $\vdash\ulcorner(\alpha_1)\ldots(\alpha_n)(\phi \equiv \phi') \supset . \eta = \eta'\urcorner$. 143

***188.** *If* $\vdash\ulcorner\phi \equiv \phi'\urcorner$, *and* η' *is formed from* η *by putting* ϕ' *for some occurrences of* ϕ, *then* $\vdash\ulcorner\eta = \eta'\urcorner$. 143

†189. $x = \hat{y}(y \in x)$ 143

†190. $x \in y . \supset . x \in V$ 144

†191. $x \in y . \equiv . x \in V . x \in y$ 144

†192. $x \bar{\in} \Lambda$ 144

†192b. $y \neq \Lambda . \equiv (\exists x)(x \in y)$ 144

***193.** $\vdash\ulcorner(\alpha)\phi \supset . \hat{\alpha}\phi = V\urcorner$. 144

PAGE

***194.** $\vdash \ulcorner (\alpha) \sim \phi \supset . \hat{\alpha}\phi = \Lambda \urcorner$. 145

***196.** *If* β *is distinct from* α, $\vdash \ulcorner (\alpha)(\alpha = \beta . \equiv \phi) \supset . \beta = (\iota\alpha)\phi \urcorner$. 148

***197.** *If* β *is not* α *nor free in* ϕ, $\vdash \ulcorner \sim (\exists\beta)(\alpha)(\alpha = \beta . \equiv \phi) \supset . (\iota\alpha)\phi = \Lambda \urcorner$. 148

***200.** *If* ϕ *has no free variables beyond* $\alpha, \beta_1, \ldots, \beta_n$, *and is formed from a stratified formula by restricting all bound variables to elements, then* $\vdash \ulcorner \beta_1, \beta_2, \ldots, \beta_n \in V . \supset . \hat{\alpha}\phi \in V \urcorner$. 162

***201.** *If* ϕ *is atomic, and* ϕ' *is formed from* ϕ *by putting* α' *for an occurrence of* α, *then* $\vdash \ulcorner \alpha = \alpha' . \supset . \phi \supset \phi' \urcorner$. 162

†210. $V \in V$ 163

†211. $\Lambda \in V$ 163

***223.** *If* ϕ' *is like* ϕ *except for containing free occurrences of* ζ' *in place of some free occurrences of* ζ, $\vdash \ulcorner \zeta = \zeta' . \supset . \phi \equiv \phi' \urcorner$. 170

***224.** *If* $\vdash \ulcorner \zeta = \zeta' \urcorner$, *and* ϕ' *is formed from* ϕ *by putting* ζ' *for some occurrences of* ζ, *then* $\vdash \ulcorner \phi \equiv \phi' \urcorner$. 170

***226.** *If* η' *is like* η *except for containing free occurrences of* ζ' *in place of some free occurrences of* ζ, $\vdash \ulcorner \zeta = \zeta' . \supset . \eta = \eta' \urcorner$. 170

***227.** *If* $\vdash \ulcorner \zeta = \zeta' \urcorner$, *and* η' *is formed from* η *by putting* ζ' *for some occurrences of* ζ, *then* $\vdash \ulcorner \eta = \eta' \urcorner$. 170

***228.** *If* [etc. as in *223] *then* $\vdash \ulcorner \zeta = \zeta' . \phi . \equiv . \zeta = \zeta' . \phi' \urcorner$. 170

***230.** $\vdash \ulcorner \alpha \in \hat{\alpha}\phi . \equiv . \alpha \in V . \phi \urcorner$. 171

***231.** *If* ψ *is like* ϕ *except for containing free occurrences of* ζ *wherever* ϕ *contains free occurrences of* α, $\vdash \ulcorner (\alpha)\phi \supset \psi \urcorner$. 171

***232.** *If* [etc. as in *231] *then* $\vdash \ulcorner \psi \supset (\exists\alpha)\phi \urcorner$. 172

***233.** *If* ψ *is like* ϕ *except for containing free occurrences of* $\zeta_1, \zeta_2, \ldots, \zeta_n$ *in place respectively of all free occurrences of* $\alpha_1, \alpha_2, \ldots, \alpha_n$, *then* $\vdash \ulcorner (\alpha_1)(\alpha_2) \ldots (\alpha_n)\phi \supset \psi \urcorner$. 172

***234.** *If* [etc. as in *231], *and* α *is not free in* ζ, *then* (a) $\vdash \ulcorner \psi \equiv (\alpha)(\alpha = \zeta . \supset \phi) \urcorner$ *and* (b) $\vdash \ulcorner \psi \equiv (\exists\alpha)(\alpha = \zeta . \phi) \urcorner$. 174

***235.** *If* [etc. as in *231] *then* $\vdash \ulcorner \zeta \in \hat{\alpha}\phi \equiv . \zeta \in V . \psi \urcorner$. 174

***245.** *If* χ *is formed from* ψ *by putting* ϕ *for some free occurrences of* $\ulcorner \alpha \in \hat{\alpha}\phi \urcorner$, *then* $\vdash \ulcorner \hat{\alpha}\psi = \hat{\alpha}\chi \urcorner$. 176

***246.** $\vdash \ulcorner \hat{\alpha}\phi = \hat{\alpha}(\alpha \in V . \phi) \urcorner$. 176

***249.** $\vdash \ulcorner \hat{\alpha}\phi = V . \equiv (\alpha)(\alpha \in V . \supset \phi) \urcorner$. 176

		PAGE
†270.	$z \in x \cap y . \equiv . z \in x . z \in y$	180
†271.	$z \in x \cup y . \equiv : z \in x . \vee . z \in y$	180
†272.	$x, y \in \mathrm{V} . \supset . x \cap y \in \mathrm{V}$	181
†273.	$x, y \in \mathrm{V} . \supset . x \cup y \in \mathrm{V}$	181
†274.	$x \in \mathrm{V} . \equiv . \bar{x} \in \mathrm{V}$	181
†275.	$x = \bar{\bar{x}}$	181
†277.	$x = x \cup x$	181
†286.	$(x \cap y) \cap z = x \cap (y \cap z)$	181
†288.	$x \cap (y \cup z) = (x \cap y) \cup (x \cap z)$	182
†293.	$\overline{\mathrm{V}} = \Lambda$	184
†294.	$x \cap \bar{x} = \Lambda$	184
†298.	$x \cap \Lambda = \Lambda$	184
†311.	$x \subset y . y \subset z . \supset . x \subset z$	185
†312.	$x = y . \equiv . x \subset y . y \subset x$	185
†316.	$x \cap y \subset x$	186
†321.	$x \subset \bar{y} . \equiv . x \cap y = \Lambda$	187
†323.	$x \subset y . \equiv . x = x \cap y$	187
†326.	$x \subset y . \equiv . y = (\bar{x} \cap y) \cup x$	187
†332.	$x \cup y \subset z . \equiv . x \subset z . y \subset z$	188
†333.	$x \subset \mathrm{V}$	188
†335.	$\mathrm{V} \subset x . \equiv . x = \mathrm{V}$	188
†342.	$x \in y . \equiv . x \in \mathrm{V} . \iota x \subset y$	190
†344.	$x \in \mathrm{V} . \equiv . x \in \iota x$	190
†345.	$y \in \iota x . \equiv . x \in \mathrm{V} . x = y$	190
†355.	$x \in \mathrm{V} . \iota x \cup y = \iota z . \supset . x = z$	192
†358.	$x \in \mathrm{V} . \supset : \iota x = \iota y . \equiv . x = y$	193
†359.	$\iota x \in \mathrm{V}$	193
†360.	$\iota x \cup \iota y \in \mathrm{V}$	193
†362.	$z \in y \cap \overline{\iota x} . \equiv . z \in y . z \neq x$	194
†411.	$x;y \in \mathrm{V}$	199
†417.	$x, y, w \in \mathrm{V} . \supset : x;y = z;w . \equiv . x = z . y = w$	200
†420.	$\mathrm{V}(x, y) \equiv . x, y \in \mathrm{V}$	200
†422.	$x \subset y . \supset . x(z, w) \supset y(z, w)$	201
†424.	$(x \cup y)(z, w) \equiv . x(z, w) \vee y(z, w)$	201
*433.	$\vdash \ulcorner \hat{\alpha}\hat{\beta}\phi\ (\alpha, \beta) \equiv . \alpha, \beta \in \mathrm{V} . \phi \urcorner$.	203
*434.	*If* ψ *is like* ϕ *except for containing free occurrences of* ζ *and*	

		PAGE
	η *wherever* ϕ *contains free occurrences respectively of* α *and* β, *then* $\vdash \ulcorner \hat{\alpha}\hat{\beta}\phi\,(\zeta, \eta) \equiv .\, \zeta, \eta \in V \,.\, \psi \urcorner$.	204
***435.**	$\vdash \ulcorner (\alpha)(\beta)(\phi \supset \psi) \supset .\, \hat{\alpha}\hat{\beta}\phi \subset \hat{\alpha}\hat{\beta}\psi \urcorner$.	204
***437.**	$\vdash \ulcorner \dot{\ }\hat{\alpha}\hat{\beta}\phi = \hat{\alpha}\hat{\beta}\phi \urcorner$.	205
†443.	$\dot{x} \subset x$	206
†446.	$\dot{x} \subset y \,. \equiv (z)(w)(x(z, w) \supset y(z, w))$	207
†461.	$\breve{\breve{x}} = \dot{x}$	209
†467.	$x(z, w) \,.\, w \in y \,. \supset .\, z \in x``y$	210
†468.	$x``y \subset z \,. \equiv (w)(w')(x(w, w') \,.\, w' \in y \,. \supset .\, w \in z)$	210
†469.	$x \subset y \,. \supset .\, x``z \subset y``z$	210
†479.	$y \in x``\iota z \,. \equiv x(y, z)$	212
†486.	$(x \mid x')(y, w) \equiv (\exists z)(x(y, z) \,.\, x'(z, w))$	213
†487.	$x(y, z) \,.\, x'(z, w) \,. \supset (x \mid x')(y, w)$	213
†491.	$(x \mid y) \mid z = x \mid (y \mid z)$	213
†495.	$(x \mid y)``z = x``(y``z)$	214
†511.	$*x(y, y) \equiv .\, y \in V$	216
***521.**	*If* α *and* α' *are distinct and not free in* ζ, *and* α' *is not free in* ϕ, *and* ϕ', ψ, *and* ψ' *are like* ϕ *except for containing free occurrences respectively of* α', η, *and* η' *wherever* ϕ *contains free occurrences of* α, *then* $\vdash \ulcorner (\alpha')(\alpha)(\zeta(\alpha', \alpha) \,.\, \phi \,. \supset \phi') \,.\, *\zeta\,(\eta', \eta) \,.\, \psi \,. \supset \psi' \urcorner$.	218
†524.	$*x \mid x = x \mid *x$	220
†533.	$y \,\tilde{\epsilon}\, \text{r}x \,. \supset .\, x`y = \Lambda$	223
†534.	$y \,\tilde{\epsilon}\, V \,. \supset .\, x`y = \Lambda$	223
†536.	$x`y \in V$	224
†538.	$z \in \text{r}y \,. \supset .\, (x \mid y)`z = x`(y`z)$	224
†539.	$\text{r}x = V \,. \supset .\, \text{r}y \subset \text{r}(x \mid y)$	224
***541.**	$\vdash \ulcorner \alpha, \zeta \in V \,. \supset .\, \lambda_{\alpha}\zeta\,`\alpha = \zeta \urcorner$.	227
***542.**	*If* θ *is like* ζ *except for containing free occurrences of* η *in place of all free occurrences of* α, $\vdash \ulcorner \eta, \theta \in V \,. \supset \lambda_{\alpha}\zeta\,`\eta = \theta \urcorner$.	228
***543.**	$\vdash \ulcorner \text{r}\,\lambda_{\alpha}\zeta = \hat{\alpha}(\zeta \in V) \urcorner$.	228
***545.**	$\vdash \ulcorner \lambda_{\alpha}\zeta\,(\theta, \eta) \supset .\, \lambda_{\alpha}\zeta\,`\eta = \theta \urcorner$.	229
†550.	$x \in V \,. \equiv .\, I`x = x$	229
†560.	$x \mid I = \dot{x}$	231
†564.	$\mathfrak{E}(x, y) \equiv .\, x \in y \,.\, y \in V$	232
†611.	$0 \in V$	238

PAGE

†615. $1 = \mathrm{S}‘0$ 239

†619. $x \in \mathrm{S}‘z \,.\!\supset (\exists y)(y \in x \,.\, x \cap \overline{\iota y} \in z)$ 239

†620. $y \in x \,.\, x \cap \overline{\iota y} \in z \,.\, z \in \mathrm{V} \,.\!\supset\!.\, x \in \mathrm{S}‘z$ 239

***637.** *If* ψ, ϕ', *and* ϕ_0 *are like* ϕ *except for containing free occurrences respectively of* ζ, $\ulcorner \mathrm{S}‘\alpha \urcorner$, *and* ‘0’ *wherever* ϕ *contains free occurrences of* α, $\vdash \ulcorner (\alpha)(\alpha \in \mathrm{Nn} \,.\, \phi \,.\!\supset \phi') \,.\, \phi_0 \,.\, \zeta \in \mathrm{Nn} \,.\!\supset \psi \urcorner$. 244

†664. $y \in \mathrm{Nn} \,.\!\supset (\exists x)(x \in \mathrm{Cs} \,.\, x \in y)$ 250

†682. $x^0 = I$ 256

†685. $y \in \mathrm{Nn} \,.\!\supset\!.\, x^{\mathrm{S}‘y} = x \mid (x^y)$ 258

†686. $y \in \mathrm{Nn} \,.\, \mathfrak{r}x = \mathrm{V} \,.\!\supset\!.\, \mathfrak{r}(x^y) = \mathrm{V}$ 258

†690. $x \in \mathrm{Nn} \,.\!\supset\!.\, x + 0 = x$ 259

†691. $y \in \mathrm{Nn} \,.\!\supset\!.\, x + (\mathrm{S}‘y) = \mathrm{S}‘(x + y)$ 259

†692. $x \times 0 = 0$ 260

†693. $y \in \mathrm{Nn} \,.\!\supset\!.\, x \times (\mathrm{S}‘y) = x + (x \times y)$ 260

†694. $x \wedge 0 = 1$ 261

†695. $y \in \mathrm{Nn} \,.\!\supset\!.\, x \wedge (\mathrm{S}‘y) = x \times (x \wedge y)$ 261

†696. $x, y \in \mathrm{Nn} \,.\!\supset\!.\, x + y \in \mathrm{Nn}$ 261

BIBLIOGRAPHICAL REFERENCES

This list includes only such works as happen to have been alluded to, by title or otherwise, in the course of the book. For a comprehensive register of the literature of mathematical logic to the end of 1935 see Church's *Bibliography*. Subsequent literature is covered by the Reviews section of the *Journal of Symbolic Logic*, which is indexed biennially.

ACKERMANN, Wilhelm. "Mengentheoretische Begründung der Logik." *Math. Annalen*, vol. 115 (1937), pp. 1–22.

——*Grundzüge*. See Hilbert.

BERKELEY, E. C. "Boolean Algebra and Applications to Insurance." *Record* (Amer. Inst. of Actuaries), vol. 26 (1937), Part III, pp. 373–414.

BERNAYS, Paul. "A System of Axiomatic Set-Theory, Part I." *J. of Symbolic Logic*, vol. 2 (1937), pp. 65–77.

——*Grundlagen*. See Hilbert.

BERRY, G. D. W. "On Quine's Axioms of Quantification." *J. of Symbolic Logic*, vol. 6 (1941), pp. 23–27.

BOOLE, George. *The Mathematical Analysis of Logic*. London and Cambridge, England, 1847. Reprinted in *Collected Logical Works*, Chicago and London, 1916.

——*An Investigation of the Laws of Thought*. London, 1854. Reprinted ibid.

CANTOR, Georg. "Ueber eine elementare Frage der Mannigfaltigkeitslehre." *Jahresb. der deutschen Math.-Vereinigungen*, vol. 1 (1890–1), pp. 75–78. Reprinted in *Gesammelte Abh.*, Berlin, 1932.

CARNAP, Rudolf. *Physikalische Begriffsbildung*. Karlsruhe, 1926.

——*Der logische Aufbau der Welt*. Berlin, 1928.

——*Abriss der Logistik*. Vienna, 1929.

——*The Logical Syntax of Language*. London and New York, 1937. Translation of *Logische Syntax der Sprache* (Vienna, 1934) with additions.

——"Testability and Meaning." *Phil. of Sci.*, vol. 3 (1936), pp.

419–471; vol. 4 (1937), pp. 1–40.

CAYLEY, Arthur. "On the Theory of Groups as Depending on the Symbolical Equation $\theta^n = 1$." *Phil. Mag.*, vol. 7 (1854), pp. 40–47, 408–409. Reprinted in *Collected Math. Papers*, Cambridge, Eng., 1895.

CHURCH, Alonzo. *A Bibliography of Symbolic Logic.* Providence, 1938. Reprinted from *J. of Symbolic Logic* (1936, 1938).

——"A Set of Postulates for the Foundation of Logic." *Annals of Math.*, vol. 33 (1932), pp. 346–366; vol. 34 (1933), pp. 839–864.

——"An Unsolvable Problem of Elementary Number Theory." *Amer. J. of Math.*, vol. 58 (1936), pp. 345–363.

——"A Note on the Entscheidungsproblem." *J. of Symbolic Logic*, vol. 1 (1936), pp. 40–41, 101–102.

——Review of Chwistek. Ibid., vol. 2 (1937), pp. 168–170.

CHWISTEK, Leon. "Antynomje logiki formalnej." *Przegl. Fil.*, vol. 24 (1921), pp. 164–171.

CURRY, H. B. "Grundlagen der kombinatorischen Logik." *Amer. J. of Math.*, vol. 52 (1930), pp. 509–536, 789–834.

——"Apparent Variables from the Standpoint of Combinatory Logic." *Annals of Math.*, vol. 34 (1933), pp. 381–404.

——"First Properties of Functionality in Combinatory Logic." *Tôhoku Math. J.*, vol. 41 (1936), pp. 371–401.

——"On the Use of Dots as Brackets in Logical Expressions." *J. of Symbolic Logic*, vol. 2 (1937), pp. 26–28.

DEDEKIND, Richard. *Stetigkeit und irrationale Zahlen.* Braunschweig, 1872. 2d edn., 1892; 3d edn., 1905; 4th edn., 1912.

DE MORGAN, Augustus. "On the Syllogism, No. IV, and on the Logic of Relations." *Trans. Camb. Phil. Soc.*, vol. 10 (1864), pp. 331–358. (Read 1860).

DITTRICH, Arnošt. "Objasnění základů užší relativistiky pomocí Russellova kalkulu relačního. (Explication des éléments de la relativité proprement dite à l'aide du calcul de relations de Russell.)" Bohemian with French abstract. *Časopis pro Pěstování Mat. a Fys.*, vol. 59 (1929–30), pp. 35–44.

FERMAT, P. de. *Œuvres.* 4 vols. and supplement. Paris, 1891–1924.

FITCH, F. B. "The Consistency of the Ramified *Principia*." *J. of Symbolic Logic*, vol. 3 (1938), pp. 140–149.

FRAENKEL, Adolf. *Einleitung in die Mengenlehre*. 3d edn., Berlin, 1928. (Earlier edns. 1919, 1923.)

——"Untersuchungen über die Grundlagen der Mengenlehre." *Math. Zeitschr.*, vol. 22 (1925), pp. 250–273.

FREGE, Gottlob. *Begriffsschrift*. Halle, 1879.

——*Die Grundlagen der Arithmetik*. Breslau, 1884. Reprinted 1934.

——*Grundgesetze der Arithmetik*. Vol. 1, 1893; vol. 2, 1903. Jena.

GERGONNE, J. D. "Essai de dialectique rationelle." *Annales de math. pures et appliquées*, vol. 7 (1816–17), pp. 189–228.

GÖDEL, Kurt. *On Undecidable Propositions of Formal Mathematical Systems*. Mimeographed. Princeton, 1934.

——"Die Vollständigkeit der Axiome des logischen Funktionenkalküls." *Monatsh. für Math. u. Phys.*, vol. 37 (1930), pp. 349–360.

——"Ueber formal unentscheidbare Sätze der *Principia Mathematica* und verwandter Systeme." Ibid., vol. 38 (1931), pp. 173–198.

GRASSMANN, Hermann. *Lehrbuch der Arithmetik für höhere Lehranstalten*. (Also under the title: *Lehrbuch der Math. für höhere Lehranstalten*. Erster Theil: Arithmetik.) Berlin, 1861. Reprinted in part in *Gesammelte math. u. phys. Werke*, Leipzig, 1894–1911.

HEMPEL, C. G., and P. OPPENHEIM. *Der Typusbegriff im Lichte der neuen Logik*. Leyden, 1936.

HENKIN, Leon. "The Completeness of the First-Order Functional Calculus." *J. of Symbolic Logic*, vol. 14 (1949), pp. 159–166.

HILBERT, David. "Neubegründung der Mathematik." *Abh. aus dem Math. Seminar der Hamburg Univ.*, vol. 1 (1922), pp. 157–177. Reprinted in *Gesammelte Abh.*, Berlin, 1935.

——and W. ACKERMANN. *Grundzüge der theoretischen Logik*. Berlin, 1928. 2d edn. 1938; 3d edn. 1949. Translation: *Principles of Mathematical Logic*, New York, 1950.

—— and P. BERNAYS. *Grundlagen der Mathematik*. Vol. 1, 1934; vol. 2, 1939. Berlin. Reprinted, Ann Arbor, 1944.

HULL, C. L., and others. *Mathematico-Deductive Theory of Rote Learning*. Lithographed. New Haven, 1940.

HUNTINGTON, E. V. "Sets of Independent Postulates for the Algebra of Logic." *Trans. Amer. Math. Soc.*, vol. 5 (1904), pp. 288–309.

——"New Sets of Independent Postulates for the Algebra of Logic." Ibid., vol. 35 (1933), pp. 274–304. Corrections pp. 557–558, 971.

——"Note on a Recent Set of Postulates for the Calculus of Propositions." *J. of Symbolic Logic*, vol. 4 (1939), pp. 10–14.

JEFFREYS, Harold. *Scientific Inference*. Cambridge, Eng., 1931. 2d edn. 1937.

JEVONS, W. S. *Pure Logic*. London, 1864. Reprinted, London and New York, 1890.

KLEENE, S. C. "λ-Definability and Recursiveness." *Duke Math. J.*, vol. 2, (1936), pp. 340–353.

——"Recursive Predicates and Quantifiers." *Trans. Amer. Math. Soc.*, vol. 53 (1943), pp. 41–73.

KURATOWSKI, Casimir. "Sur la notion de l'ordre dans la théorie des ensembles." *Fundamenta Math.*, vol. 2 (1921), pp. 161–171.

LANGFORD, C. H. See Lewis.

LEŚNIEWSKI, Stanisław. "Grundzüge eines neuen Systems der Grundlagen der Mathematik I." *Fundamenta Math.*, vol. 14 (1929), pp. 1–81.

LEWIS, C. I. *A Survey of Symbolic Logic*. Berkeley, 1918.

——and C. H. LANGFORD. *Symbolic Logic*. New York, 1932.

ŁUKASIEWICZ, Jan. "O logice trójwartościowej." *Ruch Fil.*, vol. 5 (1920), pp. 169–171.

——"Zur Geschichte der Aussagenlogik." *Erkenntnis*, vol. 5 (1935–6), pp. 111–131. Translation of "Z historji logiki zdań" (*Przegl. Fil.*, 1934).

——and A. TARSKI. "Untersuchungen über den Aussagenkalkül." *C. r. Soc. Sci. Lett. Varsovie*, Classe III, vol. 23 (1930), pp. 30–50.

NEUMANN, J. v. "Eine Axiomatisierung der Mengenlehre." *Jour. für reine u. angew. Math.*, vol. 154 (1925), pp. 219–240.

Correction in vol. 155, p. 128.
——"Zur Hilbertschen Beweistheorie." *Math. Zeitschr.*, vol. 26 (1927), pp. 1–46.
OPPENHEIM, Paul. See Hempel.
PASCAL, Blaise. *Œuvres*. Paris, 1889.
PEANO, Giuseppe. *Arithmetices Principia*. Turin, 1889.
——*Formulaire de Mathématiques*. Introduction, 1894; vol. 1, 1895; vol. 2, 1897–9. Turin. Vol. 3, 1901, Paris. Vol. 4, 1902–3; vol. 5 (s. v. *Formulario Mathematico*), 1905–8. Turin.
——"Sui numeri irrazionale." *Revue de Math.* (*Riv. di Mat.*), vol. 6 (1896–99), pp. 126–140.
——"Sulla definizione di funzione." *Atti della Reale Accad. dei Lincei*, Rendiconti, classe di sci. fis., mat., e nat., vol. 20 (1911), pp. 3–5.
PEIRCE, C. S. *Collected Papers*. Ed. by C. Hartshorne and P. Weiss. 6 vols. Cambridge, Mass., 1931–5.
POST, E. L. "Introduction to a General Theory of Elementary Propositions." *Amer. J. of Math.*, vol. 43 (1921), pp. 163–185.
QUINE, W. V. *A System of Logistic*. Cambridge, Mass., 1934.
——*Methods of Logic*. New York, 1950.
——"Ontological Remarks on the Propositional Calculus." *Mind*, vol. 43 (1934), pp. 472–476.
——"On the Axiom of Reducibility." Ibid., vol. 45 (1936), pp. 498–500.
——"A Reinterpretation of Schönfinkel's Logical Operators." *Bull. Amer. Math. Soc.*, vol. 42 (1936), pp. 87–89.
——"Definition of Substitution." Ibid., pp. 561–569.
——"A Theory of Classes Presupposing No Canons of Type." *Proc. Nat. Acad. of Sci.*, vol. 22 (1936), pp. 320–326.
——"Truth by Convention." *Philosophical Essays for A. N. Whitehead* (New York, 1936), pp. 90–124.
——"Set-Theoretic Foundations for Logic." *J. of Symbolic Logic*, vol. 1 (1936), pp. 45–57.
——"On Derivability." Ibid., vol. 2 (1937), pp. 113–119.
——"Logic Based on Inclusion and Abstraction." Ibid., pp. 145–152.

——"New Foundations for Mathematical Logic." *Amer. Math. Monthly*, vol. 44 (1937), pp. 70–80.

——"On the Theory of Types." *J. of Symbolic Logic*, vol. 3 (1938), pp. 125–139.

——"Relations and Reason." *Technology Rev.*, vol. 41 (1939), pp. 299–301, 324–327.

——"Designation and Existence." *J. of Phil.*, vol. 36 (1939), pp. 701–709.

——"A Logistical Approach to the Ontological Problem." *J. of Unified Sci.*, vol. 9 (1940); preprinted 1939 for 5th International Congress for Unity of Sci.

Ramsey, F. P. *The Foundations of Mathematics and Other Logical Essays.* New York and London, 1931.

Rosser, J. B. "A Mathematical Logic Without Variables." *Annals of Math.*, vol. 36 (1935), pp. 127–150; *Duke Math. J.*, vol. 1 (1935), pp. 328–355.

——"On the Consistency of Quine's 'New Foundations for Mathematical Logic.'" *J. of Symbolic Logic*, vol. 4 (1939), pp. 15–24.

——"An Informal Exposition of Proofs of Gödel's Theorems and Church's Theorem." Ibid., pp. 53–60.

——"Definition by Induction in Quine's 'New Foundations for Mathematical Logic.'" Ibid., pp. 80–81.

——"The Burali-Forti Paradox." Ibid., vol. 7 (1942), pp. 1–17.

Russell, Bertrand. *The Principles of Mathematics.* Cambridge, Eng., 1903. 2d edn., New York, 1938.

——"On Denoting." *Mind*, vol. 14 (1905), pp. 479–493.

——"Mathematical Logic as Based on the Theory of Types." *Amer. J. of Math.*, vol. 30 (1908), pp. 222–262.

——"On Order in Time." *Proc. Camb. Phil. Soc.*, vol. 32 (1936), pp. 216–228.

——*Principia.* See Whitehead.

Schönfinkel, Moses. "Ueber die Bausteine der mathematischen Logik." *Math. Annalen*, vol. 92 (1924), pp. 305–316.

Schröder, Ernst. *Vorlesungen über die Algebra der Logik.* Vol. 1, 1890; vol. 2, 1891–1905; vol. 3, 1895. Leipzig.

Shannon, C. E. "A Symbolic Analysis of Relay and Switching

Circuits." *Trans. Amer. Inst. of Electrical Engineers*, vol. 57 (1938), pp. 713–723.

SHEFFER, H. M. "A Set of Five Independent Postulates for Boolean Algebras." *Trans. Amer. Math. Soc.*, vol. 14 (1913), pp. 481–488.

SKOLEM, Thoralf. "Einige Bemerkungen zu der Abhandlung von E. Zermelo: 'Ueber die Definitheit in der Axiomatik.'" *Fundamenta Math.*, vol. 15 (1930), pp. 337–341.

SMITH, H. B. "The Algebra of Propositions." *Phil. of Sci.*, vol. 3 (1936), pp. 551–578.

STAMM, Edward. "Zastosowanie algebry logiki do teorji szyfrów (Application of the Algebra of Logic to Cryptography)." *Rozprawy Polskiego Tow. Mat.*, vol. 1 (1921), pp. 40–52.

STONE, M. H. "Postulates for Boolean Algebras and Generalized Boolean Algebras." *Amer. J. of Math.*, vol. 57 (1935), pp. 703–732.

STUDY, Eduard. *Die realistische Weltansicht und die Lehre vom Raume.* Braunschweig, 1914.

TARSKI, Alfred. "Sur les ensembles définissables de nombres réels I." *Fundamenta Math.*, vol. 17 (1931), pp. 210–239.

——"Einige Betrachtungen über die Begriffe der ω-Widerspruchsfreiheit und der ω-Vollständigkeit." *Monatsh. für Math. u. Phys.*, vol. 40 (1933), pp. 97–112.

——"Grundzüge des Systemenkalküls." *Fundamenta Math.*, vol. 25 (1935), pp. 177–198; vol. 26 (1936), pp. 283–301.

——"Der Wahrheitsbegriff in den formalisierten Sprachen." *Studia Phil.*, vol. 1 (1936), pp. 261–405. Translation of *Pojęcie Prawdy w Językach Nauk Dedukcyjnych* (Warsaw, 1933) with additions.

——"On Undecidable Statements in Enlarged Systems of Logic and the Concept of Truth." *J. of Symbolic Logic*, vol. 4 (1939), pp. 105–112.

——"Untersuchungen." See Łukasiewicz.

TURING, A. M. "On Computable Numbers." *Proc. London Math. Soc.*, ser. 2, vol. 42 (1937), pp. 230–266. Correction in vol. 43, pp. 544ff.

WANG, Hao. "A Formal System of Logic." *J. of Symbolic Logic*, vol. 15 (1950), pp. 25–32.

WHITEHEAD, A. N. "On Mathematical Concepts of the Material World." *Phil. Trans. Royal Soc. of London*, ser. A, vol. 205 (1906), pp. 465–525.

——"La théorie relationniste de l'espace." *Revue de Métaph. et Morale*, vol. 23 (1916), pp. 423–545.

——and B. RUSSELL. *Principia Mathematica.* Vol. 1, 1910; vol. 2, 1912; vol. 3, 1913. Cambridge, England. 2d edn. 1925–7.

WIENER, Norbert. "A Simplification of the Logic of Relations." *Proc. Camb. Phil. Soc.*, vol. 17 (1912–14), pp. 387–390.

——"Studies in Synthetic Logic." Ibid., vol. 18 (1914–16), pp. 14–28.

——"A New Theory of Measurement." *Proc. London Math. Soc.*, vol. 19 (1919–20), pp. 181–205.

WITTGENSTEIN, Ludwig. *Tractatus Logico-Philosophicus.* New York and London, 1922. Reprint of "Logisch-philosophische Abhandlung" (*Annalen der Naturphil.*, 1921) with English translation in parallel.

WOODGER, J. H. *The Axiomatic Method in Biology.* Cambridge, Eng., 1937.

——*The Technique of Theory Construction.* (Int. Encyc. of Unified Sci., vol. 2, no. 5.) Chicago, 1939.

——"The 'Concept of Organism' and the Relation between Embryology and Genetics." *Qtly. Rev. of Biology*, vol. 5 (1930), pp. 1–22, 438–463; vol. 6 (1931), pp. 178–207.

——"The Formulation of a Psychological Theory." *Erkenntnis*, vol. 7 (1937), pp. 195–198.

ZERMELO, Ernst. "Untersuchungen über die Grundlagen der Mengenlehre I." *Math. Annalen*, vol. 65 (1908), pp. 261–281.

ZWICKY, F. "On a New Type of Reasoning and Some of Its Possible Consequences." *Phys. Rev.*, vol. 43 (1933), pp. 1031–1033.

ŻYLIŃSKI, Eustachy. "Some Remarks concerning the Theory of Deduction." *Fundamenta Math.*, vol. 7 (1925), pp. 203–209.

INDEX OF PROPER NAMES

Ackermann, W. 75, 100, 164.
Aristotle 1.
Berkeley, E. C. 8.
Bernays, P. 81, 165.
Berry, G. D. W. 89.
Boole, G. 1, 5, 14, 55, 61, 71, 145.
Cantor, G. 274.
Carnap, R. 8, 26, 33, 55, 75, 80, 138, 279.
Cauchy, A. 276.
Cayley, A. 214.
Church, A. 81, 163, 229, 292, 331.
Chwistek, L. 163.
Curry, H. B. 41, 71.
Dedekind, R. 271.
De Morgan, A. 210, 214.
Dittrich, A. 8.
Duns Scotus 61.
Fermat, P. de 243.
Fitch, F. B. 89.
Fraenkel, A. 122, 136, 164.
Frege, G. 1, 18, 26, 61, 71, 75, 88f, 102, 126, 128, 132, 136, 149, 189, 202, 214, 221, 229, 238, 243.
Gautam, N. D. 191.
Gergonne, J. D. 18, 189.
Gödel, K. 81, 126, 287, 292, 318.
Grassmann, H. 266.
Hempel, C. G. 8.
Henkin, L. 81.
Herbrand, J. 292.
Hilbert, D. 75, 80f, 88, 100.
Hull, C. L. 8.
Huntington, E. V. 33, 184.
Jeffreys, H. 279.
Jevons, W. S. 14.
Kant, I. 55, 151.
Kaplansky, I. 144.
Kleene, S. C. 292.
Kuratowski, C. 202.
Leibniz, G. W. v. 1, 136.
Leśniewski, S. 60.
Lewis, C. I. 32, 136.
Łukasiewicz, J. 14, 18, 42, 61.
MacLaurin, C. 276.
McKinsey, J. C. C. 60.
Neumann, J. v. 75, 80, 89, 130, 165f, 247.
Norris, M. J. 144.
Oppenheim, P. 8.
Pascal, B. 243.
Peano, G. 1, 41, 61, 71, 80, 102, 119, 136, 145, 149, 184, 189, 202, 222, 259, 266, 271, 276.
Peirce, C. S. 1, 14, 18, 49, 55, 61, 89, 136, 189, 202, 214.
Petrus Hispanus 61.
Philo of Megara 18.
Post, E. L. 14, 43, 88.
Quine, W. V. 2, 8, 32, 55, 71, 119, 121, 126, 145, 151, 163ff, 184.
Ramsey, F. P. 163.
Ravven, R. M. 247.
Rosser, J. B. 71, 81, 159, 165.
Russell, B. 1, 8, 14, 31f, 61, 71, 75, 80, 88f, 102, 121, 126, 128, 131f, 145, 149, 151, 163ff, 189, 197, 202, 214, 221f, 258, 267, 271.
Schönfinkel, M. 71.
Schröder, E. 1, 14, 61, 89, 106, 189.
Shannon, C. E. 8.
Sheffer, H. M. 49, 126.
Skolem, T. 164.
Smith, H. B. 32,
Stamm, E. 8.

Steinhardt, L. D. 95, 137.
Stoics 61.
Stone, M. H. 184.
Study, E. 279.
Symon, K. R. 124.
Tarski, A. 4, 26, 33, 55, 89, 126, 287, 318.
Turing, A. M. 292.

Wang, H. 159, 166.
Whitehead, A. N. 1, 8, 14, 31f, 61, 71, 75, 80, 88f, 102, 126, 163, 221, 258, 267.
Wiener, N. 8, 126, 163, 202.
Wittgenstein, L. 14, 55.
Woodger, J. H. 8.
Zermelo, E. 163ff.
Zwicky, F. 8.
Zyliński, E. 49.

INDEX OF SUBJECTS

Abbreviation 7, 47, 75, 100, 126, 133, 143, 314ff. See also Definition.
Abstract 132, 142f, 149. See also Abstraction.
Abstract entities 32, 120f.
Abstraction 126, 128ff, 140ff, 177f, 304; of functions 225ff, 229; of relations 202ff, 208.
Accents 35, 69, 110, 284, 298.
Addition, see Sum.
Aggregate 120, 127. See also Class.
Algebra 34, 61, 279; of logic 14, 184.
Alphabet 69, 300, 306.
Alphabetic order 79f, 89, 288ff, 300.
Alphabetic variance 109ff, 159, 177.
Alternation 12ff, 41, 47, 301.
Alternative denial 48f.
Analysis 278f.
Analytic 55. See also Truth, logical.
Analytic geometry 279.
Ancestral 215ff, 221, 241f, 285.
Antecedent 14.
Apodasis 14.
Apparent variable, see Bondage.
Application: of functions 221f; of logic 7f; principle of 82ff, 88, 151, 171f.
Arabic notation 265f, 313.
Arch 283ff. See also Concatenation.
Argument 222.
Arithmetic 127f, 259ff, 317; of ratios 268ff; of real numbers 274ff; of signed reals 277f.
Arithmetization of syntax 313.
Arrow projection 303.
Assertion sign 88, 162. See also Theorem; Closure.
Associativity 41, 59, 61, 181ff, 213f, 232, 262, 264, 284.
Asterisk 88, 145, 216. See also Metatheorem; Ancestral.
Aussonderungsaxiom 164.
Atomic: context 140f; formula 73, 89f, 124, 127, 283ff, 296f.
Axioms: of membership 155ff, 162, 305; of quantification 80ff, 85ff, 300f.

Begin 284.
Biconditional 20, 30f, 48, 303; stacked 94; substitutivity of 96ff.
Binary connectives 13, 37, 180.
Bondage 76ff, 142f, 298.
Bonds 70, 76.
Boolean algebra, see Class algebra.
Bound, upper 272f.
Bound variable, occurrence, etc. 76ff, 142f, 298.
Brackets 91, 101, 103f.
Breve 208. See also Converse.

Calculus: differential 278f; of classes 14, 179ff, 184.
Classes, 119ff, 128ff, 135ff, 148f, 163ff, 180ff; algebra of 14, 179ff, 184; class of 122, 237; theory of 127f.
Closed 215, 217, 242f.
Closure 76ff, 89, 299f.
Combinatory logic 71.
Commutativity 57, 61, 138, 181, 183, 213, 232, 263f.
Compatibility 30.
Complement 164f, 180ff.

Completeness 6, 86, 307, 312ff.
Complex numbers 278.
Component, truth-functional 52ff, 299.
Composition, truth-functional 11ff, 29, 33, 42ff.
Computability 292.
Concatenation 283ff, 288f, 294.
Concrete objects 121f. See also Individuals.
Conditional 14ff, 18, 29, 31f, 48, 301; stacked 92f; subjunctive 16, 29, 33.
Conjunction 11ff, 30, 41, 47, 303.
Connectives 11ff, 119, 180.
Consequent 14.
Consistency 166, 292.
Constant function 226, 279.
Constants, logical 1.
Constructivity 291f, 316ff.
Context 25, 140ff, 167, 294.
Contextual definition 121, 126, 132, 140, 145.
Continuity 273.
Contradiction 128ff, 157, 163ff, 197; law of 51.
Converse 208ff, 212ff.
Converse domain 213.
Corners 33ff, 110, 287, 290.
Counter set 247ff.
Counting 247, 254.
Couple, ordered 126, 198ff, 202, 278.
Cyclic formulae 157.

Dagger 88, 137, 139, 145, 173f.
Decidability 5f, 86ff, 156, 291f.
Decimal notation 265f, 313f.
Definability 286, 292f, 307, 312f. 316ff.
Definition 47, 103, 126, 133f, 293; contextual 121, 126, 132, 140, 145; protosyntactical 292ff, 306f, 312f, 316ff; syntactical 317f. See also Abbreviation.
DeMorgan's law 57, 61, 181, 183.
Denial 13f, 47, 301; scope of 16, 40.
Density 273.
Derivative 279.
Description 145ff, 292ff.
Designation 24ff, 32ff, 146ff, 226.
Development, laws of 58.
Differential 25; calculus 279.
Disjunction, see Alternation.
Distributivity 60f, 182f, 210f, 232, 263f.
Diversity 137.
Division 269f, 275ff.
Domain 212.
Dots 37ff, 41; relational 205ff, 209.
Double denial 56; negation 181, 183.
Dyadic relations 201, 225.

Element 131f, 155ff, 165f, 197f, 200, 251f.
Elementhood: conditions for 158f; theorems on 144, 178f, 181, 193f, 199, 224, 238, 246f.
Element-identity 190, 230.
End 284.
Epsilon 119ff, 155, 284, 293, 313f.
Epsilon abbreviate 314f.
Equivalence 30f, 56. See also Biconditional.
Essential 1f, 28, 50.
Excluded middle 51.
Existence 150f, 162.
Existential quantification 101ff, 303.
Exponents 254, 259; laws of 264f. See also Power.
Expression 5f, 283ff, 287.
Extensionality 11, 16, 29, 33, 73f, 120f.
Extra-logical discourse 127, 141, 149ff, 161, 168, 292, 296f.

False 27ff.

Final segment 284.
Finite classes 194, 250ff.
Foreign variable 133f, 304.
Formality 283ff, 286f, 317ff.
Formalization, importance of 6ff.
Formula 71ff, 79, 127f, 286f, 297; atomic 73, 89f, 123f, 127, 283ff, 296f; logical 124ff, 127f, 283ff, 295f; mathematical 128; protosyntactical 292, 306f; syntactical 290.
Fraction 25. See also Ratio; Quotient.
Framed ingredient 295f.
Free variable, occurrence, etc. 76ff, 143, 298.
Function 214, 221ff; abstraction of 225ff, 229.

Generalization, see Quantification.
Geometry 279.
Greater 265ff, 272, 277.
Greek letters 34ff, 68, 75, 110, 135, 287, 290.
Group theory 214.

Hieroglyph 26.
Hypothesis 93.
Hypothetical, see Conditional.

Idempotence 56, 61, 181, 183, 232.
Identities of arithmetic 262ff.
Identity 122f, 134ff, 185; in protosyntax 293, 303f, 307; of indiscernibles 136, 175f; stacked 182; substitutivity of 160f, 167ff, 174.
Identity function 214, 229ff, 253, 256.
Identity matrix 308ff.
Image 209ff, 214.
Implication 14, 16, 28f, 31f, 56. See also Conditional.
Inclusion 185ff, 204, 206f; and abstraction 126, 184; and membership 185, 189.
Incompletability 6, 307, 312ff.
Individuals 121ff, 135ff, 148f.
Induction, mathematical 243f.
Inference 7, 85; rules of 88f.
Infinite classes 194, 248, 250ff.
Initial segment 284.
Integer 276. See also Whole number.
Inverse 275f.
Irrational number 272f.
Iteration 30, 32, 41.

Joint 39.
Joint denial 45ff, 284.

Lambda 226ff.
Lambda-definability 292.
Least upper bound 272f.
Less 265ff, 272, 277.
Limit 278f.
Line 136f, 233, 279.
Logic: subject matter of 1ff, 127f; symbolic 5; traditional 1, 7, 189, 210; universality of 2; vocabulary of 1ff. See also Mathematical logic.
Logical: formula 124ff, 127f, 283ff, 295f; statement 125.
Looseness 39.

Mathematical formula 128.
Mathematical induction 243f.
Mathematical logic 5, 7f; applications of 7f; history of 1, 14, also footnotes and small print *passim*.
Mathematics: extensionality of 73, 121; its derivability from logic 5, 73, 126ff, 279.
Matrix: name 152, 226; statement 71ff, 79, 298.
Meaningfulness 146f, 150, 163ff.
Measurement 7, 279.
Membership 119ff, 131ff, 155ff, 185, 316; as relation 232f, 248; axioms of 155ff, 162, 305.

Mengenlehre 127, 164ff.
Metamathematics 126, 286. See also Syntax.
Metatheorem 89ff, 319ff.
Model 271, 317f.
Modus ponens 85, 88f.
Multiplication, see Product.

Name matrix 152, 226.
Names 32ff, 83f, 119, 125, 149ff, 171, 226.
Natural number 237ff, 247, 250ff.
Negation, see Complement; Denial.
Negative number 276f.
Nominalism 121.
Non-element 131, 177, 179, 228, 251, 316.
Notation 34f, 37, 47, 75, 132, 283ff, 287ff.
Null class 122, 144f, 163, 183f, 188.
Number: complex 278; natural 237ff, 247, 250ff; rational 25, 266ff, 271ff; real 271ff.
Numbering of theorems and metatheorems 88, 90, 137.

Occurrence 297f.
Official notation 40f. See also Primitive notation.
One 237ff, 267, 275.
Ontology 121f, 150f, 162.
Ordinary usage 12ff, 67ff.

Pair, ordered 126, 198ff, 202, 278.
Paradox, see Contradiction.
Parentheses 5, 15f, 37ff, 124, 180, 221, 296; restoration of 40f.
Part 295.
Particular quantification 101ff, 303.
Perpositive 314.
Pi 273.
Plane 136f, 233.
Point 136f, 233.
Polyadic relations 201, 208, 225, 227.
Ponential 85ff.
Positive number 276f.
Potentiality 16.
Power: arithmetical 259, 261f, 264f, 270, 274f, 277f; relative 253ff, 259, 265.
Predecessor 252.
Predicate 27ff, 119, 149, 151.
Predication, relational 200f, 227.
Prefix, see Quantifier.
Primitive notation 40f, 47, 75, 100, 133, 143, 149ff.
Primitives, reduction of 2f, 5ff, 43ff, 49, 72f, 126, 288ff.
Priority 296. See also Alphabetic order.
Product: arithmetical 259ff, 268ff, 274, 277; logical 179ff; relative 213f.
Projection, see Arrow projection; also Image.
Proof 5f, 319ff.
Proof notation 91ff, 101ff, 129, 138f, 145, 173f, 176f, 182, 204, 262f.
Proper ancestral 220f.
Proper names 149ff.
Property 119f, 237.
Proposition 32. See also Statement.
Propositional function 121. See also Property; Statement matrix.
Protasis 14.
Protosyntax 291ff, 306ff, 314f.

Quantification 18, 55, 67ff, 285; axioms of 80ff, 87ff, 300f; existential 101ff, 303; universal 102; vacuous 74, 82, 88, 104.
Quantificational diagrams 70, 76.
Quantifier 67ff, 142; confinement of 108f; distribution of 82, 88, 105ff; insertion of 139f; permutation of 89, 95, 104; scope of 76f.

Quasi-atomic context 140ff, 167, 294.
Quasi-quotation 33ff, 110, 287, 290.
Quasi-theorem 315.
Quotation 23ff, 33ff, 290.
Quotient 269f, 275ff.

Range of functionality 222ff, 228.
Ratio 25, 266ff, 271ff.
Rational 272. See also Ratio.
Real number 271ff.
Real variable, see Free variable.
Recursive 86, 292, 318.
Reduction sentences 33.
Reflexivity 56, 61, 138, 185, 232.
Relational part 205ff.
Relations 197ff; abstraction of 202ff, 208; predication of 200ff, 227; theory of 61, 127, 279.
Relative: power 253ff, 259, 265; product 213f.
Relettering 109ff, 159, 177.
Reverse, see Converse.
Root 275f.
Russell's class 128ff, 157, 179, 249.

Scope 16, 40, 190, 217, 227.
Self-augment 246ff.
Self-membership 125, 128ff, 157, 175, 179, 249.
Semantic 24f.
Sequence 295f, 299.
Set 127. See also Class; Counter set; Truth set.
Sign 283ff, 288; and object 23ff, 283ff.
Signed real numbers 276f.
Single-valuedness 222.
Singulary 13; scope of such operators 16, 190, 217, 227.
Specification 82ff, 88, 151, 171f.
Spelling 284.
Star, see Asterisk.
Statement 11ff, 71ff, 79, 298; logical 125.
Statement connective 27ff.
Statement matrix 71ff, 79, 298.
Statement predicate 27ff.
Stratification 155ff, 164, 302f; of matrices 159f; test of 158, 238, 302f.
Strict implication 32.
Subjunctive 16, 29, 33.
Subscripts 35, 67, 254, 269, 275, 278.
Substitution 170ff, 301ff.
Substitutivity: of biconditional 96ff; of identity 160f, 167ff, 174.
Subtraction 276f.
Successor 237ff, 252f, 259ff. See also Alphabetic order.
Sum, arithmetical 259ff, 268f, 274f, 277; logical 180ff.
Symmetry 138, 232. See also Commutativity.
Syntax 3ff, 283ff.

Tautology 50ff, 55, 86ff, 93, 199, 299.
Term 119ff, 135f, 145, 152, 294.
Testability 5f, 81, 86ff, 156, 291f.
Theorem 6, 85ff, 127, 155f, 162f, 291f, 318ff; notation of 137, 162; protosyntactical 306f; protosyntactical definition of 305, 315f.
Tilde 14, 140.
Total reflexivity 138, 232.
Traditional logic 1, 7, 189, 210.
Transitivity 58, 61, 138, 185, 232.
Transposition 57, 186f.
Truth 1f, 4, 27ff; logical 1ff, 28, 30, 50, 80ff, 86, 313, 315, 318; protosyntactical 313, 317; syntactical 317.
Truth-functional: component 52ff, 299; composition 11ff, 29, 33, 42ff.
Truth sequence 299.
Truth set 53ff, 299.
Truth table 11ff, 19ff, 42ff, 51ff.

Truth value 11ff.
Two 237ff, 267, 275.
Types, theory of 163ff.

Unit class 135f, 189ff, 241.
Universal class 144f, 163, 183f, 188, 304f; infinitude of 248, 251ff.
Universal quantification 102. See also Quantification.
Universal relation 205f.
Unnamable numbers 273f.
Unsigned real numbers 276.
Upper bound 272f.
Use and mention 23ff, 283ff.

Vacuous: occurrence 1f, 28, 50; quantification 74, 82, 88, 104.
Value: of a function 222; of a variable 34f, 287.
Variables 34f, 68ff, 75, 109f, 283ff, 287, 300; as pronouns 3, 5, 68, 71; collision of 74f, 110ff; elimination of 71; rewriting 109ff, 133f, 159, 177; values of 34f, 287.

Whole number 267, 270, 275f. See also Natural number.

Zero 237ff, 267ff, 275f.